SALVAGE

UNIFICATION OF FINITE ELEMENT METHODS

NORTH-HOLLAND
MATHEMATICS STUDIES

94

UNIFICATION OF FINITE ELEMENT METHODS

Edited by

H. KARDESTUNCER

University of Connecticut
Storrs
Connecticut
U.S.A.

1984

NORTH-HOLLAND–AMSTERDAM · NEW YORK · OXFORD

ISBN: 0 444 87519 0

Publishers:

ELSEVIER SCIENCE PUBLISHERS B.V.
P.O. Box 1991
1000 BZ Amsterdam
The Netherlands

Sole distributors for the U.S.A. and Canada:

ELSEVIER SCIENCE PUBLISHING COMPANY, INC.
52 Vanderbilt Avenue
New York, N.Y. 10017
U.S.A.

Library of Congress Cataloging in Publication Data
Main entry under title:

Unification of finite element methods.

  (North-Holland mathematics studies)
  Bibliography: p.
  1. Finite element method. 2. Argyris, J. H. (John
H.), 1916-   . I. Kardestuncer, Hayrettin.
II. Series.
TA347.F5U55 1984   620'.001'515353   84-6006
ISBN 0-444-87519-0 (U.S.)

PRINTED IN THE NETHERLANDS

7th

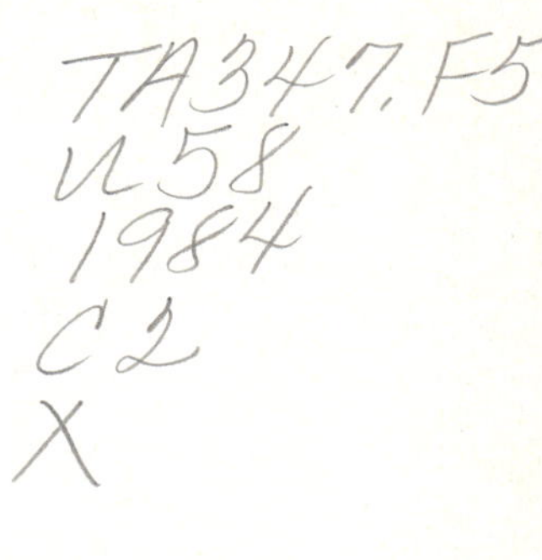

Dedicated to

Professor John H. Argyris

for his pioneering and continuing contributions
to the finite element methods

Alliance of Industry and Academe

PROFESSOR JOHN H. ARGYRIS

A man who unifies engineering and mathematics with elegance

My first encounter with Professor John H. Argyris' work occurred during my graduate studies at MIT in the mid fifties. His elegant treatment of Bernoulli's virtual work and energy principles mounted on Menabrea's *il minimo lavoro* with Castigliano's two theorems, St. Venant's theories of torsion, Maxwell's reciprocity principles, Lord Rayleigh's variational principles, Müller-Breslau's and Otto Mohr's unit load ideas, etc. gave me the impression that this man belonged to the last century. Yet the methodology presented (stiffness and flexibility methods in structural analysis) was so new that it was unknown to my fellow students and did not even exist in the curriculum. A few years later, I learned that he was the holder of the prestigious Chair of Aeronautical Structures at the University of London where he was also Professor of Aerospace Sciences and at the same time was Director of the Institute of Statics and Dynamics and Director of the Computer Center at the University of Stuttgart. I began to wonder if perhaps there were more than one J.H. Argyris, and whose work was I studying?

The more I studied his work and the more I learned of his accomplishments the more convinced I was that the man must be older than I thought; perhaps he was born a century *before* the last. However, when I finally met him in 1961, I was sure that he must be the grandson of the man whose work was so inspiring me and guiding my doctoral dissertation at the Sorbonne.

A citizen of Great Britain, a resident of West Germany, John H. Argyris was born August 19, 1916, in the *Land of the Gods*. A child prodigy who graduated from the Technical University of Athens at the age of eighteen, he received the all-German Prize of Deutscher Stahlbauverband during his postgraduate studies in Munich when he was only twenty years old. Many believe that it was not merely coincidental that Sir Isaac Newton was born on Christmas day of 1642, the same day (with acceptable approximation based on the theories presented in this volume) that another genius, Galileo Galilei, had died. I am curious to know what genius it was who died on August 19, 1916.

I had every intention here to write more about Professor Argyris and his work but the more I wrote the more I became convinced that my writing could in no way reflect the accomplishments of this great man. His life can not be told in an essay; his work can not be assessed in an article; his abundant energy can not be formulated as an energy functional. He is beyond and above *all* that most of us know of him.

H. Kardestuncer

MAIN DISTINCTIONS OF
PROFESSOR JOHN H. ARGYRIS, D.Sc., Dr.h.c.mult.

1937	Dipl.-Ing.D.E.	Munich
1954	D.Sc. (Eng.)	University of London
1955	Fellow R.Ae.S.	Royal Aeronautical Society, London
1962	Honorary Associate Hon. A.C.G.I.	City Guilds Institute, London
1970	Laura h.c.dott.Ing.	University of Genoa, Special Distinction on the 100th Anniversary of the Faculty of Applied Mechanics and Ship Building
1971	George Taylor Prize	Royal Aeronautical Society, London
1971	Silver Medal	Royal Aeronautical Society, London
1972	Principal Editor	Computer Methods of Applied Mechanics and Engineering (Journal)
1972	dr.techn.h.c. and jus docendi	University of Norway, Trondheim
1973	Corresponding member	Academy of Sciences of Athens (Positive Sciences)
1974	Honorary Fellow	Groupe pour l'Avancement des Méthodes Numeriques de l'Ingenieur (GAMNI), Paris
1975	von Kármán Medal	Highest Scientific Award, American Society of Civil Engineers, New York
1976	Honorary Fellow Hon.F.C.G.I.	City Guilds Institute, London
1979	Member A.S.C.E.	American Society of Civil Engineers, New York

1979	Copernicus Medal	Highest Award in Natural Sciences Polish Academy of Sciences, Warsaw
1980	Gold Medal	of the Land Baden-Württemberg
1980	Honorary Professor	Northwest Polytechnical University, Xian, People's Republic of China
1981	Timoshenko Medal	Highest Scientific Award, American Society of Mechanical Engineers, New York
1981	Life Member A.S.M.E.	American Society of Mechanical Engineers
1981	Member	The New York Academy of Sciences, New York
1982	I.B. Laskowitz Award with Gold Medal	Highest Astronautical Award of the New York Academy of Sciences
1983	Fellow of the AIAA	Highest Grade of Membership, American Institute of Aeronautics and Astronautics, New York
1983	Dr.Ing.E.h.	University of Hanover, Honorary Doctorate
1983	Honorary Professor	Technical University of Peking (Beijing)
1983	Honorary Life Member	New York Academy of Sciences, New York
1983	World Prize in Culture and Election as Personality of the Year 1984	Centro Studi e Ricerche delle Nazioni Accademia Italia, Salsomaggiore Terme
1984	Honorary Professor	Qinghua University, Beijing
1940–	340 scientific publications	and continuing

ACKNOWLEDGMENTS

The UFEM series could not take place without the generous help of the following friends, organizations, and societies. Their encouragement, support, and sharing of the ideals of the conference are sincerely appreciated and gratefully acknowledged.

Organizing Committee Members
H. Clark, Hon. Chairman (UConn)
H. Kardestuncer, Chairman (UConn)
W.W. Bowley (UConn)
J.J. Connor (MIT)
H.A. Koenig (UConn)
A. Phillips (Yale)
R.J. Pryputniewicz (WPI)

Advisory Board Members
I. Babuška (Maryland)
L. Collatz (Hamburg, Germany)
A.C. Eringen (Princeton)
R.H. Gallagher (Arizona)
J.T. Oden (Texas)
T.H.H. Pian (MIT)
O.C. Zienkiewicz (Swansea, U.K.)

Session Chairmen

H. Allik (BBN)
W.W. Bowley (UConn)
F. Camaretta (Sikorsky)
A.D. Carlson (NUSC)
M.K.V. Chari (General Electric)
L. Collatz (Hamburg, Germany)
J.H. Connor (MIT)
A.C. Eringen (Princeton)
S. Gordon (Electric Boat)

H.A. Koenig (UConn)
R. Lalkaka (United Nations)
H. Mayer (Hamilton Standard)
D.H. Norrie (Calgary)
T. Onat (Yale)
A. Phillips (Yale)
T.H.H. Pian (MIT)
J.A. Roulier (UConn)

Local Arrangements
J.J. Farling (UConn, Conf. & Inst.)
G.D. Smith (UConn)
G.M. Wallace (UConn)

Participating Societies
AIAA
ASME
CC–ASCE

Sponsoring Organizations

Analysis & Technology, Inc.
AVCO Lycoming Corp.
Bolt Beranek & Newman, Inc.
Control Data Corp.
Electric Boat
General Electric
Hamilton Standard
Conf. & Inst. (UConn)

Kaman Aerospace Corp.
Naval Underwater Systems Center
Northeast Utilities
Perkin Elmer
Pratt & Whitney Aircraft
Sikorsky Aircraft
UConn Foundation
UConn Research Foundation

LIST OF CONTRIBUTORS

J.H. Argyris (1), Institut für Statik und Dynamik der Luft- und Raumfahrtkonstruktionen, University of Stuttgart, Stuttgart, Fed. Rep. Germany.

J.F. Abel (47), Department of Structural Engineering, Cornell University, Ithica, New York, U.S.A.

S.N. Atluri (65), CACM, Georgia Institute of Technology, Atlanta, Georgia, U.S.A.

I. Babuška (97), Institute of Physical Science and Technology, University of Maryland, College Park, Maryland, U.S.A.

K.-J. Bathe (123), Department of Mechanical Engineering, Massachusetts Institute of Technology, Cambridge, Massachusetts, U.S.A.

J. Bielak (149), Department of Civil Engineering, Carnegie-Mellon University, Pittsburgh, Pennsylvania, U.S.A.

J.H. Bramble (167), Department of Mathematics, Cornell University, Ithica, New York, U.S.A.

C.A. Brebbia (185), The Institute of Computational Mechanics, Ashurst Lodge, Southampton, England

M.A. Celia (303), Civil Engineering Department, Princeton University, Princeton, New Jersey, U.S.A.

A. Chaudhary (123), Department of Mechanical Engineering, Massachusetts Institute of Technology, Cambridge, Massachusetts, U.S.A.

J. St. Doltsinis (1), Institut für Statik und Dynamik der Luft- und Raumfahrtkonstruktionen, University of Stuttgart, Stuttgart, Fed. Rep. Germany.

J.F. Hajjar (47), Department of Structural Engineering, Cornell University, Ithica, New York, U.S.A.

T.-Y. Han (47), Department of Structural Engineering, Cornell University, Ithica, New York, U.S.A.

A.R. Ingraffea (47), Department of Structural Engineering, Cornell University, Ithica, New York, U.S.A.

K. Izadpanah (97), Computational Mechanics Center, Washington University, St. Louis, Missouri, U.S.A.

H. Kardestuncer (207), Department of Civil Engineering, University of Connecticut, Storrs, Connecticut, U.S.A.

R.C. MacCamy (149), Department of Civil Engineering, Carnegie-Mellon University, Pittsburgh, Pennsylvania, U.S.A.

D.S. Malkus (235), Mathematics Department, Illinois Institute of Technology, Chicago, Illinois, U.S.A.

A. Needleman (249), School of Engineering, Brown University, Providence, Rhode Island, U.S.A.

T. Nishioka (65), CACM, Georgia Institute of Technology, Altlanta, Georgia, U.S.A.

A.K. Noor (275), NASA Langley Research Center, The George Washington University, Hampton, Virginia, U.S.A.

E.T. Olsen (235), Mathematics Department, Illinois Institute of Technology, Chicago, Illinois, U.S.A.

J.E. Pasciak (167), Brookhaven National Laboratory, Upton, New York, U.S.A.

R. Perucchio (47), Department of Structural Engineering, Cornell University, Ithica, New York, U.S.A.

A. Philpott (321), Department of Mathematics, Massachusetts Institute of Technology, Cambridge, Massachusetts, U.S.A.

G.F. Pinder (303), Civil Engineering Department, Princeton University, Princeton, New Jersey, U.S.A.

R.J. Pryputniewicz (207), Department of Mechanical Engineering, Worcester Polytechnic Institute, Worcester, Massachusetts, U.S.A.

G. Strang (321), Department of Mathematics, Massachusetts Institute of Technology, Cambridge, Massachusetts, U.S.A.

B. Szabo (97), Computational Mechanics Center, Washington University, St. Louis, Missouri, U.S.A.

PREFACE

The 7th UFEM gathering, like its predecessors, advances further toward its goal of accomplishing "a unified method" in computational mechanics. No matter how powerful a methodology might be for a certain class of problems, it often presents shortcomings for others. Since engineering problems today are very complex and contain subregions with completely different physical and geometrical characteristics, certainly no single method is capable of handling the entirety of the problem. Consequently, the identification of various methodologies, each suitable for a particular subregion, and their unification have recently been in the minds of many researchers.

The flow chart in Fig. 1 indicates three stages of such a unification: *unified formulation, unified means,* and *unified methods.*

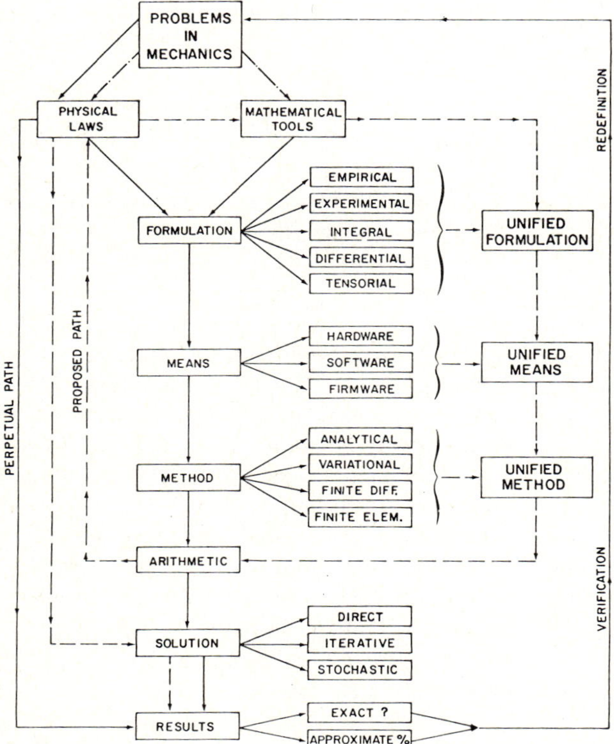

Fig. 1. Flowchart for the Unification of Methods in Mechanics.

The components of the first stage of this unification are illustrated in Fig. 2.

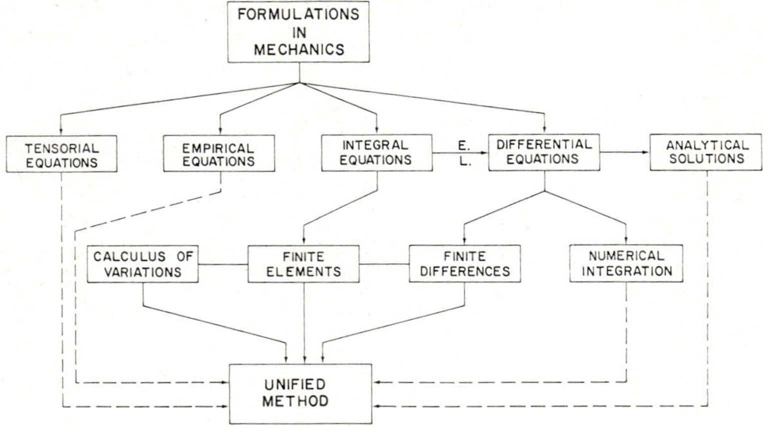

Fig. 2. Unification of Formulations on Mechanics.

Many of the papers presented here address various stages of unification, and we believe that in the near future commercial or in-house codes will be developed to accomplish this task.

The possibility of unifying various numerical methodologies using interactive graphics has been investigated by John ABEL and his co-workers. Their work is fostering the unification concept with a *unified means* which interconnects analysis methods and design parameters. They are not only improving man—machine communication but communication between methodologies employed in different regions of the domain and stages of processing.

ATLURI and NISHIOKA emphasize the unification (hybridization) of various methodologies (numerical, analytical, and experimental) in engineering for the solution of complex problems (e.g. crack propagation in 3-D domain with irregular geometry and material properties) for which none of the existing methodologies *alone* is sufficient. The authors have, in fact, been unifying these methodologies in their earlier work and they advocate the necessity of unification. The problems in this presentation, drawn from the field of fracture mechanics, demonstrate the use of more than one methodology (in time and space) for their solution. Undoubtedly, one can easily apply concepts presented in this paper to other problems. Inter-method compatibilities and error bounds, however, remain to be explored. In the opinion of the editor, the concepts presented here are firm enough ground to stand on when reaching for further goals in unification.

Dealing primarily with problems for which energy functionals exist, BABUŠKA and his co-workers present h-, p-, and h-p versions of the finite element methods.

They claim that error measures in stresses often do not follow monotonic behavior of the error measure in the energy norm. To overcome this difficulty, they introduce an extraction function and demonstrate the selection of such a function during adaptive post-processing. A numerical example accompanying the presentation uses the extraction technique.

Contact problems, in particular between nonlinear deformable bodies subject to large deformation with sticking, sliding and separating, are among the most difficult problems in solid mechanics. BATHE and CHAUDHARY present a solution algorithm that they have developed for two-dimensional contact problems. They believe that alongside finite differences, finite elements, and surface integral techniques, there is still room for more reliable and effective algorithms to analyze general problems in this field. Numerical results for two problems — a pipe buried in soil and a traction of a rubber sheet embedded in a rigid channel — accompany the paper.

BIELAK and MACCAMY unify variational finite element methods with the boundary integral equation method using the former in the interior of the domain and the latter at the exterior. They apply the methodology to a two dimensional electromagnetic interface problem: the interaction between air and a dielectric obstacle subject to two different sets of Maxwell's equations. In this problem, a homogeneous differential equation defined over an infinite domain interfaces with a non-homogeneous differential equation defined over a finite domain.

After reviewing the fundamental principles behind various approximate methods, BREBBIA embarks on the unification of finite elements and boundary elements. While acknowledging the power and potential of the former, he points out certain advantages of the latter and maintains that the complexity of the problems at hand necessitates combining (unifying) many methodologies. He refers to these as "the discrete element methods" and cites some recent attempts coinciding with the philosophy behind UFEM gatherings.

KARDESTUNCER and PRYPUTNIEWICZ explore the possibility of unifying finite element modeling with laser experimentation in two different stages of the procedure. The first part deals with evaluation of the stiffness and/or flexibility matrix coefficients for irregular (geometrically as well as physically) elements by experiment. The second part deals with determining the unknown values of the function by lasers. This, in turn, leads to a reduction in the order of the stiffness matrix and to an increase in the accuracy of the results. Measurement techniques and numerical examples accompany the presentation.

The main theme of the paper by MALKUS and OLSEN centers on the question of whether the NCR (Nagtegaal redundant constraint) element which fails to satisfy the LBB condition can be used for incompressible media. The NRC element is a quadrilateral *macroelement* with four triangles and has been used successfully by others for problems involving inelastic deformation. Here, the authors discuss why NRC elements violate the LBB condition for convergence and how this condition can be removed so that the NRC element can be used for plane and axisymmetric incompressible flows. In order to demonstrate this, they present error estimates for the element when it is used in Stokesian flow.

In his work, NEEDLEMAN applies finite element techniques to necking instabilities subject to classical and nonclassical constitutive relationships. The presentation is accompanied by a numerical analysis of tension tests using constitutive descriptions for polycrystalline metal. He presents remarkably good agreement between the FEM analysis and experimental results and claims that this is the result of incorporating into FEM modeling the constitutive relations for polycrystalline metals arising due to crystallographic texture. He also points out that localized shear stresses play a significant role in texture development.

Two recent advances toward the unification of various methodologies in one physical problem are presented by A.K. NOOR. They are (i) a hybrid method based on the combination of the direct variational techniques with perturbation methods, and (ii) a two-stage direct variational technique. Advantages of both forms of unification are illustrated for nonlinear steady-state thermal and structural problems. The author also points out other combinations of various methodologies and research areas for more effective solution of nonlinear problems. Comparative numerical studies accompany the presentation.

PHILPOTT and STRANG idealize the internal fiber of a human patella as a plane truss and then, using linear programming techniques, they try to optimize the system to accomplish minimum weight. After presenting the standard procedure for a fixed geometry problem, they develop an algorithm for problems with variable geometry, indeed an interesting and difficult task. Most truss problems consist of members with zero loads which in turn introduce degeneracy during the optimization procedure. The authors attempt in particular to deal with this difficulty of optimization.

The alternating-direction collocation (ADC) method is presented by CELIA and PINDER with particular application to multi-dimensional transport equations. These authors enhance the ADC procedure by adding a small number of quintic elements along the principle direction of flow governed by the convection-dominated transport equation. A numerical example confined to a rectangular region and a flow chart for the enhanced ADC procedure accompany the presentation.

All of these invited presentations have been written specifically to honor Professor Argyris with the understanding that they also follow as much as possible the spirit of the conference.

On behalf of the Organizers and Advisory Board members, I would like to thank our distinguished speakers, session chairmen, and participants for their kind cooperation. Some prepared their papers under a severe deadline, some traveled long distances, and some took time off from their demanding tasks to be here today. The result is most gratifying.

The Editor

CONTENTS

CHAPTER 1

ON THE NATURAL APPROACH TO FLOW PROBLEMS

J.H. Argyris & J. St.Doltsinis

The paper surveys recent work on fluid dynamics performed at
the ISD, University of Stuttgart. It is in particular directed to
a natural description of the flow phenomena and includes also
a consideration of thermally coupled problems. The derivation
of the relevant finite element equations when referred to natural quantities is outlined and examples of application are
given. For a discussion on the associated modern developments
in numerical solution techniques the reader may consult [28].

1. INTRODUCTION

The present paper surveys recent work on fluid dynamics performed at the ISD, University
of Stuttgart. The paper serves in the main as a survey on modern developments in finite
element methods for fluid motion, and is particularly devoted to a natural description of
the relevant phenomena. Its main attention is focused on incompressible media. First
draft of the theory has been presented at a lecture given at the Conference on Finite
Elements in Water Resources in Hanover in 1982.

In section 2, the natural terminology [1, 2] is introduced and methodically applied to
the formulation of field quantities characteristic of fluid motion, such as the scalar
pressure field and the vectorial velocity field. The condition of conservation of mass is
derived in natural terms and natural measures for the stress and the rate of deformation
are connected by the appropriate constitutive relations. Aiming at the analysis of fluid
motion coupled with thermal phenomena, the natural approach is subsequently extended
to the consideration of the temperature field and the heat flow [3].

Section 3 indicates the transition to finite domains as a foundation for the development
of the finite element theory of the flow problem. The streamline upwind/Petrov-
Galerkin formulation of [4] may be used for the discretisation technique in connection
with either the strict fulfilment of the incompressible statement or with the penalty approach to the condition of incompressibility. In a subsequent step the finite element
discretisation of the thermally coupled fluid flow problem is considered and the governing equations are established.

For typographical brevity, we omit in this paper a discussion of numerical integration
schemes in the time domain. Also the important task of an effective solution of the

equations governing the flow problem is not handled in the present contribution. For this purpose the reader is referred to the presentation in [28].

The theory presented in the paper is applied in section 4 to the numerical analysis of some typical examples of viscous fluid motion. Thus, the convection dominated flow over a step is considered for the two- and the three-dimensional case, and the solution of thermally coupled flows is demonstrated on the Bénard instability phenomenon in a fluid between two planes of different temperatures. The interested reader may consult [28] for an analysis of cavity flows involving free and forced heat convection with a change from liquid to solid phase of the material.

2. ON THE NATURAL APPROACH TO FLUID MOTION

2.1 Natural approach

In the natural methodology of continuum mechanics, all considerations are established on or derived from an infinitesimal tetrahedron element which replaces the classical parallelopiped applied in the traditional cartesian point of view. For comparison purposes both elements are shown in fig. 2.1 together with the associated coordinate systems. An elegant application of the tetrahedron element demands the use of supernumerary or homogeneous reference systems. One of these may be defined by the directions of the six edges of the tetrahedron. The natural formulation of the mechanics and thermomechanics of solids [5, 3, 2] may be based on the Lagrangean approach in the sense that the tetrahedron constitutes then a moving and deforming material element. In our present considerations of fluid motion, however, we prefer to adopt the Eulerian description [6, 7] in which the tetrahedron represents a fixed geometrical element in space.

Before developing the natural concept we first review alternative representations of a vector in three-dimensional space [3, 2] and illustrate then the argument on the two-dimensional case depicted in fig. 3.2. Consider the vector $\mathbf{r}$ defined by cartesian entries

$$\mathbf{r} = \{ r_x \ r_y \ r_z \} = \{ r^1 \ r^2 \ r^3 \} = \{ r^i \} \qquad i = 1, 2, 3 \qquad (2.1)$$

In the natural terminology the vector $\mathbf{r}$ may, on the one hand, be composed from non-unique independent vectorial contributions taken along the tetrahedral edges

$$\mathbf{r}_c = \{ r_c^\alpha \ r_c^\beta \ r_c^\gamma \ r_c^\delta \ r_c^\varepsilon \ r_c^\zeta \} = \{ r_c^\vartheta \} \qquad \vartheta = \alpha, \dots, \zeta \qquad (2.2)$$

This forms the so-called component description of a vector. On the other hand, we may introduce as measures the unique orthogonal projections of the vector $\mathbf{r}$ onto the natural coordinate axes,

$$\mathbf{r}_t = \{ r_t^\alpha \ r_t^\beta \ r_t^\gamma \ r_t^\delta \ r_t^\varepsilon \ r_t^\zeta \} = \{ r_t^\vartheta \} \qquad \vartheta = \alpha, \dots, \zeta \qquad (2.3)$$

This forms the total description of a vector.

Considering fig. 2.2, we observe that for a given component representation r_c of the vector its cartesian form r is deduced by the transformation

$$r = B_{ND}^{t} \, r_c \qquad (2.4)$$

with the matrix

$$B_{ND} = \left[\cos(\vartheta, i)\right] \qquad \vartheta = \alpha, \dots, \zeta \, ; \ i = 1, 2, 3 \qquad (2.5)$$

The total natural entries r_t of (2.3) are then obtained through the relation

$$r_t = B_{ND} \, r = B_{ND} B_{ND}^{t} \, r_c = B_{tc} \, r_c \qquad (2.6)$$

where (2.4) has been used. The symmetric matrix

$$B_{tc} = B_{ND} B_{ND}^{t} = \left[\cos(\vartheta, \psi)\right] \qquad \vartheta = \alpha, \dots, \zeta \, ; \ \psi = \alpha, \dots, \zeta \qquad (2.7)$$

establishes the direct connection between total and component definitions. It is evident that due to the redundancy of the natural quantities, the above transformations are not invertible.

Finally, we note that the scalar product of two arbitrary vectors a and b may be given in one of the equivalent forms

$$a^{t}b = a_c^{t} b_t = a_t^{t} b_c \qquad (2.8)$$

as is easily confirmed with the aid of (2.4) and (2.6)

2.2 Pressure field

We proceed next to the description of a scalar field e.g. the pressure p in the fluid which for arbitrary non-steady conditions is a function of the time t and the position vector x . We express this by

$$p = p(t, x) \qquad (2.9)$$

and examine the consequences of the different representations of the vector x on the description of the scalar field.

Positions may be defined by component coordinates x_c , cf. (2.2). From the associated dependence in (2.9) the pressure gradient then reads

$$g_t(p) = \left\{\frac{\partial p}{\partial x_c^{\vartheta}}\right\} = \frac{\partial p}{\partial x_c^{t}} = \left[\frac{\partial p}{\partial x_c}\right]^{t} = \left[\frac{\partial p}{\partial x} \frac{dx}{dx_c}\right]^{t} = B_{ND} \, g(p) \qquad (2.10)$$

where the chain rule confirms the transformation (2.6) between total natural and cartesian specification of a gradient vector and justifies the total notation of (2.10).

4   *J.H. Argyris & J. St. Doltsinis*

Actually, $g_t(p)$ comprises the rates of change of p in the natural directions (and hence the corresponding orthogonal projections of the gradient vector).

Consider next the transformation rule (2.4) leading from the component to the cartesian definition of the gradient vector,

$$g(p) = \left\{\frac{\partial p}{\partial x^i}\right\} = \frac{\partial p}{\partial x^t} = \left[\frac{\partial p}{\partial x}\right]^t = \left[\frac{\partial p}{\partial x_t}\frac{dx_t}{dx}\right]^t = B_{Nb}^t \, g_c(p) \qquad (2.11)$$

Here the component vector $g_c(p)$ is clearly

$$g_c(p) = \left\{\frac{\partial p}{\partial x_t^{\,\partial}}\right\} = \frac{\partial p}{\partial x_t^t} = \left[\frac{\partial p}{\partial x_t}\right]^t \qquad (2.12)$$

In fact, $g_c(p)$ merely comprises the formal derivatives of p with respect to a non-unique dependence on x_t and represents component contributions to the gradient vector. Applying next the chain rule to the gradient of (2.10) we obtain the relation

$$g_t(p) = \left[\frac{\partial p}{\partial x_c}\right]^t = \left[\frac{\partial p}{\partial x_t}\frac{dx_t}{dx_c}\right]^t = B_{tc} \, g_c(p) \qquad (2.13)$$

which agrees with the transformation of (2.6) between component natural and total natural quantities. In conclusion we list the invariance of the expression

$$g_t^t(p)\,dx_c = g^t(p)\,dx = g_c^t(p)\,dx_t \qquad (2.14)$$

which furnishes the increment of the scalar p associated with a change of spatial location, and may be verified by the chain rule, or via an appropriate interpretation of (2.8).

2.3 Velocity field

The extension of the above terminology to the description of a vector field is straightforward. Consider for instance the velocity field,

$$v = v(t, x) \qquad (2.15)$$

which is in general unsteady. The acceleration of a certain particle may be obtained by the so-called material differentiation of the velocity vector with respect to time as

$$\dot{v} = \frac{dv}{dt} = \frac{\partial v}{\partial t} + \frac{\partial v}{\partial x}\,v \qquad (2.16)$$

The first term of the expression in (2.16) represents the local derivative of the velocity with respect to time and is to be evaluated at a fixed location. The second term represents the contribution of convection and is dependent on the gradient of the velocity field.

Disregarding for the time being a particular representation of the velocity vector $\mathbf{v}$, its gradient may be measured with respect to one of the different specifications of the location vector $\mathbf{x}$. Taking component natural coordinates $\mathbf{x}_c$ we obtain in analogy to (2.10) the total natural gradient

$$\mathcal{G}_t(\mathbf{v}) = \left\{ \frac{\partial \mathbf{v}}{\partial x_c^\vartheta} \right\} = \frac{\partial \mathbf{v}}{\partial x_c^t} = \left[\frac{\partial \mathbf{v}}{\partial \mathbf{x}_c} \right]^t = \left[\frac{\partial \mathbf{v}}{\partial \mathbf{x}} \frac{d\mathbf{x}}{d\mathbf{x}_c} \right]^t = B_{ND}\, \mathcal{G}(\mathbf{v}) \tag{2.17}$$

In (2.17) the cartesian gradient

$$\mathcal{G}(\mathbf{v}) = \left\{ \frac{\partial \mathbf{v}}{\partial x^i} \right\} = \frac{\partial \mathbf{v}}{\partial \mathbf{x}^t} = \left[\frac{\partial \mathbf{v}}{\partial \mathbf{x}} \right]^t = \left[\frac{\partial \mathbf{v}}{\partial \mathbf{x}_t} \frac{d\mathbf{x}_t}{d\mathbf{x}} \right]^t = B_{ND}^t\, \mathcal{G}_c(\mathbf{v}) \tag{2.18}$$

may be related in analogy to (2.11) to the component gradient of $\mathbf{v}$,

$$\mathcal{G}_c(\mathbf{v}) = \left\{ \frac{\partial \mathbf{v}}{\partial x_t^\vartheta} \right\} = \frac{\partial \mathbf{v}}{\partial \mathbf{x}_t^t} = \left[\frac{\partial \mathbf{v}}{\partial \mathbf{x}_t} \right]^t \tag{2.19}$$

which represents an extension of (2.12) and is derived from a functional dependence of $\mathbf{v}$ on $\mathbf{x}_t$. Applying once more the chain rule to (2.17) we deduce

$$\mathcal{G}_t(\mathbf{v}) = \left[\frac{\partial \mathbf{v}}{\partial \mathbf{x}_c} \right]^t = \left[\frac{\partial \mathbf{v}}{\partial \mathbf{x}_t} \frac{d\mathbf{x}_t}{d\mathbf{x}_c} \right]^t = B_{tc}\, \mathcal{G}_c(\mathbf{v}) \tag{2.20}$$

which relates directly the total to the component natural gradient and represents an extension of (2.13) to a vector field $\mathbf{v}$. We also note the invariance of the expression

$$\mathcal{G}_c^t(\mathbf{v})\, \mathbf{v}_c = \mathcal{G}^t(\mathbf{v})\, \mathbf{v} = \mathcal{G}_t^t(\mathbf{v})\, \mathbf{v}_c \tag{2.21}$$

which is analogous to (2.14) and represents the convective acceleration term of (2.16). Here $\mathbf{v}$ symbolises one of the three differently defined representations of the velocity vector. The invariance of (2.21) may be confirmed by the chain rule.

We next turn our attention to the particle acceleration of (2.16) and observe that it may be represented by one of the alternative forms adopted for the description of the velocity vector $\mathbf{v}$. Thus, in component natural terms we have

$$\dot{\mathbf{v}}_c = \frac{\partial \mathbf{v}_c}{\partial t} + \mathcal{G}_t^t(\mathbf{v}_c)\, \mathbf{v}_c \tag{2.22}$$

with the total gradient matrix of $\mathbf{v}_c$

$$\mathcal{G}_t(\mathbf{v}_c) = \frac{\partial \mathbf{v}_c}{\partial x_c^t} = \left[\frac{\partial \mathbf{v}_c}{\partial \mathbf{x}_c} \right]^t \tag{2.23}$$

The cartesian form of the acceleration on the other hand may be expressed as

$$\dot{v} = \frac{\partial v}{\partial t} + \mathcal{G}^t(v)\, v = B_{ND}^t\, \dot{v}_c \tag{2.24}$$

Here v comprises the cartesian components of the velocity, cf. (2.1). Here the cartesian gradient matrix

$$\mathcal{G}(v) = \frac{\partial v}{\partial x^t} = \left[\frac{\partial v}{\partial x}\right]^t \tag{2.25}$$

should not be confused with the expression of (2.18), which is not limited to a particular representation of the vector v. The total natural formulation of the acceleration is given by

$$\dot{v}_t = \frac{\partial v_t}{\partial t} + \mathcal{G}_c^t(v_t)\, v_t = B_{ND}\, \dot{v} = B_{tc}\, \dot{v}_c \tag{2.26}$$

with the associated component gradient matrix of v_t

$$\mathcal{G}_c(v_t) = \frac{\partial v_t}{\partial x_t^t} = \left[\frac{\partial v_t}{\partial x_t}\right]^t \tag{2.27}$$

We observe that relations (2.4) and (2.6) between the alternative vector specifications also apply to the acceleration, as confirmed by (2.24) and (2.26).

2.4 Continuity condition

We proceed to the natural formulation of the continuity condition. To this end, consider in fig. 2.3 the infinitesimal tetrahedron element defined by the lengths of the six edges

$$l = \lceil l^\alpha \; l^\beta \; l^\gamma \; l^\delta \; l^\varepsilon \; l^\zeta \rceil = \lceil l^\vartheta \rceil \qquad \vartheta = \alpha, \ldots, \zeta \tag{2.28}$$

with a volume V. When determining the flow of mass through the element as induced by component natural velocities we stipulate the column matrix

$$w_c = \{w_c^\vartheta\} = \left\{\frac{\partial v_c^\vartheta}{\partial x_c^\vartheta}\right\} \qquad \vartheta = \alpha, \ldots, \zeta \tag{2.29}$$

containing the rate of change of all component velocities along the edges ϑ.

Consider next a component natural velocity characterised by the intensity v_c^α and assume for the time being an incompressible fluid with density ρ. Under these conditions mass permeates at a rate $\rho v_c^\alpha A^\alpha$ through the centre of the face A^α not containing l^α into the element and is discharged at a rate $\rho(v+dv)_c^\alpha A^\alpha$ through the opposite face at a distance $l^\alpha/3$ from the point of input. The balance of input-output of fluid mass due to the component natural velocity v_c^α is seen to be simply

$$-\rho \, dv_c^\alpha A^\alpha = -\rho \, w_c^\alpha \frac{l^\alpha}{3} A^\alpha = -\rho \, w_c^\alpha V \tag{2.30}$$

where

$$w_c^\alpha = \frac{\partial v_c^\alpha}{\partial x_c^\alpha} \tag{2.31}$$

denotes the rate of change of v_c^α along α. Generalising to all natural directions ϑ and applying the column matrix w_c of (2.29) we obtain the rate of mass supply by summation of the individual contributions defined by (2.30) as arising for all component natural velocities of v_c . Hence the condition of conservation of mass for an incompressible fluid is given by

$$e_6^t w_c = e_3^t w = div\, v = o \tag{2.32}$$

where

$$w = \{\,w^i\,\} = \left\{\frac{\partial v^i}{\partial x^i}\right\} \qquad\qquad i = 1, 2, 3 \tag{2.33}$$

is the cartesian counterpart of w_c. Note also the summation vectors

$$e_6 = \{\,1\ 1\ 1\ 1\ 1\ 1\,\} \quad, \quad e_3 = \{\,1\ 1\ 1\,\} \tag{2.34}$$

In the case of a compressible fluid, expression (2.30) for a typical rate of the component mass supply must account for a change of the density ρ along α . Consequently, the condition of conservation of mass (2.31) is modified into

$$\rho\, e_6^t w_c + g_t^t(\rho) v_c = \rho\, e_3^t w + g^t(\rho) v = div\,(\rho v) = o \tag{2.35}$$

Here the density gradients

$$g_t(\rho) = \frac{\partial \rho}{\partial x_c^t} = \left[\frac{\partial \rho}{\partial x_c}\right]^t = B_{ND}\, g(\rho) \tag{2.36}$$

and

$$g(\rho) = \frac{\partial \rho}{\partial x^t} = \left[\frac{\partial \rho}{\partial x}\right]^t \tag{2.37}$$

correspond to the definitions of (2.10) and (2.11).

2.5 Rate of deformation and stress

A specification of the behaviour of fluid flow demands the introduction of suitable stress
and deformation measures. In classical continuum mechanics (see e.g. [8]), the rate
of deformation is defined by the symmetric part of the cartesian gradient of the instantaneous velocity field. With reference to (2.18) we may thus write

$$\frac{\partial v}{\partial x} = \left[\frac{\partial v}{\partial x}\right]_s + \left[\frac{\partial v}{\partial x}\right]_A = \mathcal{G}^t(v) \tag{2.38}$$

The associated instantaneous material spin is then

$$W = \frac{1}{2}\left\{\frac{\partial v}{\partial x} - \left[\frac{\partial v}{\partial x}\right]^t\right\} = \left[\frac{\partial v}{\partial x}\right]_A = -\mathcal{G}_A(v) \tag{2.39}$$

and is defined by the antisymmetric part. The rate of deformation of the material

$$D = \begin{bmatrix} D^{11} & D^{12} & D^{13} \\ D^{21} & D^{22} & D^{23} \\ D^{31} & D^{32} & D^{33} \end{bmatrix} = \frac{1}{2}\left\{\frac{\partial v}{\partial x} + \left[\frac{\partial v}{\partial x}\right]^t\right\} = \left[\frac{\partial v}{\partial x}\right]_s = \mathcal{G}_s(v) \tag{2.40}$$

is correspondingly specified by the symmetric part of the cartesian velocity gradient
matrix. In what follows we refer to the column matrix

$$\delta = \left\{ D^{11} \quad D^{22} \quad D^{33} \quad \sqrt{2}\,D^{12} \quad \sqrt{2}\,D^{23} \quad \sqrt{2}\,D^{13} \right\} \tag{2.41}$$

as the cartesian rate of deformation.

Natural measures of the rate of deformation were originally defined by reference to the
deformation of the fluid material instantaneously occupying the tetrahedron element
[9, 7]. They may be expressed in terms of the natural definitions of the velocity gradient [10]. Thus, the total natural rate of deformation is given by

$$\delta_t = \left\{ \frac{\partial v_t^\alpha}{\partial x_c^\alpha} \quad \frac{\partial v_t^\beta}{\partial x_c^\beta} \quad \frac{\partial v_t^\gamma}{\partial x_c^\gamma} \quad \frac{\partial v_t^\delta}{\partial x_c^\delta} \quad \frac{\partial v_t^\varepsilon}{\partial x_c^\varepsilon} \quad \frac{\partial v_t^\zeta}{\partial x_c^\zeta} \right\} \tag{2.42}$$

The column matrix δ_t comprises the rates of extension of the material along the six
natural directions [5]. It may be related to the cartesian definition of the rate of deformation via

$$\delta_t = C^t \delta \tag{2.43}$$

where the detailed structure of the transformation matrix C may be found in [9].
The component natural rate of deformation is defined in analogy to the total one in
(2.42) as

$$\delta_c = \left\{ \frac{\partial v_c^\alpha}{\partial x_t^\alpha} \quad \frac{\partial v_c^\beta}{\partial x_t^\beta} \quad \frac{\partial v_c^\gamma}{\partial x_t^\gamma} \quad \frac{\partial v_c^\delta}{\partial x_t^\delta} \quad \frac{\partial v_c^\varepsilon}{\partial x_t^\varepsilon} \quad \frac{\partial v_c^\zeta}{\partial x_t^\zeta} \right\} \tag{2.44}$$

The column matrix δ_c may be related uniquely to the total natural rate of deformation
by

$$\delta_t = \mathcal{A} \, \delta_c \tag{2.45}$$

where the transformation matrix

$$\mathcal{A} = C^t C = [\cos^2(\vartheta, \varphi)] \qquad \vartheta = \alpha, \dots, \zeta \; ; \; \varphi = \alpha, \dots, \zeta \tag{2.46}$$

is also given in [9] and presumes that component velocities vary only along the direction of their action [10]. In this case, δ_c of (2.44) and w_c of (2.29) are identical.

For stresses we must adopt a corresponding definition to the rate of deformation so that
their scalar product satisfies the condition of invariance for the virtual rate of work.
Thus the column matrix

$$\sigma = \left\{ \sigma^{11} \quad \sigma^{22} \quad \sigma^{33} \quad \sqrt{2}\,\sigma^{12} \quad \sqrt{2}\,\sigma^{23} \quad \sqrt{2}\,\sigma^{13} \right\} \tag{2.47}$$

comprises the Cauchy stresses in their cartesian form and corresponds to the rate of deformation δ of (2.41). The natural component stresses

$$\sigma_c = \left\{ \sigma_c^\alpha \quad \sigma_c^\beta \quad \sigma_c^\gamma \quad \sigma_c^\delta \quad \sigma_c^\varepsilon \quad \sigma_c^\zeta \right\} \tag{2.48}$$

correspond following [9, 5] to the total natural rate of deformation δ_t of (2.42),
while the total natural stresses

$$\sigma_t = \left\{ \sigma_t^\alpha \quad \sigma_t^\beta \quad \sigma_t^\gamma \quad \sigma_t^\delta \quad \sigma_t^\varepsilon \quad \sigma_t^\zeta \right\} \tag{2.49}$$

correspond to the component natural rate of deformation δ_c of (2.44) (c.f. fig. 2.4).
The invariance of the virtual rate of work may now be expressed as

$$\delta_t^t \sigma_c = \delta^t \sigma = \delta_c^t \sigma_t \tag{2.50}$$

Bearing in mind (2.43) and (2.45) we easily confirm the relation

$$\sigma_c = C^{-1} \sigma = \mathcal{A}^{-1} \sigma_t \tag{2.51}$$

connecting the different representations of the stress state.

2.6 Constitutive relations for incompressible viscous fluids

In formulating the stress-strain relations appertaining to the fluid motion, it is convenient to split the stress state into hydrostatic and deviatoric contributions. We may ignore here an account of the standard cartesian approach (see e.g. [11]) and apply instead the natural approach to this subject as developed in [7, 9]. Considering first total stresses we write

$$\sigma_t = \sigma_{tH} + \sigma_{tD} \tag{2.52}$$

and obtain the hydrostatic part of the total stress in the form

$$\sigma_{tH} = \frac{1}{3} E_c \, \sigma_c \tag{2.53}$$

where the matrix

$$E_c = e_c \, e_c^t \tag{2.54}$$

performs the summation of the component stresses in each row and yields the total hydrostatic stress in each of the natural directions. The deviatoric part of the total stress follows then from (2.52) as

$$\sigma_{tD} = \left[\mathbf{1} - \frac{1}{3} E_c \right] \sigma_c \tag{2.55}$$

in which relation (2.51) between total and component definitions is used. Partitioning next the component stress as

$$\sigma_c = \sigma_{cH} + \sigma_{cD} \tag{2.56}$$

we may derive the hydrostatic and the deviatoric part by application of (2.51) to the total quantities of (2.53) and (2.55), respectively. A decomposition of the total natural rate of deformation

$$\delta_t = \delta_{tV} + \delta_{tD} \tag{2.57}$$

into volumetric parts

$$\delta_{tV} = \frac{1}{3} E_c \, \delta_c \tag{2.58}$$

and deviatoric parts

$$\delta_{tD} = \left[A - \tfrac{1}{3} E_6 \right] \delta_c \tag{2.59}$$

proceeds along the same argument. Also the component natural rate of deformation

$$\delta_c = \delta_{cV} + \delta_{cD} \tag{2.60}$$

may be partitioned analogously.

Consider next an incompressible fluid, i.e. a fluid undergoing only isochoric deformations. In this case the volumetric rate of deformation must vanish. This yields,

$$\delta_V = \tfrac{1}{3} e_6^t \delta_c = \tfrac{1}{3} e_6^t A^{-1} \delta_t = 0 \tag{2.61}$$

which is equivalent to (2.32). In the absence of viscous effects the incompressible fluid is described as an ideal one for which deviatoric stresses are absent. Then the stress field derives simply from a static pressure p . Consequently, the total stresses reduce to

$$\sigma_t \equiv \sigma_{tH} = -p\, e_6 \tag{2.62}$$

and by (2.51) the component stresses become

$$\sigma_c \equiv \sigma_{cH} = -p\, A^{-1} e_6 \tag{2.63}$$

In a viscous incompressible fluid on the other hand, a rate of deformation – which is exclusively deviatoric because of (2.61) – leads to deviatoric total stresses of the form

$$\sigma_{tD} = 2\mu\, \delta_t = 2\mu\, A\, \delta_c \tag{2.64}$$

or to deviatoric component stresses,

$$\sigma_{cD} = A^{-1} \sigma_{tD} = 2\mu\, \delta_c = 2\mu\, A^{-1} \delta_t \tag{2.65}$$

where μ denotes the viscosity coefficient of the fluid.

For the viscous case the stress is ultimately obtained by a superposition of a hydrostatic contribution arising from (2.62), (2.63) and the deviatoric contribution of (2.64), (2.65). Thus, the total natural stress reads

$$\sigma_t = 2\mu\, A\, \delta_c - p\, e_6 \tag{2.66}$$

and the component one

$$\sigma_c = \mathbf{A}^{-1}\left[2\mu\,\delta_t - p\,e_6\right] \tag{2.67}$$

We observe that the constitutive relations in (2.66) and (2.67) are expressed in terms of corresponding stress and rate of deformation measures.

If standard computational procedures are to be applied to the analysis of the isochoric motion of an incompressible fluid, one may use the so-called penalty approach. The isochoric condition (2.61) can then be relaxed and the pressure p is related to the volumetric rate of deformation as follows

$$p = -3\bar{\kappa}\,\delta_V \tag{2.68}$$

where

$$\bar{\kappa} = \frac{2\mu}{3}\,\frac{1+\bar{\nu}}{1-2\bar{\nu}} \;\rightarrow\; \infty \tag{2.69}$$

represents the penalty parameter. In (2.68), $\bar{\kappa}$ may be interpreted as a modulus of viscous compressibility and is expressed in (2.69) in an analogous manner to the well-known elastic bulk modulus. The strictly incompressible constitutive relations (2.66), (2.67) may now be modified accordingly. For instance, (2.67) assumes in the penalty approach the form

$$\sigma_c = 2\mu\left[\mathbf{A}^{-1} + \frac{\bar{\nu}}{1-2\bar{\nu}}\mathbf{A}^{-1}E_6\mathbf{A}^{-1}\right]\delta_t \tag{2.70}$$

in which

$$\bar{\nu} \;\rightarrow\; \frac{1}{2} \tag{2.71}$$

may be used as an alternative penalty parameter.

2.7 Fluid motion coupled with thermal phenomena

In this subsection we consider fluid motion coupled to thermal phenomena. To this purpose we assume the following unsteady temperature field

$$T = T(t,\mathbf{x}) \tag{2.72}$$

where $\mathbf{x}$ denotes the position vector. In extension of the argument in subsection 2.2 the time rate of the temperature of a particle may be expressed in the alternative forms

$$\dot{T} = \frac{\partial T}{\partial t} + \underset{t}{g}^t(T)\, v_c$$

$$= \frac{\partial T}{\partial t} + g^t(T)\, v$$

$$= \frac{\partial T}{\partial t} + \underset{c}{g}^t(T)\, v_t \tag{2.73}$$

The different formulations of the temperature gradient in (2.73) may be compared to the definitions in (2.10), (2.11) and (2.12), respectively. In the present case the invariance condition of (2.14) becomes

$$\underset{t}{g}^t(T)\, v_c = g^t(T)\, v = \underset{c}{g}^t(T)\, v_t \tag{2.74}$$

The time rate of the temperature is associated with a rate of heat stored in the fluid material. The latter may be expressed per unit material volume as

$$\dot{q}_m = \rho\, c\, \dot{T} = \dot{q}_s + \dot{q}_k \tag{2.75}$$

where c denotes the specific heat capacity of the fluid. In accordance with (2.73) the rate of heat stored in the material may be composed in the Eulerian approach of two parts. Thus, the rate of heat stored in a unit volume when fixed in space reads

$$\dot{q}_s = \rho\, c\, \frac{\partial T}{\partial t} \tag{2.76}$$

and is associated with the temperature rate obtained at a fixed location. The contribution

$$\dot{q}_k = \rho\, c\, \underset{t}{g}^t(T)\, v_c = \rho\, c\, g^t(T)\, v = \rho\, c\, \underset{c}{g}^t(T)\, v_t \tag{2.77}$$

is the heat convection term due to the motion of the fluid and may be presented in one of the alternative formulations, natural or cartesian, as shown in (2.77).

We proceed next to the specification of the heat supply to a unit volume of space due to a directed heat flow, i.e. conduction. Following subsection 2.1, the heat flow vector $\dot{q}$ with cartesian entries

$$\dot{q} = \{\, \dot{q}_x \quad \dot{q}_y \quad \dot{q}_z \,\} = \{\, \dot{q}^1 \quad \dot{q}^2 \quad \dot{q}^3 \,\} = \{\, \dot{q}^i \,\} \qquad i = 1, 2, 3 \tag{2.78}$$

may alternatively be represented by the component natural contributions

$$\dot{q}_c = \{\, \dot{q}_c^\alpha \quad \dot{q}_c^\beta \quad \dot{q}_c^\gamma \quad \dot{q}_c^\delta \quad \dot{q}_c^\varepsilon \quad \dot{q}_c^\zeta \,\} = \{\, \dot{q}_c^\vartheta \,\} \qquad \vartheta = \alpha, \dots, \zeta \tag{2.79}$$

or by the total natural quantities

$$\dot{q}_t = \{\, \dot{q}_t^\alpha \quad \dot{q}_t^\beta \quad \dot{q}_t^\gamma \quad \dot{q}_t^\delta \quad \dot{q}_t^\varepsilon \quad \dot{q}_t^\zeta \,\} = \{\, \dot{q}_t^\vartheta \,\} \qquad \vartheta = \alpha, \dots, \zeta \tag{2.80}$$

The reader is reminded of the interrelations between the alternative representations of the heat flow vector in accordance with (2.4) and (2.6).

When determining the heat flow through an infinitesimal tetrahedron element shown in fig. 2.5, as arising from component natural heat fluxes [3], we have to introduce the column matrix

$$\dot{\tau}_c = \{\, \dot{\tau}_c^\vartheta \,\} = \left\{\, \frac{\partial \dot{q}_c^\vartheta}{\partial x_c^\vartheta} \,\right\} \qquad \vartheta = \alpha, \dots, \zeta \tag{2.81}$$

Consider now in fig. 2.5 a component natural heat flux characterised by the intensity $\dot{q}_c^\alpha$ progressing through the tetrahedron. Noting the input $\dot{q}_c^\alpha A^\alpha$ and the output $(\dot{q}+d\dot{q})_c^\alpha A^\alpha$ of the heat rate emerging at a distance $l^\alpha/3$ from the point of input we deduce for the rate of heat supply to the element as contributed by the component natural heat flux $\dot{q}_c^\alpha$

$$-d\dot{q}_c^\alpha A^\alpha = -\dot{\tau}_c^\alpha \frac{l^\alpha}{3} A^\alpha = -\dot{\tau}_c^\alpha V \tag{2.82}$$

where

$$\dot{\tau}_c^\alpha = \frac{\partial \dot{q}_c^\alpha}{\partial x_c^\alpha} \tag{2.83}$$

denotes the rate of change of $\dot{q}_c^\alpha$ along α. Generalising for all natural directions we apply the column matrix $\dot{\tau}_c$ of (2.81) and obtain the rate of heat supply by summation of all individual contributions as expressed by (2.82). Hence, we find

$$\dot{q}_t = -e_6^t \dot{\tau}_c = -e_3^t \dot{\tau} = -div\,\dot{q} \tag{2.84}$$

where $\dot{c}$ is the cartesian counterpart of the column matrix $\dot{c}_c$ and reads

$$\dot{c} = \{\dot{c}^i\} = \left\{\frac{\partial \dot{q}^i}{\partial x^i}\right\} \qquad i = 1, 2, 3 \qquad (2.85)$$

We conclude this subsection by presenting a natural counterpart to the Fourier's law relating the heat flow to the temperature gradient [3]. Starting with the cartesian form

$$\dot{q} = -\lambda \left[\frac{\partial T}{\partial x}\right]^t = -\lambda \, g\,(T) \qquad (2.86)$$

where λ denotes the thermal conductivity of the fluid, we deduce the natural relation

$$\dot{q}_t = -\lambda \, B_{ND} \, B_{ND}^t \, g_c\,(T) = -\lambda \, B_{tc} \, g_c\,(T) = -\lambda_{tc} \, g_c\,(T) \qquad (2.87)$$

by an appropriate application of (2.6) and (2.4) or (2.11). We note that

$$\lambda_{tc} = \lambda \, B_{tc} \qquad (2.88)$$

symbolises the natural thermal conductivity matrix connecting via (2.87) the total natural heat flow vector $\dot{q}_t$ with the component temperature gradient $g_c\,(T)$. We also observe that the connection between $\dot{q}_t$ and the total temperature gradient $g_t\,(T)$ is simply given by the thermal conductivity λ of the material as in (2.86).

3. DISCRETISATION BY FINITE ELEMENTS

3.1 Weak form of the equations governing fluid and thermal flow

Bearing in mind our prospective application of the finite element technique to the flow problem we write in the following the basic equations in their weak form assuming a finite volume V bounded by the surface S. Thus, a weak form of the momentum balance may be expressed in natural terms as

$$\int_V (\rho \, \bar{v}_t^t \dot{v}_c + \bar{\delta}_t^t \sigma_c)\, dV = \int_V \bar{v}_t^t f_c \, dV + \int_S \bar{v}_t^t n \sigma_c \, dS \qquad (3.1)$$

where $\bar{v}$ symbolises a virtual velocity field and $\bar{\delta}$ the associated rate of deformation. Also, f denotes the body force vector acting per unit volume and n a normal operator yielding the surface tractions. Alternative formulations of (3.1) in natural or in cartesian terms are possible as outlined in section 2. The virtual rate of kinetic energy, for instance, on the left-hand side in (3.1) is given in terms of one of the expressions

$$\bar{v}_t^{\,t}\dot{v}_c = \bar{v}_c^{\,t}\dot{v}_t = \bar{v}^{\,t}\dot{v} \tag{3.2}$$

offered in (2.8) for the scalar product of two vectors. Clearly, the acceleration $\dot{v}$ consists, in the Eulerian approach adopted here of a local part and a convective part and may be specified in the component natural form $\dot{v}_c$ of (2.32), the cartesian form $\dot{v}$ of (2.24), or the total natural form of (2.26). We also observe that the component natural stress σ_c in (3.1) obeys the constitutive laws of subsection 2.6.

For a weak formulation of the isochoric condition we rely on expression (2.61) and write in natural terms

$$\int_V \bar{p}\, e_6^t\, \delta_c\, dV = \int_V \bar{p}\, e_6^t\, \mathcal{A}^{-1}\delta_t\, dV = 0 \tag{3.3}$$

where $\bar{p}$ represents the virtual pressure field.

We next turn our attention to the heat flow as occurring concurrently with the fluid motion. The heat balance of the volume in question may be expressed in natural terms as

$$\int_V \bar{T}\rho c\left[\frac{\partial T}{\partial t} + g_c^t(T)v_t\right]dV + \int_V \bar{T}e_6^t\dot{c}_c\, dV = \int_V \bar{T}\delta_t^t\sigma_c\, dV \tag{3.4}$$

where $\bar{T}$ denotes a virtual temperature field. In (3.4), the first integral on the left-hand side is due to the rate of heat stored in the material, in accordance with (2.75). It is specified through the local term of (2.76) and the convective term of (2.77). The second integral reproduces the rate of heat supply (2.84) by heat conduction. It balances the stored heat expression with due consideration of the rate of dissipation in the material as given by the right-hand side of (3.4). The second integral in (3.4) associated with the heat flux may be transformed as follows (cf. [3])

$$\int_V \bar{T}e_6^t\dot{c}_c\, dV = -\int_V g_t^t(\bar{T})\dot{q}_c\, dV + \int_S \bar{T}\dot{q}_n\, dS \tag{3.5}$$

where due to (2.8) and (2.87)

$$g_t^t(\bar{T})\dot{q}_c = g_c^t(\bar{T})\dot{q}_t = -g_c^t(\bar{T})\lambda_{tc}g_c(T) \tag{3.6}$$

Furthermore, the boundary condition

$$\dot{q}_n = \alpha \left(T - T_\infty \right) \tag{3.7}$$

expresses the local heat exchange between the surface S under the temperature T and the surrounding medium under a temperature T_∞ ; the associated heat transfer coefficient is denoted by α . Thus (3.5) may be brought into the final form

$$\int_V \bar{T} e_6^t \dot{c}_c \, dV = \int_V \lambda \, q_t^t(\bar{T}) \, q_c(T) \, dV + \int_S \bar{T} \alpha \left(T - T_\infty \right) dS \tag{3.8}$$

By substitution of (3.8) in (3.4) one obtains the fundamental expression for the derivation of the relevant finite element relations.

3.2 Natural finite element equations for fluids

To set up a finite element formulation of the flow problem consider first the weak momentum equation (3.1) in conjunction with an approximate representation of the velocity field within each finite element expressed by

$$v_c = \omega_N(x_t) \, V_c(t) \tag{3.9}$$

The column matrix

$$V_c = \left\{ V_{c1} \quad V_{c2} \quad \cdots \quad V_{cn} \right\} = \left\{ V_{cj} \right\} \qquad j = 1, \ldots, n \tag{3.10}$$

comprises the component natural contributions to the velocity vector at any one of the n nodes of an element

$$V_{cj} = \left\{ V_c^\alpha \quad V_c^\beta \quad V_c^\gamma \quad V_c^\delta \quad V_c^\varepsilon \quad V_c^\varsigma \right\}_j = \left\{ V_c^\partial \right\}_j \quad \partial = \alpha, \ldots, \varsigma \tag{3.11}$$

Correspondingly, the matrix

$$\omega_N = \left[\omega_{N1} \quad \omega_{N2} \quad \cdots \quad \omega_{Nn} \right] = \left[\omega_{Nj} \right] \qquad j = 1, \ldots, n \tag{3.12}$$

contains the diagonal matrices,

$$\omega_{Nj} = \lceil \omega_j \rfloor \tag{3.13}$$

of dimensions 6 x 6 which interpolate the velocities V_{cj} . Note also that the ω_j's depend only on the total natural coordinates x_t .

The local part of the acceleration within the element may now be established immediately via (3.9) as

$$\frac{\partial v_c}{\partial t} = \omega_N \frac{d V_c}{dt} = \omega_N \dot{V}_c \tag{3.14}$$

Before entering into the derivation of the convective part of the acceleration we observe that the velocity field (3.9) may alternatively be described by

$$v_c = \underline{V}_c(t)\,\underline{\omega}(x_t) \tag{3.15}$$

where $\underline{V}_c$ is here the super row matrix of the component nodal velocities,

$$\underline{V}_c = \begin{bmatrix} V_{c1} & V_{c2} & \cdots & V_{cn} \end{bmatrix} = \begin{bmatrix} V_{cj} \end{bmatrix} \qquad j = 1,\dots,n \tag{3.16}$$

and $\underline{\omega}$ the column matrix,

$$\underline{\omega} = \{ \omega_1 \quad \omega_2 \cdots \omega_n \} = \{ \omega_j \} \qquad j = 1,\dots,n \tag{3.17}$$

Hence, the velocity gradient may be written as

$$\mathcal{G}_c(v_c) = \left[\frac{\partial v_c}{\partial x_t}\right]^t = \left[\underline{V}_c \frac{d\underline{\omega}}{dx_t}\right]^t = \mathcal{G}_c(\underline{\omega})\,\underline{V}_c^{\,t} \tag{3.18}$$

with

$$\mathcal{G}_c(\underline{\omega}) = \left[\frac{d\underline{\omega}}{dx_t}\right]^t \tag{3.19}$$

The convective term of the acceleration (cf. (2.21)) may now be expressed as,

$$\mathcal{G}_c^{\,t}(v_c)\,v_t = \underline{V}_c \mathcal{G}_c^{\,t}(\underline{\omega})\,v_t \tag{3.20}$$

in which the velocity gradient is represented by (3.18). The total natural velocity appearing in (3.20) obeys the interpolation rule of (3.9) in the form

$$v_t = \omega_N V_t \tag{3.21}$$

with the column matrix (cf. (3.10))

$$V_t = \{V_{t1} \; V_{t2} \; \cdots \; V_{tn}\} = \{V_{tj}\} \qquad j = 1, \ldots, n \qquad (3.22)$$

Here V_t comprises the field of total natural velocities at each of the n element nodes

$$V_{tj} = \{V_t^{\alpha} \; V_t^{\beta} \; V_t^{\gamma} \; V_t^{\delta} \; V_t^{\varepsilon} \; V_t^{\zeta}\}_j = \{V_t^{\vartheta}\}_j, \qquad \vartheta = \alpha, \ldots, \zeta \qquad (3.23)$$

Applying next expression (2.22), we obtain the acceleration by a summation of the local part (3.14) and the convective part (3.20) in the form,

$$\dot{v}_c = \frac{\partial v_c}{\partial t} + \mathop{\mathcal{G}}_c^t(v_c)v_t = \omega_N \dot{V}_c + \underline{V}_c \mathop{\mathcal{G}}_c^t(\omega)\,\omega_N V_t \qquad (3.24)$$

We now proceed to the rate of deformation within the finite element. To this purpose we consider the total natural rate of deformation δ_t of (2.42) and rewrite it in the form

$$\delta_t = \left\lceil \frac{\partial}{\partial x_c^{\vartheta}} \right\rceil v_t = \mathcal{D}_c v_t \qquad \vartheta = \alpha, \ldots, \zeta \qquad (3.25)$$

where the operator $\mathcal{D}_c$ is the (6×6) diagonal matrix

$$\mathcal{D}_c = \left\lceil \frac{\partial}{\partial x_c^{\alpha}} \; \frac{\partial}{\partial x_c^{\beta}} \; \frac{\partial}{\partial x_c^{\gamma}} \; \frac{\partial}{\partial x_c^{\delta}} \; \frac{\partial}{\partial x_c^{\varepsilon}} \; \frac{\partial}{\partial x_c^{\zeta}} \right\rfloor$$

$$= \left\lceil \frac{\partial}{\partial x_c^{\vartheta}} \right\rfloor \qquad \vartheta = \alpha, \ldots, \zeta \qquad (3.26)$$

and,

$$\frac{\partial}{\partial x_c^{\vartheta}} = \frac{\partial}{\partial x_t} \frac{\partial x_t}{\partial x_c^{\vartheta}} = \frac{\partial}{\partial x_t} b_{t\vartheta} \qquad (3.27)$$

Here $b_{t\vartheta}$ symbolises the ϑ-th column of the matrix B_{tc} in (2.6), respectively in (2.7). Application of the interpolation rule (3.21) furnishes the total natural rate of deformation within the element as,

$$\delta_t = \mathcal{D}_c \,\omega_N\, V_t = a_t\, V_t \tag{3.28}$$

Turning our attention to the virtual velocity field $\bar{v}_t$ introduced in (3.1) we set,

$$\bar{v}_t = \bar{\omega}_N\, \overline{V}_t \tag{3.29}$$

The definition of the column matrix $\overline{V}_t$ is in line with that of V_t in (3.22). As to $\bar{\omega}_N$ its formation is that of ω_N of (3.12) but may be based on different interpolation functions $\bar{\omega}_j \neq \omega_j$. The associated virtual rate of deformation reads then in analogy to (3.28)

$$\bar{\delta}_t = \mathcal{D}_c\, \bar{\omega}_N\, \overline{V}_t = \bar{a}_t\, \overline{V}_t \tag{3.30}$$

In finite element theory, forces are assumed to be transmitted exclusively through the element nodes. Let the column matrix

$$P_c = \{\, P_{c_1}\ P_{c_2}\ \cdots\ P_{c_n}\,\} = \{\, P_{c_j}\,\} \qquad j = 1,\ldots, n \tag{3.31}$$

comprise the component natural element contribution to the force vector at each of the n element nodes,

$$P_{c_j} = \{\, P_c^{\alpha}\ P_c^{\beta}\ P_c^{\gamma}\ P_c^{\delta}\ P_c^{\varepsilon}\ P_c^{\zeta}\,\}_j = \{\, P_c^{\vartheta}\,\}_j \qquad \vartheta = 1,\ldots, 5 \tag{3.32}$$

In accordance with the invariance rule (2.8), the component natural representation of the nodal force vector P_{c_j} of (3.32) corresponds to the total natural definition of the nodal velocity vector V_{t_j} of (3.23). Disregarding for simplicity the volume forces on the right-hand side of (3.1) and expressing the surface integral through the nodal quantities the virtual work expression (3.1) assumes for a finite element of volume V the form

$$\int_V (\rho\, \bar{v}_t^{\,t}\, \dot{v}_c + \bar{\delta}_t^{\,t}\, \sigma_c)\, dV = \overline{V}_t^{\,t}\, P_c \tag{3.33}$$

Introducing the kinematic relations (3.29), (3.30) in (3.33) we obtain the component natural forces at the element nodes as

$$P_c = \int_V (\rho \, \bar{\omega}_N^t \, \dot{v}_c + \bar{a}_t^t \, \sigma_c) \, dV \tag{3.34}$$

The acceleration term on the right-hand side of (3.34) may be transformed with the aid of (3.24) into,

$$\int_V \rho \, \bar{\omega}_N^t \, \dot{v}_c \, dV = \left[\int_V \rho \, \bar{\omega}_N^t \, \omega_N \, dV \right] \dot{V}_c + \left[\int_V \rho \, \bar{\omega}_N^t \, \underline{V}_c \, \mathcal{G}_c^t(\underline{\omega}) \, \omega_N \, dV \right] V_t$$

$$= m_N \, \dot{V}_c + n_N \, V_t \tag{3.35}$$

where

$$m_N = \int_V \rho \, \bar{\omega}_N^t \, \omega_N \, dV \tag{3.36}$$

corresponds to the Lagrangean mass matrix while

$$n_N = \int_V \rho \, \bar{\omega}_N^t \, \underline{V}_c \, \mathcal{G}_c^t(\underline{\omega}) \, \omega_N \, dV \tag{3.37}$$

accounts for the nonlinear convective contribution inherent to the present Eulerian approach.

To specify the stress term on the right-hand side of (3.34) we call upon expression (2.67) for the component natural stresses and obtain

$$\int_V \bar{a}_t^t \, \sigma_c \, dV = \int_V 2\mu \, \bar{a}_t^t \, A^{-1} \, \delta_t \, dV - \int_V \bar{a}_t^t \, A^{-1} \, e_6 \, p \, dV \tag{3.38}$$

Using the kinematics as prescribed in (3.28), the first integral on the right-hand side of (3.38) is transformed into,

$$\int_V 2\mu \, \bar{a}_t^t \, A^{-1} \, \delta_t \, dV = \left[\int_V 2\mu \, \bar{a}_t^t \, A^{-1} \, a \, dV \right] V_t = d_N \, V_t \tag{3.39}$$

22   *J.H. Argyris & J. St. Doltsinis*

where

$$d_N = \int_V 2\mu\, \bar{a}_t^t\, A^{-1}\, a\; dV \tag{3.40}$$

represents the viscosity matrix of the element and reflects the deviatoric response of the isochoric fluid. With respect to the second integral on the right-hand side of (3.38) we introduce the approximation

$$p = \pi\, \mathbf{p} \tag{3.41}$$

to describe the pressure field within the element. In (3.41) $\mathbf{p}$ is the column matrix

$$\mathbf{p} = \{\, p_1,\ p_2\ \cdots\ p_m\,\} = \{\, p_k\,\} \qquad k = 1,\ldots, m \tag{3.42}$$

and contains the pivotal values of the pressure and π the interpolation functions within the row matrix

$$\pi = [\, \pi_1,\ \pi_2\ \cdots\ \pi_m\,] = \{\, \pi_k\,\} \qquad k = 1,\ldots, m \tag{3.43}$$

Introducing (3.41) into (3.38) one obtains,

$$\int_V \bar{a}_t^t\, A^{-1}\, e_6\, p\; dV = \left[\int_V \bar{a}_t^t\, A^{-1}\, e_6\, \pi\; dV\right] \mathbf{p} = h_N\, \mathbf{p} \tag{3.44}$$

where

$$h_N = \int_V \bar{a}_t^t\, A^{-1}\, e_6\, \pi\; dV \tag{3.45}$$

is the hydrostatic element matrix. Using (3.44), (3.39) and (3.35), the component natural forces (3.34) of the element may ultimately be expressed as

$$P_c = m_N\, \dot{V}_c + n_N\, V_t + d_N\, V_t - h_N\, \mathbf{p} \tag{3.46}$$

3.3 Transition to cartesian definitions; discretised Navier-Stokes equations

Before proceeding to the assembly of the element contributions (3.46) within the region considered, we transform (3.46) into a global cartesian system of reference. Denoting the respective cartesian element nodal forces by

$$P = \{P_1\ P_2\ \ldots\ P_n\} = \{P_j\} \qquad j = 1, \ldots, n \qquad (3.47)$$

and the corresponding velocities by

$$V = \{V_1\ V_2\ \ldots\ V_n\} = \{V_j\} \qquad j = 1, \ldots, n \qquad (3.48)$$

we may apply relation (2.4) connecting natural and cartesian definitions of vectors to obtain on the element level,

$$P = \lceil B_{ND}^t \rfloor P_c \qquad (3.49)$$

and

$$V = \lceil B_{ND}^t \rfloor V_c \qquad (3.50)$$

We note also that in

$$v = \omega V \qquad (3.51)$$

the interpolation matrix ω corresponds to the definition of ω_N in (3.12) but with entries ω_j (cf. (3.13)) of dimension 3×3, in order to maintain consistency with the cartesian definition. One may now substitute (3.50) in (3.51) to express v in terms of V_c. Relating on the other hand v to v_c via (2.4) and expressing the latter through the interpolation (3.9) we deduce a second expression for v. Thus,

$$v = \omega \lceil B_{ND}^t \rfloor V_c = B_{ND}^t \omega_N V_c \qquad (3.52)$$

and hence

$$\omega \lceil B_{ND}^t \rfloor = B_{ND}^t \omega_N \qquad (3.53)$$

Applying next the transformation to the velocities (2.6) we obtain for the total elemental velocities

$$V_t = \lceil B_{ND} \rfloor \, V \tag{3.54}$$

An analogous argument to that used in (3.52) yields in the present case

$$v_t = \omega_N \lceil B_{ND} \rfloor \, V = B_{ND}\, \omega \, V \tag{3.55}$$

and hence

$$\omega_N \lceil B_{ND} \rfloor = B_{ND}\, \omega \tag{3.56}$$

Substituting in (3.49) P_c , as given in (3.46), and V_t as defined in (3.54) furnishes the cartesian forces

$$P = \lceil B_{ND}^t \rfloor \, P_c$$

$$= \lceil B_{ND}^t \rfloor m_N \dot{V}_c + \lceil B_{ND}^t \rfloor n_N \lceil B_{ND} \rfloor V + \lceil B_{ND}^t \rfloor d_N \lceil B_{ND} \rfloor V$$

$$- \lceil B_{ND}^t \rfloor h_N \, p$$

$$= m\dot{V} + nV + dV - hp \tag{3.57}$$

Using (3.56) and (3.53) as well as (3.36) and (3.50) we may verify that the first term in the second expression of (3.57) reduces to

$$\lceil B_{ND}^t \rfloor m_N \dot{V}_c = \lceil B_{ND}^t \rfloor \, [\, \int_V \rho \, \bar{\omega}_N^t \, \omega_N \, dV\,] \, \dot{V}_c$$

$$= [\, \int_V \rho \, \bar{\omega}^t \omega \, dV\,] \lceil B_{ND}^t \rfloor \, \dot{V}_c$$

$$= [\, \int_V \rho \, \bar{\omega}^t \omega \, dV\,] \, \dot{V} = m \dot{V} \tag{3.58}$$

Note the expression for the elemental mass matrix

$$m = \int_V \rho \, \bar{\omega}^t \omega \, dV \tag{3.59}$$

Consider next the second term on the right-hand side of (3.58). Application of (3.37) for the natural convectivity matrix yields the cartesian counterpart

$$n = \lceil B_{ND}^t \rfloor \, n_N \, \lceil B_{ND} \rfloor$$

$$= \lceil B_{ND}^t \rfloor \, [\int_V \rho \, \bar{\omega}_N^t \, \underline{V}_c \, \mathcal{G}_c^t(\underline{\omega}) \, \omega_N \, dV] \, \lceil B_{ND} \rfloor$$

$$= \int_V \rho \, \bar{\omega}^t \, \underline{V} \, \mathcal{G}^t(\underline{\omega}) \, \omega \, dV \tag{3.60}$$

in which use is made of the relation (2.18) connecting $\mathcal{G}$ and $\mathcal{G}_c$. Also, $\underline{V}$ denotes the super row matrix of the cartesian nodal velocities.

Finally, the cartesian viscosity and hydrostatic elemental matrix

$$d = \lceil B_{ND}^t \rfloor \, d_N \, \lceil B_{ND} \rfloor \tag{3.61}$$

and

$$h = \lceil B_{ND}^t \rfloor \, h_N \tag{3.62}$$

represent standard transformations and do not require further elaboration.

We observe in the above finite element idealisation that identity of $\bar{\omega}_j$, the weighting functions, with ω_j , the interpolation functions, reduces the discretisation procedure to that of Galerkin. In most structural applications, this method leads to symmetric matrices and the associated solutions are known to possess the property of best approximation. In convection dominated flow problems, however, we adopt a suggestion of [4] and prefer to apply the streamline upwind/Petrov-Galerkin technique. In this case $\bar{\omega}_j$ and ω_j are taken to be different. Bearing in mind the aforementioned publication in which a detailed description of the method is given we restrict our present account to an elaboration of the alternative natural formulation. Following [4], the weighting functions $\bar{\omega}_j$ are formed as

$$\bar{\omega}_j = \omega_j + s_j \tag{3.63}$$

where ω_j is the standard interpolation function at the j-th element node and s_j a perturbation defined by

$$s_j = k \, \frac{v^t g(\omega_j)}{v^t v} = k \, \frac{v_t^t g_c(\omega_j)}{v_t^t v_c} \tag{3.64}$$

which induces an upwinding in the streamline direction. The scalar coefficient k is specified in [4] as a function of the velocity and the element dimensions. The natural expression for S_j in (3.64) may be seen to simply rely on the invariance of alternative expressions of scalar products as shown in (2.8). In (3.64) ω_j is assumed to be a function of the total natural coordinates x_t. The associated gradient g_c follows then the definition of (2.12) with ω_j in place of the pressure. In conclusion we note that the upwind technique introduces an additional dependence on the velocity into the finite element characteristics. As outlined in [4] under certain conditions the up-wind scheme affects merely the weighting of the acceleration term in (3.34) but not that of the stress term. In this case the element viscosity matrix is symmetric.

Turning next our attention to the entire flow domain, the element contributions to the nodal forces as given by (3.57) may be summed up and yield the global relation

$$R = M\dot{V} + NV + DV - Hp \tag{3.65}$$

which represents the discretised form of the Navier-Stokes equations. In (3.65) R denotes the column matrix of the nodal forces applied to the flow domain, V and $\dot{V}$ are the corresponding velocities and accelerations, and the column matrix P defines the pressure field in the entire flow domain. The matrices M, N, D and H may be deduced by a straightforward assembly procedure from the matrices m, n, d and h of the individual elements.

3.4 Isochoric condition. Exact analysis and approximate penalty formulation

We now proceed to the discretisation of the isochoric condition using the natural methodology and consider to this end the last expression in (3.3). Introducing a relation analogous to that of (3.41) for the variation of $\bar{p}$ and expressing δ_t as in (3.28) we deduce for a finite element the condition

$$\int_V \bar{p}\, e_6^t A^{-1} \delta_t\, dV = \bar{p}^t [\int_V \bar{n}^t e_6^t A^{-1} a_t\, dV]\, V_t = \bar{p}^t g_N^t V_t = 0 \tag{3.66}$$

where the matrix

$$g_N = [\int_V \bar{n}^t e_6^t A^{-1} a_t\, dV]^t = \int_V a_t^t A^{-1} e_6 \bar{n}\, dV \tag{3.67}$$

coincides with the matrix h_N in (3.45) for the case when $\bar{\omega}_j = \omega_j$ and $\bar{\pi}_j = \pi_j$. To obtain the cartesian form of (3.66) we refer to (3.54) and deduce

$$g_N^t V_t = g_N^t \lceil B_{ND} \rceil V = g^t V = 0 \tag{3.68}$$

Hence the cartesian counterpart of the natural matrix g is

$$g = \lceil B_{ND}^{t} \rceil g_{N} \tag{3.69}$$

The isochoric condition for the entire flow domain may now be symbolised by (cf. (3.68))

$$G^{t} V = 0 \tag{3.70}$$

where the column matrix V comprises the velocities at the nodal points of the finite element mesh, and G is composed by the individual element matrices g in (3.69).

In the penalty approach the isochoric condition is relaxed in accordance with (2.68). As a consequence the weak formulation in (3.3) is correspondingly affected. Adopting the finite element approximation in (3.66) one obtains in the penalty approach

$$g_{N}^{t} V_{t} = - \bar{X}^{-1} p \tag{3.71}$$

The matrix $\bar{X}^{-1}$ may be seen to represent the integral expression,

$$\bar{X}^{-1} = \int_{V} \frac{1}{\bar{K}} \, \bar{\pi}^{t} \pi \, dV \tag{3.72}$$

Solution of (3.71) for the pressure yields

$$p = - \bar{K} g_{N}^{t} V_{t} = - \bar{K} g_{N}^{t} \lceil B_{ND} \rceil V = - \bar{K} g^{t} V \tag{3.73}$$

where use is made of (3.54), (3.69) when forming the alternative cartesian expression on the right-hand side of (3.73). Substitution of (3.73) in (3.57) determines a pure velocity formulation. Isolating the two last terms in the final expression in (3.57) we consequently have

$$dV - hp = dV + h\bar{K} g^{t} V = [d + h\bar{K} g^{t}] V = \bar{d} V \tag{3.74}$$

The matrix

$$\bar{d} = d + h\bar{K} g^{t} \tag{3.75}$$

represents the elemental viscosity in the penalty approach and is a symmetric matrix in an ordinary Galerkin approximation. The above procedure corresponds to the mixed finite element technique of [15] in which velocity and pressure field are approximated independently. An alternative penalty formulation of the viscous incompressible problem may be obtained by substitution of (2.68) in (3.38). This leads to a pure velocity formulation in (3.46) or (3.57) without the need of a separate approximation for the pressure. On the other hand, this advantage involves necessarily a reduced integration scheme for the volumetric part of the associated viscosity matrix $\boldsymbol{d}$ [13, 14]. Summarising, the discretised Navier-Stokes equations for the entire flow domain may be written in the penalty approach as

$$R = M\dot{V} + NV + \bar{D}V \tag{3.76}$$

where the relaxed isochoric constraint in the viscosity matrix is included in $\bar{D}$ in accordance with one or the other approximation technique.

3.5 Finite element equations for heat flow

As a final item we consider the finite element approximation of the heat balance in the fluid as governed by (3.4) and (3.8). To this end we write the temperature field within the element as

$$T = \boldsymbol{\tau}(\boldsymbol{x}_t)\,T(t) \tag{3.77}$$

where the column matrix

$$T = \{\, T_1 \;\; T_2 \;\cdots\; T_n \,\} = \{\, T_j \,\} \qquad j = 1, \ldots, n \tag{3.78}$$

comprises the temperatures at the element nodes and the row matrix

$$\boldsymbol{\tau} = [\, \tau_1 \;\; \tau_2 \;\cdots\; \tau_n \,] = [\, \tau_j \,] \qquad j = 1, \ldots, n \tag{3.79}$$

the interpolation functions. Analogously, we express the virtual temperature field as

$$\bar{T} = \bar{\boldsymbol{\tau}}(\boldsymbol{x}_t)\,\bar{T}(t) \tag{3.80}$$

where the weighting functions $\bar{\tau}_j$ in $\bar{\boldsymbol{\tau}}$ may be constructed in accordance with the streamline upwind/Petrov-Galerkin concept, as detailed in (3.63) for $\bar{\omega}_j$. Applying (3.77) we may obtain the local part of the temperature rate as

$$\frac{\partial T}{\partial t} = \boldsymbol{\tau}\,\dot{T}$$

(3.81)

Correspondingly the convective part becomes

$$\boldsymbol{g}_c^t(T)\,\boldsymbol{v}_t = \boldsymbol{v}_t^t\,\boldsymbol{g}_c(T) = \boldsymbol{V}_t^t\,\boldsymbol{\omega}_N^t\,\boldsymbol{g}_c(\boldsymbol{\tau}^t)\,T$$

(3.82)

where use is made of (3.21) for $\boldsymbol{v}_t$.

With the aid of (3.80), (3.81) and (3.82) the first integral in the heat balance of (3.4) may be transformed into

$$\int_V \bar{T}\rho c\left[\frac{\partial T}{\partial t} + \boldsymbol{g}_c^t(T)\,\boldsymbol{v}_t\right]dV =$$

$$= \bar{T}^t\left[\int_V \rho c\,\bar{\boldsymbol{\tau}}^t\boldsymbol{\tau}\,dV\right]\dot{T} + \bar{T}^t\left[\int_V \rho c\,\bar{\boldsymbol{\tau}}^t\boldsymbol{V}_t^t\,\boldsymbol{\omega}_N^t\,\boldsymbol{g}_c(\boldsymbol{\tau}^t)\,dV\right]T$$

$$= \bar{T}^t\left[\boldsymbol{c}\dot{T} + \boldsymbol{k}T\right]$$

(3.83)

The matrix

$$\boldsymbol{c} = \int_V \rho c\,\bar{\boldsymbol{\tau}}^t\boldsymbol{\tau}\,dV$$

(3.84)

represents the heat capacity matrix of the element in a Lagrangean approach and must be supplemented in the present Eulerian presentation by the convective contribution associated in (3.83) with the coefficient matrix

$$\boldsymbol{k} = \int_V \rho c\,\bar{\boldsymbol{\tau}}^t\boldsymbol{V}_t^t\,\boldsymbol{\omega}_N^t\,\boldsymbol{g}_c(\boldsymbol{\tau}^t)\,dV = \int_V \rho c\,\bar{\boldsymbol{\tau}}^t\boldsymbol{V}^t\boldsymbol{\omega}^t\boldsymbol{g}(\boldsymbol{\tau}^t)\,dV$$

(3.85)

The second integral expression $\boldsymbol{k}$ in (3.85) refers to a cartesian specification, the transition from the first natural expression being a consequence of (2.74).

The second integral on the left-hand side of (3.4) may be put as a consequence of (3.8) into the finite element form

$$\int_V \boldsymbol{g}_c^t(\bar{T}) \, \boldsymbol{\lambda}_{tc} \, \boldsymbol{g}_c(T) \, dV + \int_S \bar{T}\alpha(T - T_\infty) \, dS = \bar{T}^t [\boldsymbol{l}\,T - \dot{q}_s] \tag{3.86}$$

S being the element surface. Application of (3.80) and (3.77) yields the equivalent expression

$$\int_V \boldsymbol{g}_c^t(\bar{T}) \, \boldsymbol{\lambda}_{tc} \, \boldsymbol{g}_c(T) \, dV + \int_S \bar{T}\alpha T \, dS =$$

$$= \bar{T}^t [\int_V \boldsymbol{g}_c^t(\bar{\tau}^t) \, \boldsymbol{\lambda}_{tc} \, \boldsymbol{g}_c(\tau^t) \, dV]\,T + \bar{T}^t[\int_S \alpha \, \bar{\tau}^t \tau \, dS]\,T$$

$$= \bar{T}^t \boldsymbol{l} \, T \tag{3.87}$$

The element conductivity matrix is thus given by

$$\boldsymbol{l} = \int_V \boldsymbol{g}_c^t(\bar{\tau}^t) \, \boldsymbol{\lambda}_{tc} \, \boldsymbol{g}_c(\tau^t) \, dV + \int_S \alpha \, \bar{\tau}^t \tau \, dS \tag{3.88}$$

Its transcription into the cartesian form may be established by substitution of (2.88) for $\boldsymbol{\lambda}_{tc}$ and application of the gradient relations (2.20), (2.17) and (2.18). We find

$$\int_V \boldsymbol{g}_c^t(\bar{\tau}^t) \, \boldsymbol{\lambda}_{tc} \, \boldsymbol{g}_c(\tau^t) \, dV = \int_V \lambda \, \boldsymbol{g}^t(\bar{\tau}^t) \, \boldsymbol{g}(\tau^t) \, dV \tag{3.89}$$

Furthermore, we observe in (3.86) that

$$\dot{q}_s = \int_S \alpha \, \bar{\tau}^t \, T_\infty \, dS = [\int_S \alpha \, \bar{\tau}^t \tau \, dS]\,T_\infty \tag{3.90}$$

represents a prescribed heat rate through the element surface.

Concerning the rate of dissipation defined by the integral on the right-hand side of (3.4), one may write,

$$\int_V \bar{\bar{T}}\, \delta_t^t\, \sigma_c\, dV = \bar{T}^t \left[\int_V \bar{\tau}^t \delta_t^t \sigma_c\, dV \right] = \bar{T}^t \dot{q}_d \tag{3.91}$$

and

$$\dot{q}_d = \int_V \bar{\tau}^t \delta_t^t \sigma_c\, dV = \int_V \bar{\tau}^t \delta^t \sigma\, dV \tag{3.92}$$

Here δ_t , σ_c and δ , σ may be deduced from the mechanical account of the flow problem in subsection 3.2.

Collecting the contributions (3.83), (3.86) and (3.91) into the overall heat balance of the element as expressed by (3.4) we obtain

$$c\dot{T} + kT + \ell T = \dot{q} \tag{3.93}$$

where

$$\dot{q} = \dot{q}_s + \dot{q}_d \tag{3.94}$$

is a generalised heat rate. The finite element equations for the entire flow domain assume then the form

$$C\dot{T} + KT + LT = \dot{Q} \tag{3.95}$$

in which $T, \dot{T}, \dot{Q}$ are column matrices comprising quantities at the nodes of the finite element mesh and C, K, L are the relevant global matrices deduced by assembly of the respective element matrices.

4.   NUMERICAL EXAMPLES

In this section we present some examples illustrating the application of the preceding
theory on the solution of pure and thermally coupled flow problems. Details of the
numerical solution methods, omitted in this paper, may be studied in [28]. There, the
numerical aspects are discussed taking account of the pertinent literature on the subject [16 - 25], which include recent developments. We should stress here that the
streamline upwind/Petrov-Galerkin scheme is applied to all our examples. The capability of this method is demonstrated in what follows for convection dominated flow in two
and three dimensions. The solution of thermally coupled flows is illustrated on the
Bénard type instability. Cavity flows with free and forced convection including a change
of phase are treated in [28].

4.1 Flow over a step

The transient incompressible flow over a step demonstrates the applicability of the independent $p\text{-}\mathbf{v}$ formulation and a two stage solution strategy as described in [4, 28].
Due to the high Reynolds number a turbulent flow field develops necessitating the use
of upwind techniques.

The geometry of the flow domain and the boundary conditions used in the calculation
are sketched in fig. 4.1 together with the material data of the medium (air). At the
inlet a constant velocity profile is prescribed which yields a Reynolds number of 14950
based on the step height. A zero velocity component in cross flow direction is assumed
at the upper side and zero pressure at the outlet of the channel. The flow region is discretised by a mesh of 1700 bilinear plane elements QUAP4 as indicated in fig. 4.1.
Starting from a quiescent initial condition the development of the turbulent flow is investigated up to a total duration of $t = 67.5$ ms using 900 time increments. In fig. 4.2
the onset of turbulent flow is shown in the upper two streamline plots while the other
plots depict the fully turbulent flow field. The disturbances in the flow field near the
outlet may be caused by the somewhat unrealistic pressure boundary condition. Also,
the zero cross flow condition at the upper side of the flow region seems to be not well
adjusted to the process. Despite all these shortcomings the long-time exposure of the
flow over a step (water, visualised by aluminium powder) shown in fig. 4.3 and extracted from [26] compares quite well with the streamlines at the instant $t = 56.25$ ms
of the numerical investigation.

4.2 Flow in a quadratic duct with a step

The efficiency of the upwind scheme and its three-dimensional generalisation involving
the two stage solution algorithm is demonstrated in this example. The flow region, the
boundary conditions and the data of the fictitious material are depicted in fig. 4.4. At
the inlet cross-section a constant flow velocity is assumed, the Reynolds number of 200
being based on the duct dimension H . A zero pressure condition is adopted at the
outlet. The flow domain is discretised by 1368 linear volumetric elements HEXE8 as
shown in fig. 4.5. Calculations were performed in 60 time steps from the initial conditions to a steady state at dimensionless times $t = 6$. At the final stage, projections

of the nodal point velocity vectors onto the xy- and yz-planes are shown in fig. 4.6.
The following example is concerned with the solution of a coupled fluid/thermal problem.

4.3 Bénard convection in a rectangular box

In a fluid heated from below buoyancy driven convection rolls will develop above a
critical value of the Rayleigh number (cf. fig. 4.7)

$$Ra = \frac{g \alpha T_\Delta H^3}{(\mu/\rho)(\lambda/\rho c)} \tag{4.1}$$

This process is analysed for water enclosed in a rectangular box, disregarding any three-
dimensional effects. The lower and upper plate of the box are held at a constant temperature, but the vertical side walls are assumed to be subject to an adiabatic state.
The fluid is initially set at the same temperature as the upper plate and is then heated
from below. As soon as the critical Rayleigh number $Ra = 1708$ is exceeded convection rolls begin to develop. To avoid the difficulties associated with the bifurcation
phenomenon at the critical Rayleigh number a perturbation in temperature is applied
which determines the rotational sense of the first vortex. The analysis is continued until stationary conditions are attained.

The mechanical and thermal data of the fluid (water) are quoted in fig. 4.7 together
with the discretisation by QUAP4 plane elements. The Rayleigh number is evaluated to
be 20250 which exceeds by far the critical value. This fact facilitates the generation
of an unstable process. The calculation extends over $t = 450$ s and involves 60 time
steps, varying between 2.5 s and 20 s. The small increments prove necessary in the
initial process of the formation of the convection rolls within the time interval between
100 s and 150 s. The temperature perturbation applied for the initiation of the convective flow is removed after 150 s when all vortices are formed. Fig. 4.9 exhibits iso-
therms and streamlines at different stages of the process. The development of the convection rolls and the steady state condition is in good agreement with experimental and
analytical investigations [27]. The time between the initiation of convection up to
the fully developed flow corresponds to the predictions. The series of differential interferograms reproduced in fig. 4.8 shows the formation of convection rolls in silicone oil
under similar conditions.

The calculation of the coupled fluid and thermal problems was performed by an iterative
sequential solution of the two individual problems (cf. [3]). The thermal equation,
dominant in the Bénard convection phenomenon, was solved first followed by the solution of the flow problem. All coupling quantities were taken into account, i.e. the
convective terms in the thermal problem and the buoyancy forces in the flow problem,
the latter being calculated using the Boussinesq approximation. The iterative solution
of the discretised equations leads to linear equation systems with non-symmetric coefficient matrices due to the convection terms. The equation system of the thermal problem was solved using the QR-factorisation for the non-symmetric coefficient matrix.

For the flow problem the penalty approach was applied with the convection terms on the right-hand side so that standard solution methods were eligible. Upwinding was used in both problems with an upwind parameter of 0.258. Convergence below the limit $\varepsilon = 10^{-2}$ in the heat rates and velocities respectively was required to terminate the iteration of the individual problems. The sequential solutions were continued until both the velocity and the temperature increments were reduced below the convergence limit of $\varepsilon = 10^{-2}$ between consecutive iterations.

REFERENCES

[1]   Argyris, J.H. et al., Finite element method – the natural approach, Fenomech '78, Comput. Meths. Appl. Mech. Engrg. 17/18 (1979) 1-106.

[2]   Argyris, J.H., Doltsinis, J.St., Pimenta, P.M. and Wüstenberg, H., Thermo-mechanical response of solids at high strains – natural approach, Fenomech '81, Comput. Meths. Appl. Mech. Engrg. 32 (1982) 3-57.

[3]   Argyris, J.H. and Doltsinis, J.St., On the natural formulation and analysis of large deformation coupled thermomechanical problems, Comput. Meths. Appl. Mech. Engrg. 25 (1981) 195-253.

[4]   Brooks, A.N. and Hughes, T.J.R., Streamline upwind/Petrov-Galerkin formulations for convection dominated flows with particular emphasis on the incompressible Navier-Stokes equations, Fenomech '81, Comput. Meths. Appl. Mech. Engrg. 32 (1982) 199-259.

[5]   Argyris, J.H. and Doltsinis, J.St., On the large strain inelastic analysis in natural formulation – Part I. Quasistatic problems, Comput. Meths. Appl. Mech. Engrg. 20 (1979) 213-252. – Part II. Dynamic problems, Comput. Meths. Appl. Mech. Engrg. 21 (1980) 91-128.

[6]   Argyris, J.H. et al., Eulerian and Lagrangean techniques for elastic and inelastic deformation processes, TICOM 2nd Int. Conf., Austin, Texas, 1979. In: Computational Methods in Nonlinear Mechanics (J.T. Oden, Editor), North-Holland Publishing Company (1980) 13-66.

[7]   Argyris, J.H., Doltsinis, J.St. and Wüstenberg, H., Analysis of thermo-plastic forming processes – natural approach, Computers and Structures, to appear.

[8]   Prager, W., Introduction to mechanics of continua, Ginn and Co., Boston (1961).

[9]   Argyris, J.H., Three-dimensional anisotropic and inhomogeneous elastic media, matrix analysis for small and large displacements, Ing.-Archiv 34 (1965) 33-55.

[10] Argyris, J.H. and Doltsinis, J.St., A prime on superplasticity in natural formulation, Comput. Meths. Appl. Mech. Engrg., to appear.

[11] Hohenemser, K., Prager, W., Über die Ansätze der Mechanik isotroper Kontinua, ZAMM 12 (1932) 216-226.

[12] Argyris, J.H. and Mareczek, G., Finite element analysis of slow incompressible viscous fluid motion, Ing. Archiv 43 (1974) 92-109.

[13] Malkus, D.S. and Hughes, T.J.R., Mixed finite element methods - reduced and selective integration technique: a unification of concepts, Comput. Meths. Appl. Mech. Engrg. 15 (1975) 63-81.

[14] Oden, J.T., RIP-methods for Stokesian flows, In: Finite Elements in Fluids, Vol. 4 (R.H. Gallagher et al., Editors), John Wiley and Sons Ltd., 1982.

[15] Taylor, R.L. and Zienkiewicz, O.C., Mixed finite element solution of fluid flow problems, In: Finite Elements in Fluids, Vol. 4 (R.H. Gallagher et al., Editors), John Wiley and Sons Ltd., 1982.

[16] Felippa, C.A. and Park, K.C., Direct time integration methods in nonlinear structural dynamics, Comput. Meths. Appl. Mech. Engrg. 17/18 (1979) 277-313.

[17] Glowinski, R., Dinh, Q.V. and Periaux, J., Domain decomposition methods for nonlinear problems in fluid dynamics, Fenomech '81, Comput. Meths. Appl. Mech. Engrg., to appear.

[18] Hestenes, M.R., Stiefel, E., Methods of conjugate gradients for solving linear systems, J. Res. Nat. Bur. Stand. 49 (1952) 409-436.

[19] Jennings, A. and Malik, G.M., The solution of sparse linear equations by the conjugate gradient method, Int. J. Num. Meths. Engrg. 12 (1978) 141-158.

[20] Dennis, J.E. and Moré, J.J., Quasi-Newton methods - Motivation and theory, SIAM Review 19 (1977) 46-89.

[21] Matthies, H. and Strang, G., The solution of nonlinear finite element equations, Int. J. Num. Meths. Engrg. 14 (1974) 1613-1626.

[22] Thomasset, F., Implementation of finite element methods for Navier-Stokes equations, Springer New-York, 1981.

[23] Hughes, T.J.R., Winget, J., Levit, I. and Tesduyer, T.E., New alternating direction procedures in finite element analysis based upon EBE approximate factorisation, Recent Developments in Computer Methods for Nonlinear Solid and Structural Mechanics (eds. S.N. Atluri and N. Perrone), ASME Applied Mechanics Symposium Series, New York, 1983.

[24] Marchuk, G.J., Methods of Numerical Mathematics, Springer-Verlag, New York - Heidelberg - Berlin, 1975.

[25] Hughes, T.J.R., Levit, I. and Winget, J., An element by element solution algorithm for problems of structural and solid mechanics, Comput. Meths. Appl. Mech. Engrg. 36 (1983) 241-254.

[26] Tani, I., Experimental investigation of flow separation over a step, IUTAM
      Symposium Freiburg 1957, Grenzschichtforschung/Boundary layer research,
      H. Görtler ed., Springer-Verlag, 1958, 377-386.

[27] Kirchartz, K.R., Oertel, H., Zeitabhängige Zellularkonvektion, ZAMM 62
      (1982), T 211 - T 213.

[28] Argyris, J., Doltsinis, J.St., Pimenta, P.M. and Wüstenberg, H., Natural
      finite element techniques for fluid motion, Comput. Meths. Appl. Mech. Engrg.,
      to appear.

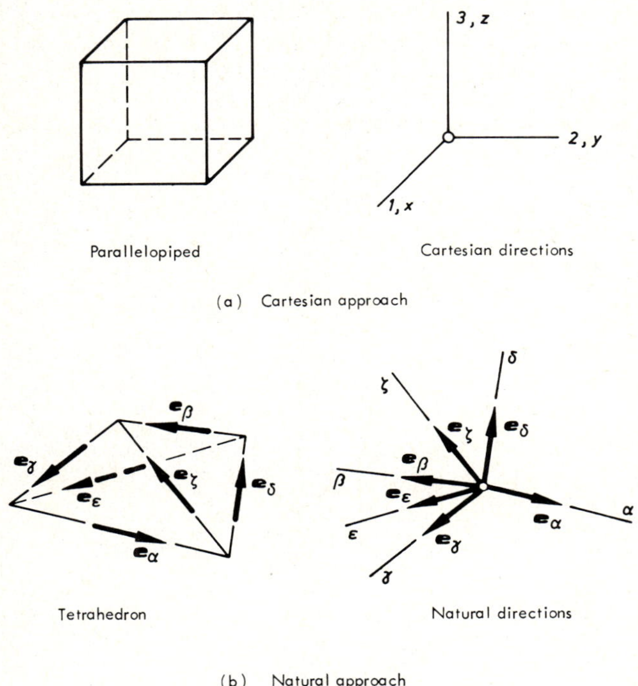

Fig. 2.1   Cartesian and natural system of reference

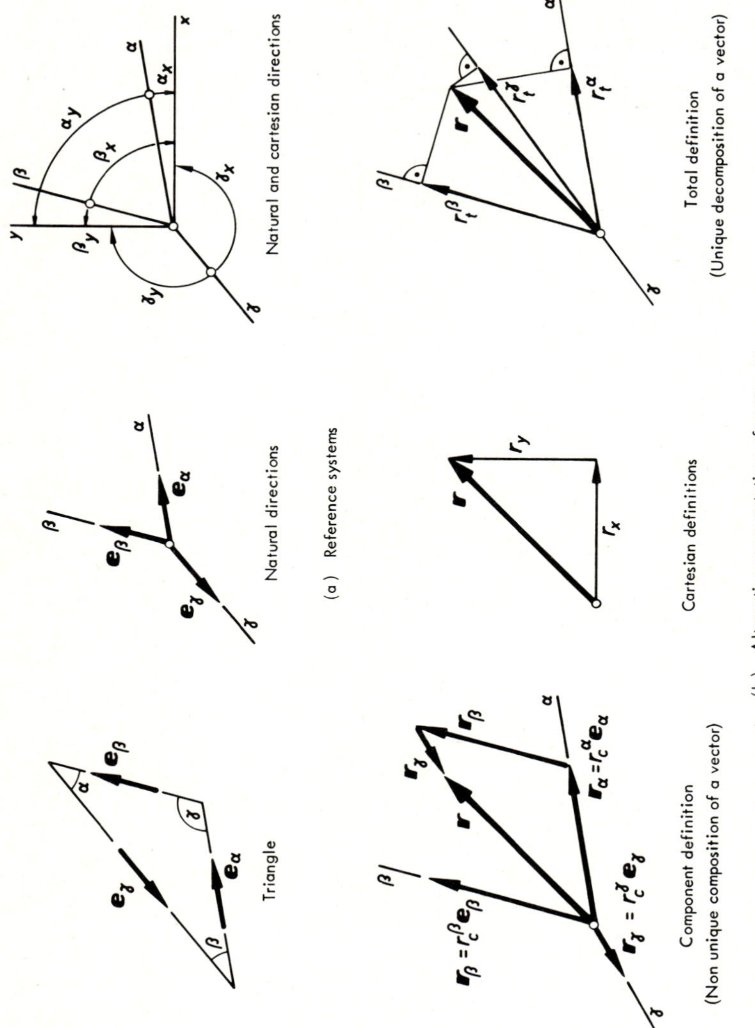

Fig. 2.2 Natural and cartesian specifications of a vector for the two-dimensional case

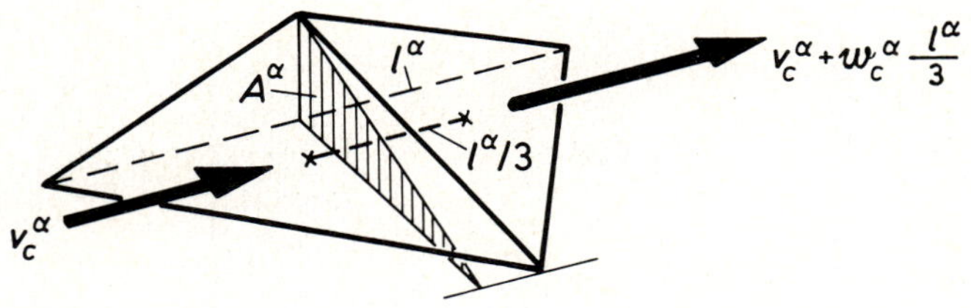

Fig. 2.3   Mass supply to a natural element due to a component velocity vector v_c^α

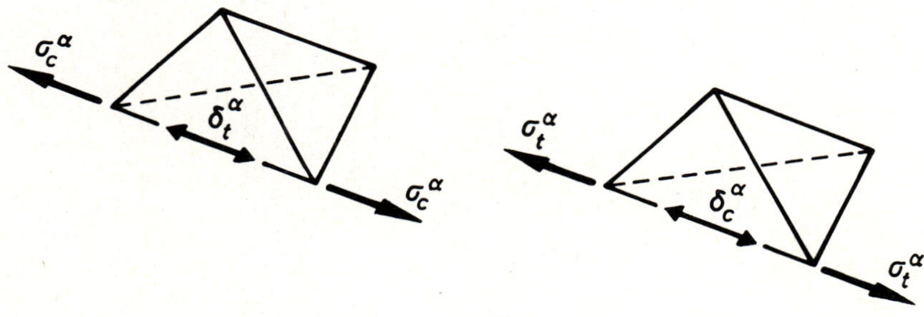

Fig. 2.4   Corresponding definitions of natural stresses and rates of deformation

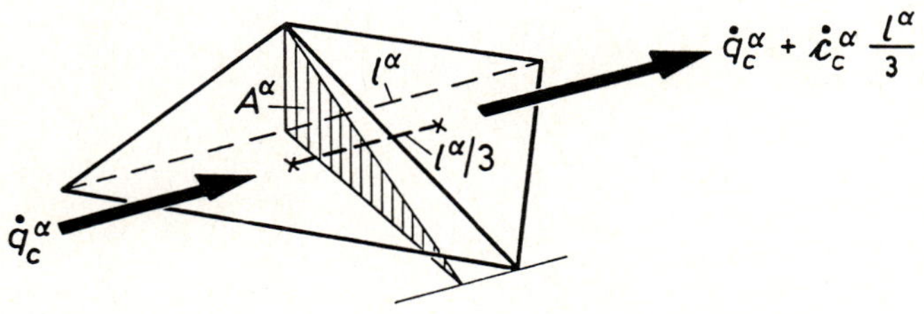

Fig. 2.5   Heat supply to a natural element due to a component heat flux vector $\dot{q}_c^\alpha$

J.H. Argyris & J. St. Doltsinis

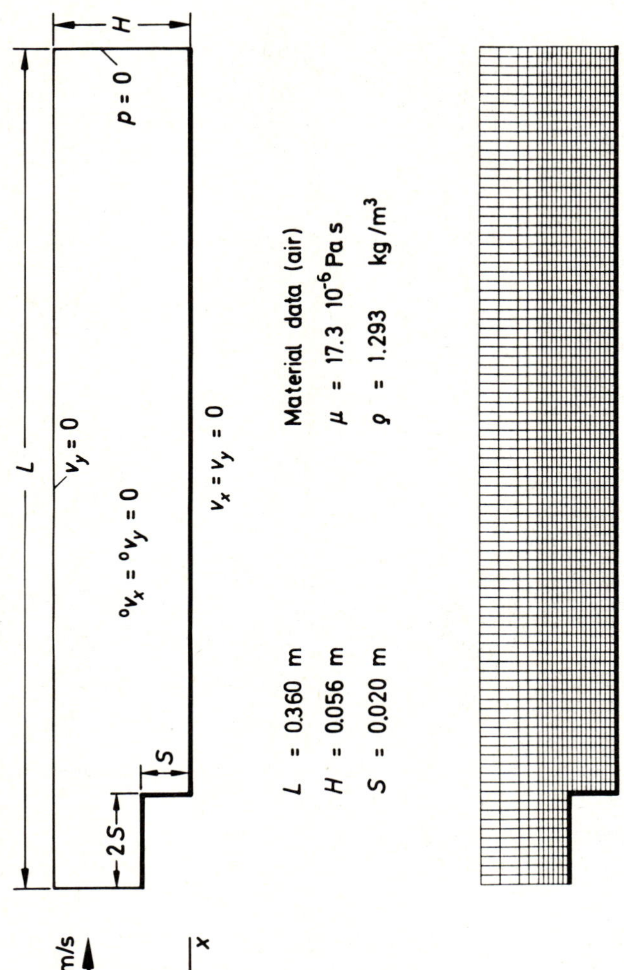

Fig. 4.1   Flow over a step. Description and finite element discretisation

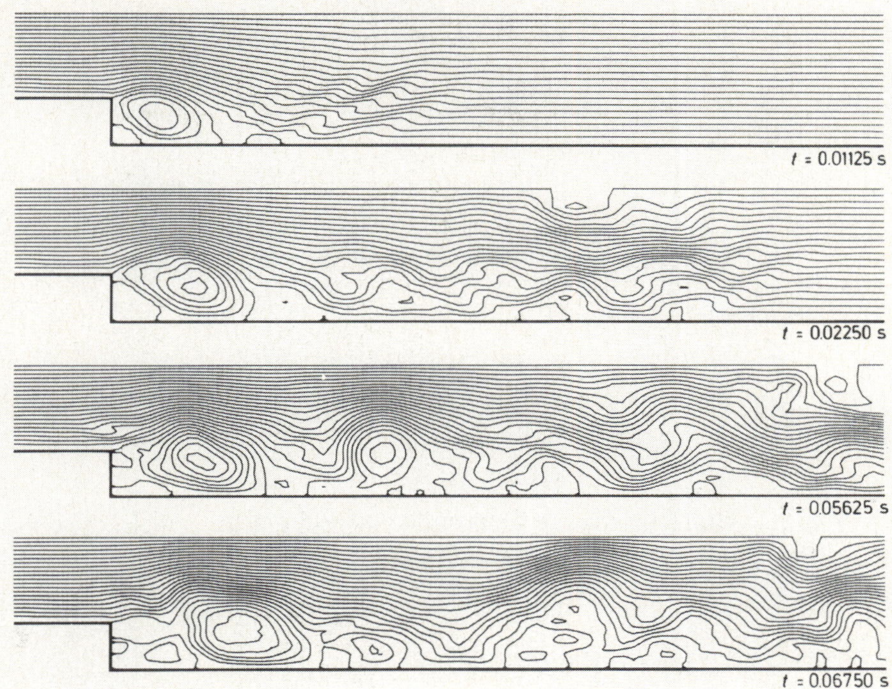

Fig. 4.2   Flow over a step. Streamlines during development of turbulent flow

Fig. 4.3   Visualisation of flow over a step by aluminium powder in water [26]

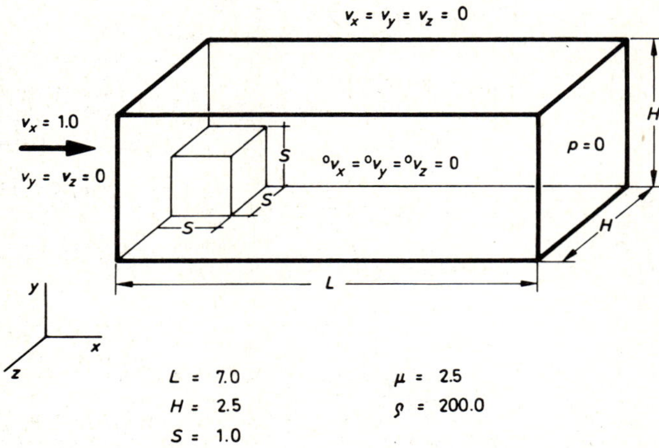

Fig. 4.4   Flow in a quadratic duct with a step

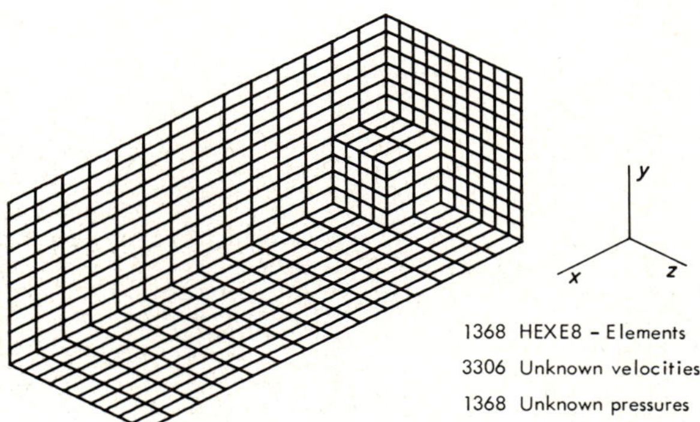

Fig. 4.5   Flow in a duct. Discretisation

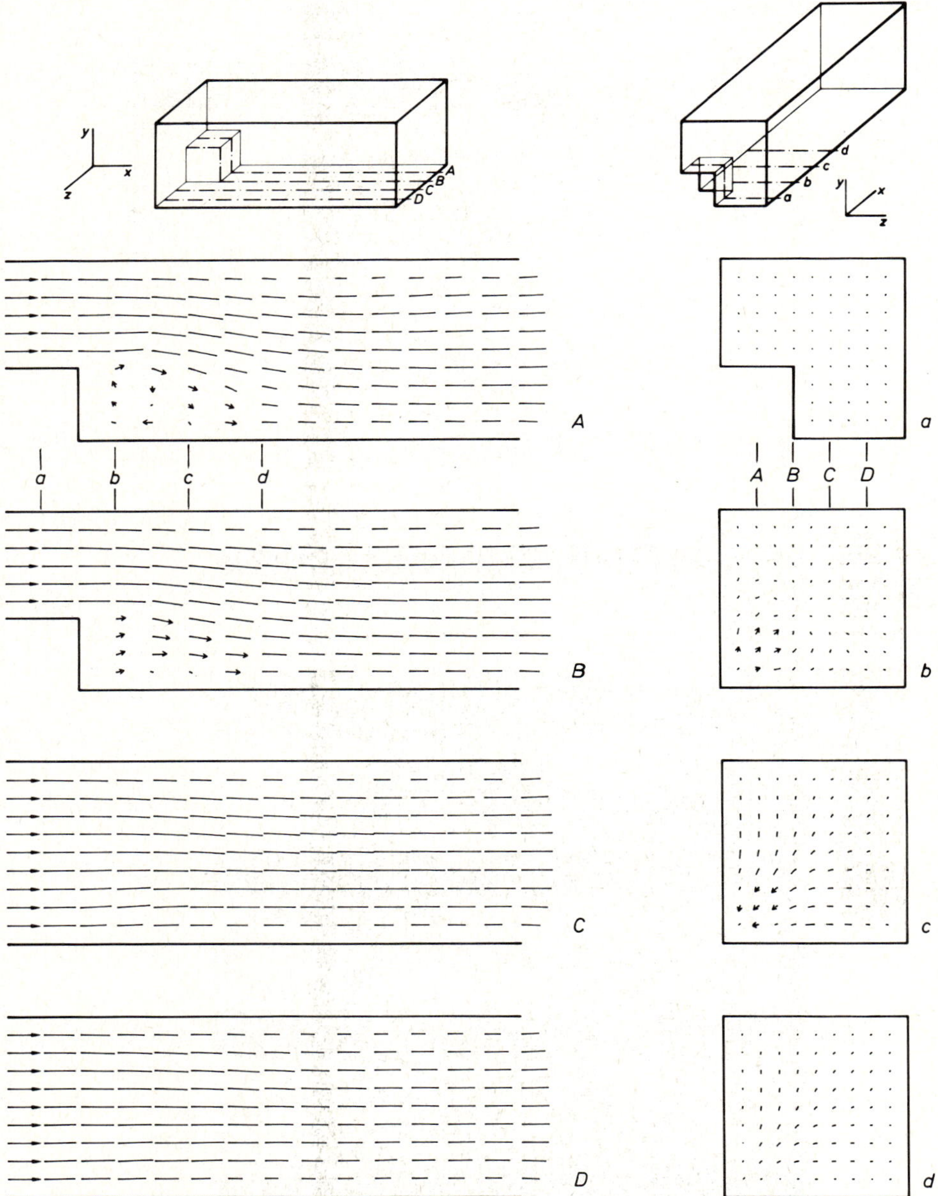

Fig. 4.6 Flow in a duct. Projections of nodal point velocity
vectors at a stationary state

   J.H. Argyris & J. St. Doltsinis

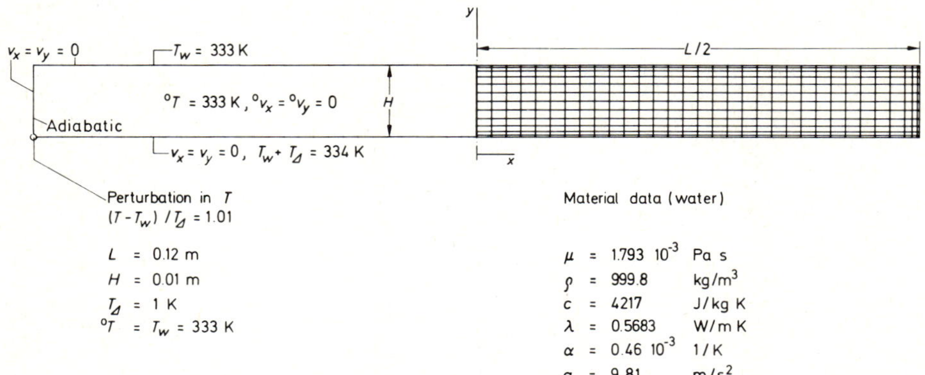

Fig. 4.7   Bénard convection in a rectangular box

Fig. 4.8   Differential interferogram of transient convection of silicone oil [27]

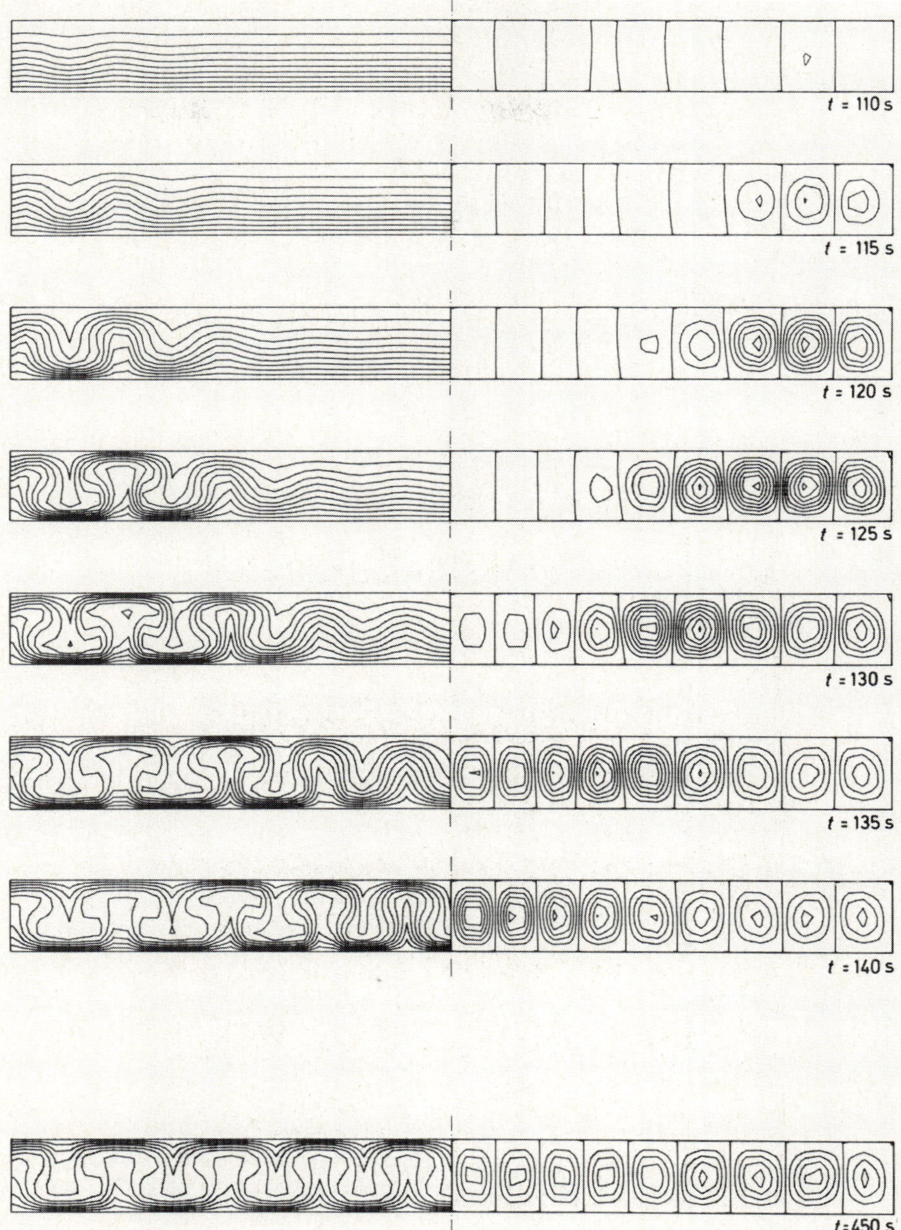

Fig. 4.9   Bénard convection. Distribution of temperature (left) and stream function (right) at transient and quasi-stationary conditions

Unification of Finite Element Methods
H. Kardestuncer (Editor)
© Elsevier Science Publishers B.V. (North-Holland), 1984

CHAPTER 2

INTERACTIVE COMPUTER GRAPHICS FOR FINITE ELEMENT, BOUNDARY ELEMENT, & FINITE DIFFERENCE METHODS

J.F. Abel, A.R. Ingraffea, R. Perucchio, T.-Y. Han, & J.F. Hajjar

The trend toward ubiquity of interactive graphics in computational engineering is helping to develop an atmosphere conducive to the combination and unification of various numerical analysis methods. A number of features of interactive computer aided design which foster unification are discussed. Examples are drawn from three principal aspects of computer graphics in engineering analysis: preprocessing, postprocessing, and interactive-adaptive analysis.

INTRODUCTION

The purpose of this paper is to present and explore some ideas about how the growing use of interactive computer graphics is both affecting the utilization of numerical methods by engineering analysts and influencing the possible unification of these methods. Recent decades have seen a dramatic development and proliferation of numerical methods. This activity has included the strengthening and diversification of established methods such as finite differences and finite elements, the rapid expansion of rediscovered techniques such as boundary-integral equations, the increasing recognition of unifying formulative notions such as variational approaches, and the combination of methods to realize diverse advantages such as the mixing of finite elements and boundary elements.

The ideas presented here represent some recent collective thoughts of a research group at Cornell University that has, for nearly a decade, been involved in developments of interactive computer graphics for their applications to both computational mechanics research and computer-aided design. Because most of this research has been limited to structural engineering and mechanics and has involved primarily finite element and boundary element approaches, this can hardly be considered a global view of disciplines and methods. Nevertheless, the emphasis in this paper on growing trends in graphics and their relation to unification of analysis methods can be considered a manifesto, albeit modest, regarding desirable directions for future developments in interactive graphics.

Clearly interactive graphics in itself is distinct from analysis and, therefore, can be viewed from one perspective as having little to do with unification of analysis techniques. However, it has been widely demonstrated that interactive graphics can be highly successful in breaking down barriers and difficulties in the performance of engineering analysis. The enhanced access to, and control of, analysis can in the same way significantly help remove obstacles to unification, as will be argued

subsequently. The provision of a proper atmosphere for amalgamation is no
less important than the recognition of the potential for unification.
Ultimately, the obvious goal is the provision of the most effective and
appropriate analysis methodologies for engineers to carry out design
responsibly and creatively.

The computational environment for analysis is changing rapidly. Computer-
aided analysis and design in both practice and industry are being tied
together with computerized information repositories and data flow capabilities, while at the same time increased computational power is being provided to the individual engineer. Three chief current developments in
hardware and software are engineering workstations which provide intimate
access to computing, networking which enables the sharing and rapid transfer of large volumes of data among workstations, and interactive graphics
capabilities integral to the workstations. The last of these is the theme
here. In this environment, within the next decade, nearly all computing
will be associated with interactive graphics. Both literally and figura-
tively, interactive computer graphics is becoming the "window" to the
world of analysis.

Interactive computer graphics was first applied to preprocessing and postprocessing only, and these are still the phases of analysis where engineering productivity is most dramatically improved by its use. However,
although analysis or processing itself has traditionally been a batch procedure, the continued increase of cheap, distributed computational capability and the growing demands of designer/analysts indicate that in the
near future some interactive processing will be effective and desirable
even for such computationally intensive procedures as finite element analysis. This notion has given rise to interactive-adaptive analysis [1],
which is defined as analysis during which the engineer can continuously
monitor what is happening and can choose to intervene at any time to
change system characteristics, models, analysis parameters, and algorithms. It is most suited to nonlinear and time-varying analyses. The
engineer is able to move backward and forward in the analysis at will and
to invoke a variety of parallel analyses at any stage. Moreover, different types of analyses can be strung together in a sequence. This analysis
control capability clearly must rely on interactive computer graphics for
its success.

In this paper all three aspects of interactive computer graphics in engineering analysis are used to illustrate the notions discussed.

UNIFYING INFLUENCES OF INTERACTIVE GRAPHICS

Experience with the development of interactive graphical software programs
for analysis applications indicates that, for well designed interactive
systems, programming is a major task that exceeds in complexity and difficulty the programming of much analysis software [2]. In addition to the
usual requirements -- a rational underlying database; modularity to foster
expandability; clear, structural programming; a system for anticipating,
handling, and correcting errors; transportability to the maximum extent
possible while maintaining necessary performance and response; and
thorough documentation -- the design for human factors or "user friendli-
ness" places extra demands on the program developer. Among these are
completeness of function, flexibility in sequence, and a humanized command
language and menus [3] [4]. The achievement of these last objectives

tends to result in a generality of the interface between the user and the computer that is conducive to flexibility, diversification, and unification of analysis methods.

In addition to the fundamental design of interactive software, a factor which is affecting the way in which particular analysis methods fit into an overall design process is the increasing use of system models independent from analysis models. In the application of, say, finite element methods, one was previously likely to describe to the computer the geometry, boundary conditions, and loadings of the problem under

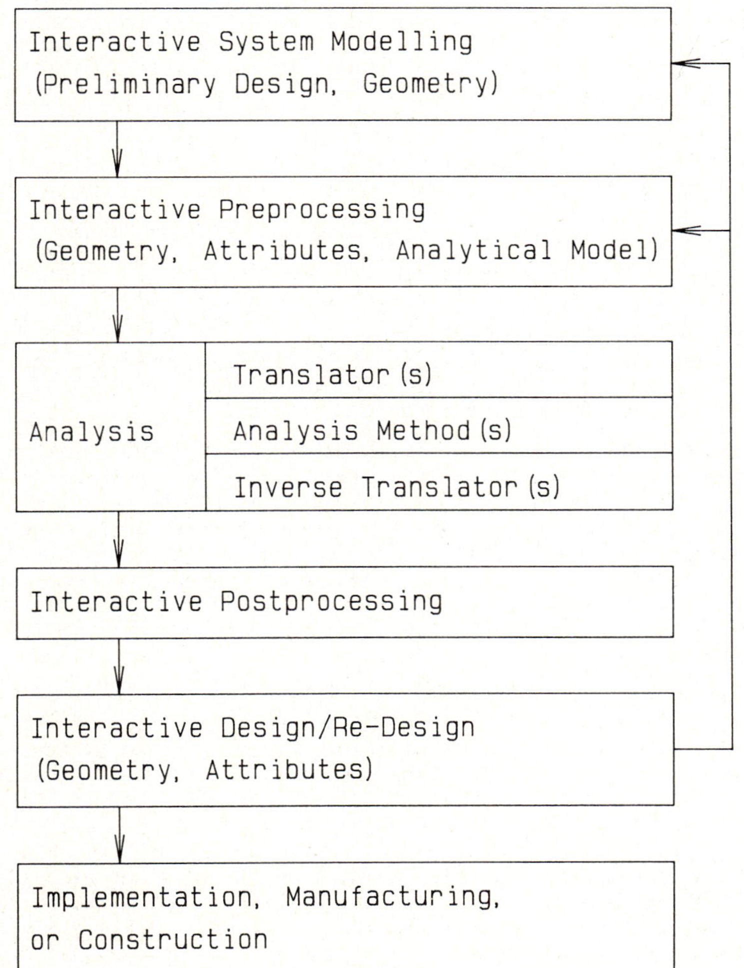

Figure 1.   The computer-aided design process. Interactive computer graphics is the principal medium for man-machine communication and control. The integration of the various stages shown is predicted on computerized databases and data flow.

consideration in terms of the elements, nodes, and connectivity of the description that was entirely linked with the analytical discretization. With the trend toward computer-aided design, analysis is now usually viewed as being integrated into an overall design process such as depicted in general form in Figure 1. The links between the various steps shown are achieved through a computerized database and data flow. Here the geometric description of the design problem is created on the computer by a modeller or, if a modeller is not available, it is created as the first step of preprocessing. Then, in preprocessing, additional attributes of the problem, such as material properties, boundary conditions, and loadings, are associated with the geometry. The result is the complete description of an analysis problem that exists within the computer prior to any meshing and, indeed, that is independent of any particular method.

In the future, the generation of the analytical model or discretization appropriate for a particular method of analysis will probably be automated [5]. This will be particularly suitable for analysis procedures that include self-adaptive mesh algorithms to achieve a specified degree of refinement or accuracy in each portion of the system. However, for at least the next decade, it is clear that the engineering analyst will need to continue to exercise strong influence over the design of the analytical model because of its direct effect on accuracy and reliability of the analysis. Interactive mesh generators are ideally suited to this task [6]. Moreover, effective interactive preprocessors provide a flexibility that is only beginning to be recognized. Not only are computer-assisted mesh generators in many cases able to be generalized for a variety of analytical approaches, but the interactive tools incorporated for, say, substructuring or partitioning also present natural opportunities for preprocessing for hybrid or combined techniques.

Another aspect of interactive software design that significantly alleviates specificity to a single analysis type is the humanization of command languages. One detectable trend is toward the use of commands more natural to the engineer. For example, in the specification of boundary conditions for finite element analysis, one may be asked to specify the type of condition (e.g., "symmetry") rather than a node-by-node list of restraint codes for each degree of freedom. The corollary is that the preprocessing software or the translator programs that are the interface to the specific analysis program (Figure 1) must include internal coding which automatically converts between analysis-specific data and the engineer's natural vocabulary.

Two factors relating to data organization for interactive systems potentially contribute to the diversification of such software. The first of these is the organization of data structures for efficient and rapid graphical display. This involves the exploitation of geometric coherence which may differ from that imposed by a particular analysis method. For instance, in displaying a finite element mesh, one may not wish to draw the mesh element by element because this will involve the redrawing of lines for the edges of adjacent elements. Instead, one would prefer to take advantage of the coherence of mesh lines by drawing each complete line in one step. The rearrangement of element edge data into complete mesh-line data in this fashion is perhaps more reminiscent of data organizations appropriate to higher order finite difference operators. Another example arises from the desire to provide display simplifications for three-dimensional geometries. Here, one technique used to permit clearer viewing of three-dimensional finite element meshes is the selected removal of lines or the selective display of some elements or groups of elements. One useful version of this is the display of only the exterior surfaces of elements [7] [8]. The distinction here in the data structure between interior and exterior element faces creates an obvious link to boundary element techniques.

The second data-related factor is the overall data transmission scheme underlying computer-aided design systems with data flows such as indicated schematically in Figure 1. The need not only to transmit information from one stage of design to another but also to exchange data between different computer-aided design systems has given rise to standardized procedures, most notably the Initial Graphics Exchange Specification or IGES [9]. Although still under development and not yet as efficient as desirable, IGES is now being successfully used to transmit geometrical data. Moreover, the specification is being expanded to include analysis information, particularly finite element data. IGES data files are rigidly formatted but are so complete and flexible in their representation of data that the files can be considered essentially system independent. The flexibility of IGES arises because its entities may be either geometric or nongeometric and because various entitites may be associated in ways appropriate to any application. The implication for analysis of the existence of such schemes as IGES is the notion that either problem descriptions or analysis models and results can be standardized in a complete, neutral form. This fosters an exchange of data that permits not only comparisons but also combinations or hybridizations of analytical

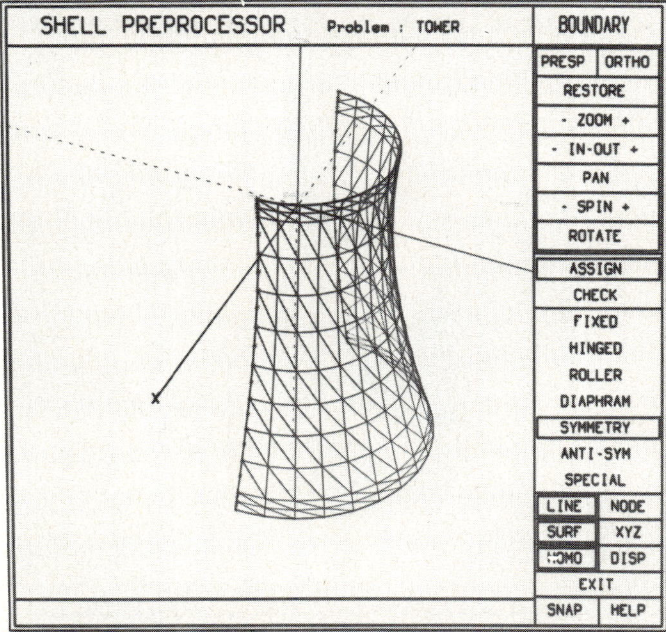

Figure 2. Assignment of boundary conditions during finite element preprocessing for shell structures [10]. Note the natural terminology for boundary condition types (middle portion of menu at right).

procedures.

Some of the ideas introduced thus far are now illustrated by discussions
and examples of interactive programs developed for research or
instructional purposes at Cornell. All of the subsequent figures are
photographs taken from graphic display devices, and these have been
reduced to the degree that some small text is not easily legible. In each
of these cases, the small text contains little pertinent information and
is retained only to illustrate the nature of the computer graphics display. In addition, the photographs of color displays are, by necessity,
reproduced here in black and white.

PREPROCESSING

A single glimpse at an interactive preprocessor for shell analysis [10]
provides an example of the humanization of commands. In Figure 2, the
menu page for the assignment of boundary conditions is shown. The commands along the right edge are divided into groups. The upper group
consists of permanent menu items for the manipulation of the main image,
while the group below these are commands and modifiers specific to the
immediate task. Here the user has activated the command ASSIGN with the
boundary condition type SYMMETRY and the modifer set to LINE. (All active
commands and modifiers are indicated by a box drawn about them.) Simply
by pointing to the meridional cuts of the hyperbolic cooling tower shell
shown, the engineer assigns the symmetry condition to this line.

By far the most significant effort in preprocessing is the creation of the

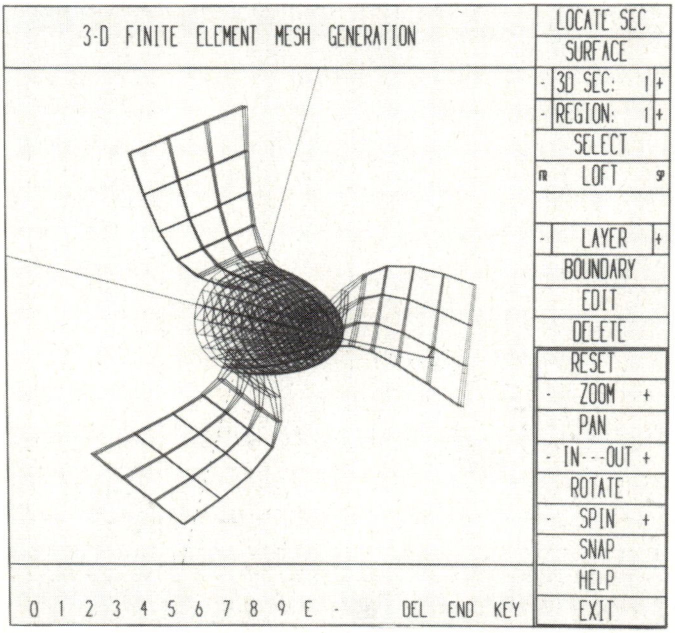

Figure 3. Complete three-dimensional finite element mesh for a propeller
generated by lofting between sections [8].

analytical model or mesh. Even with mesh generation algorithms, some
intervention by the engineer is usually necessary or desirable to control
the design of the descretization. For example, in the discrete transfinite mapping methods of mesh generation, the density and gradation of the
mesh in a region is determined by the placement and spacing of nodes
around the boundary of the region [7], and this task is best performed by
the analyst interactively. The techniques of mesh creation that have
been found successful for interactive applications have also proven to be
applicable for a variety of different analysis types. At Cornell, the
lofting specialization of discrete transfinite mapping has been used for
both finite element [7] [8] and boundary element [11] [12] preprocessing.
Figure 3 shows a complete three-dimensional finite element mesh for a pro-
peller-like solid generated interactively in this fashion. The mesh for
the hub has been created by lofting among parallel planar sectional meshes
with different circular diameters, while the lofting sections for the
blades are not parallel and, at the root of the blades, are nonplanar
sections interactively selected from the surface of the hub mesh [8]. In
Figure 4, the mesh in Figure 3 is simplified for clearer viewing by
displaying only the exterior facets of elements, and the result is clearly
suggestive of the applicability of this meshing procedure for boundary
element preprocessing. An example of a direct application of this meshing
technique for a boundary element problem is shown in Figures 5 and 6 [11]
[12].

The methods in these examples could be similarly extended to such alternatives as finite difference mesh generation. However, one obstacle remaining to the complete generalization of interactive preprocessing techniques

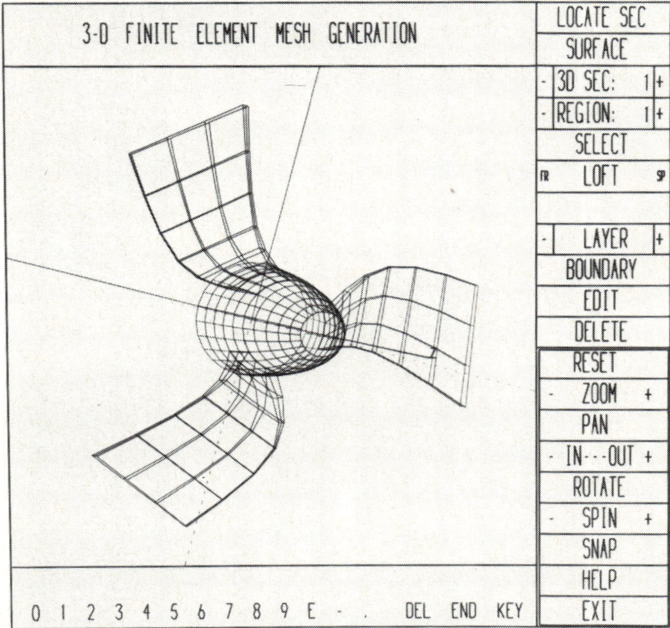

Figure 4.   The mesh in Figure 3 is simplified for clearer viewing by
displaying only exterior (boundary) facets of elements.

for analytical model generation is the nonorthogonality of the coordinate
systems employed in the mesh generation procedures. Many finite difference algorithms, for example, require a mesh which consists of isocoor-
dinate lines in an orthogonal coordinate system. The common finite element mesh generation techniques based on isoparametric or transfinite map-
pings do not automatically fulfill this requirement. Nevertheless, it is
clear that the meshing methods can be extended with minor adaptions to
less restrictive analysis methods such as finite cell methods.

In many cases, boundary element analysis requires not only a discretization of the surface of the domain but also an interior mesh. The consideration of body effects such as gravity loads and of nonlinearities are
two examples. In addition, if full field or selected interior results are
required, then interior points must be chosen as locations for the evaluation of responses. Although interior meshes for the boundary element
method need not satisfy the same strict topological and connectivity
requirements as a finite element mesh, the use of an interior finite element discretization is acceptable in most cases. The combination of
finite and boundary element preprocessing capabilities is therefore fruit-
ful. A modest two-dimensional example is shown in Figures 7 and 8 [13]
[14]. Here the motivation for use of an interior mesh is the identification of both interior points and a natural interpolation scheme for the
evaluation and visualization of the full stress field. First the boundary
element mesh in Figure 7 is created interactively by selecting the number,
spacing, and gradation of mesh points along each edge. Then, in Figure 8,
an automatic interior mesh generation scheme is used to find rapidly an
interior mesh of sampling points with a natural interpolatory basis, in
this case triangular regions. This mesh is determined solely by the

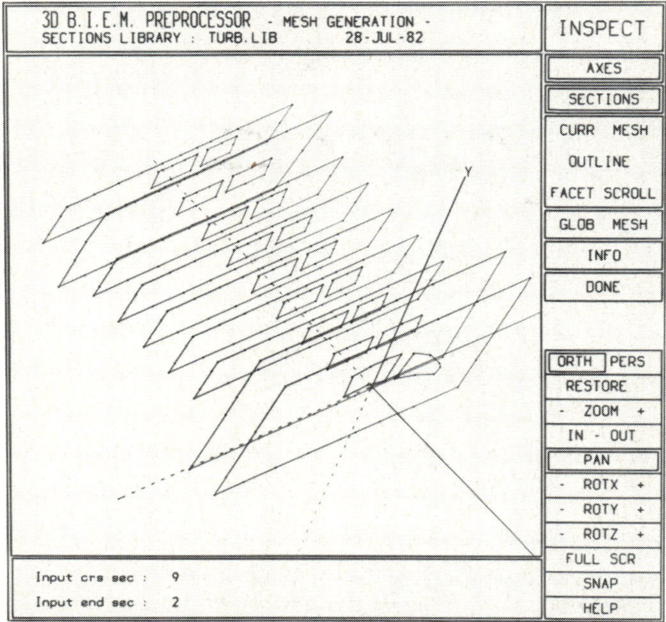

Figure 5.   Planar sections for creation by lofting methods of a boundary
             element mesh for the root of a turbine blade [11] [12].

boundary nodes and by a single interior density parameter. Although the resulting mesh could be adjusted interactively, the design considerations constraining the interior mesh are less stringent than those for a finite element analysis. It is sufficient that the mesh is fine in the region of expected stress concentration, and this refinement is achieved directly as a consequence of the design of the boundary element mesh with smaller elements in this region.

POSTPROCESSING

The combination of finite and boundary element capabilities carries over into postprocessing, particularly for an example motivated by full-field visualization such as the two-dimensional problem shown in Figures 7 and 8. Figures 9 and 10 show how an interactive two-dimensional finite post-processor [15] is used, without significant modification, to inspect the results of the boundary element analysis. The first of these figures shows the examination of the value of displacements at nodes by pointing to a particular node; the deformed mesh can also be displayed at any selected magnification. Figure 10 is a black and white photo of color stress contours throughout the domain. Here some of the stress ranges have been darkened interactively to produce a clearer view of how stresses vary near the central hole.

The general graphical problem in postprocessing is the visualization of how a scalar, vector, or tensor response parameter varies over the geometry of the system. Particularly for three-dimensional solids or surfaces, it is usually possible to display only a single parameter component over the surface of the body while using coloration to convey

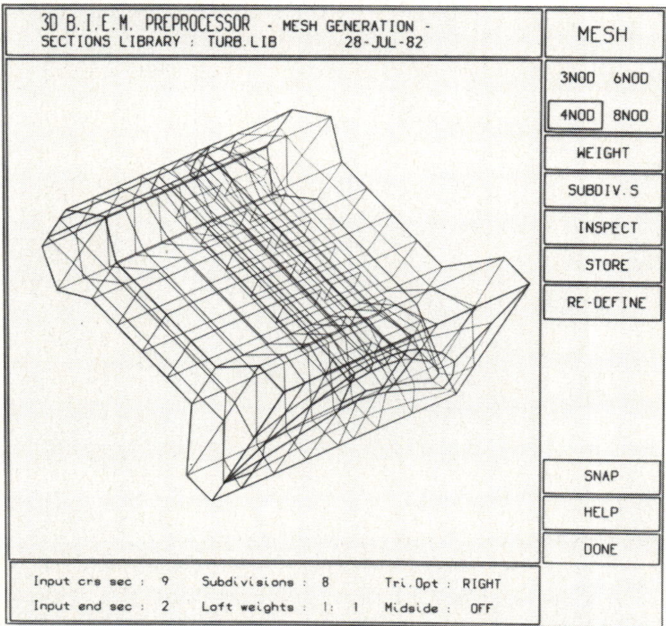

Figure 6. The complete boundary element mesh resulting from Figure 5.

simultaneously the three dimensionality of the geometry, Figure 11 [15].
As long as one has both an appropriate way of evaluating response parameters accurately and an inherent or assumed basis for spatial interpolation of these parameters, an interactive graphical postprocessor need not
be specific to a particular method of analysis and may, in fact, be used
effectively even for combined methods.

As an illustration, following are some of the major design features of a
general-purpose interactive graphical postprocessor currently under development at Cornell. The postprocessor is being designed to accommodate
three-dimensional geometries with specializations available for both
curved surfaces in three-dimensional space and two-dimensonal problems.
The postprocessor serves for finite element, finite difference, boundary
element, and combined finite/boundary element analyses. Flexibility for a
variety of technical analyses is built in, including stress analysis, heat
conduction, and compressible fluid flow. The basic display capability
consists of color contouring on the surfaces as shown in Figure 11. The
user is able to choose any viewing angle and may have single or multiple
views of the system. Moreover, the user is able to section, cut, or
"unfold" the geometry interactively in a variety of coordinate systems to
obtain views of the interior distributions of parameters. When a cut is
made, the program automatically interpolates interior results to obtain
the variation of the parameter over the exposed surface. When no interior
results are available for a boundary element analysis, the program can
treat postprocessing in much the same manner as it would handle result
displays on surfaces such as shell and membrane structures. Finally,

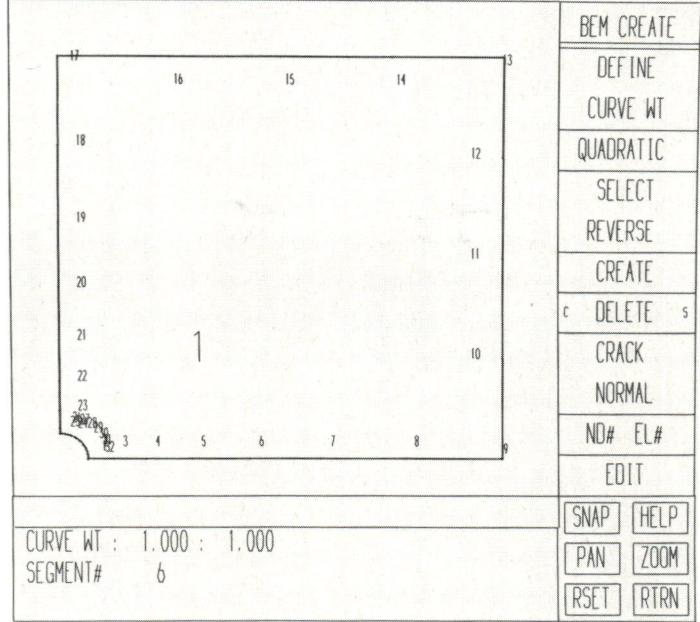

Figure 7.   Interactive boundary element preprocessing for a two-
            dimensional problem. Creation of a discretized outline for one
            quarter of a stretched plate with a circular hole [13].

although the program has its own internal data organization designed for
efficient interactive graphical manipulation, it is assumed that data from
all different types of analyses are received by the program in a standardized IGES format. This represents a finite element prejudice coinciding
with the initial directions of the IGES treatment of analysis data, but
for the most part it seems feasible to expect that the results from nearly
all analysis types can be translated into a finite-element-like format.

INTERACTIVE-ADAPTIVE ANALYSIS

Interactive-adaptive analyses has been defined earlier as continuous
graphical monitoring of the progress of analysis by the engineer with the
opportunity to interrupt and to change either the design or the analysis
[1]. In a sense, this is a combination of preprocessing and postprocessing with additional facilities of user control over analysis. Obviously
if analysis results are to be monitored in real time, the computation must
be very rapid or the user will be wasting his time sitting before a very
slowly changing graphic display. Currently, this limits the use of interactive-adaptive techniques to relatively small problems, to efficient
special-purpose analysis and design systems, to teaching applications, and
to research. However, the trends toward increasingly cost-effective computing and memory and increasingly rapid data transfer indicate that these
constraints will be alleviated. An example of the use of the approach to
monitor nonlinear structural behavior as it is being calculated is shown
in Figure 12. Here four different portions of the display are used to

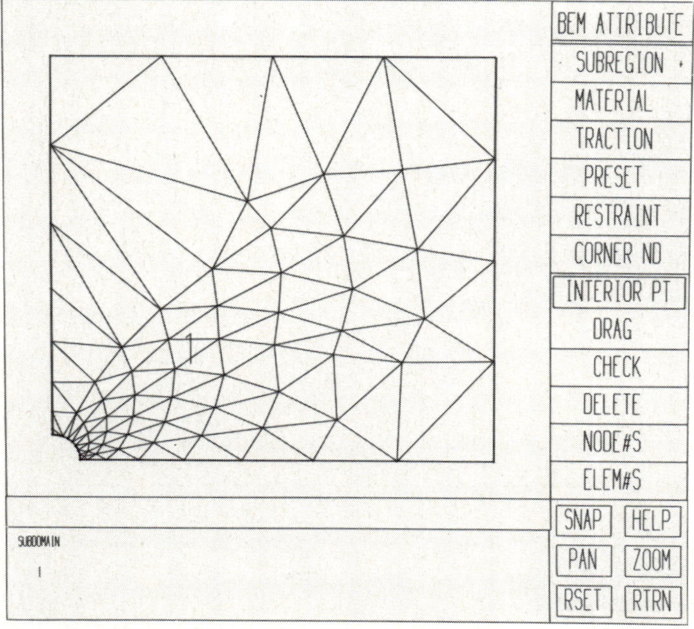

Figure 8.   For the problem of Figure 7, creation of a mesh of interior
            points used to obtain full-field information from a boundary
            element analysis [13] [14].

trace analysis progress. The main viewport shows the deformation behavior
of the structure, while the upper left and upper center viewports include
graphs of response selected interactively by the user; lastly, the upper
right monitor summarizes analysis information such as load level, load
step, number of iterations, etc.

The analysis diversification opportunities presented by interactive-
adaptive analysis are several. First is the ability to perform analyses
which aid in establishing analysis parameters; for example, the ability to
perform rapid natural period calculations assists in determining appropriate time increments for marching algorithms. Second, one role for interactive-adaptive approaches is to change analysis strategies after analysis
has begun. For example, in Figure 13 the analyst has, partly on the basis
of an initial analysis, identified a group of elements to constitute an
elastoplastic nonlinear substructure, while the remaining elements are
grouped into a linear substructure [8]. In addition, a growing number of
self-adaptive algorithms are becoming available for such aspects as fracture propagation, substructuring, marching schemes, and mesh improvements.
Although many of these self-adaptive algorithms are stable, some have
heuristic bases and are best monitored to ensure the continued validity of
computations. For example, the finite element mesh changes that are
necessary to trace an arbitrarily propagating crack may lead to meshes
with excessively distorted elements; such distortions can be corrected by
the user under an interactive-adaptive approach. Finally, the implementation of interactive-adaptive procedures presents two other possible computing strategies that are particularly significant for nonlinear analysis.

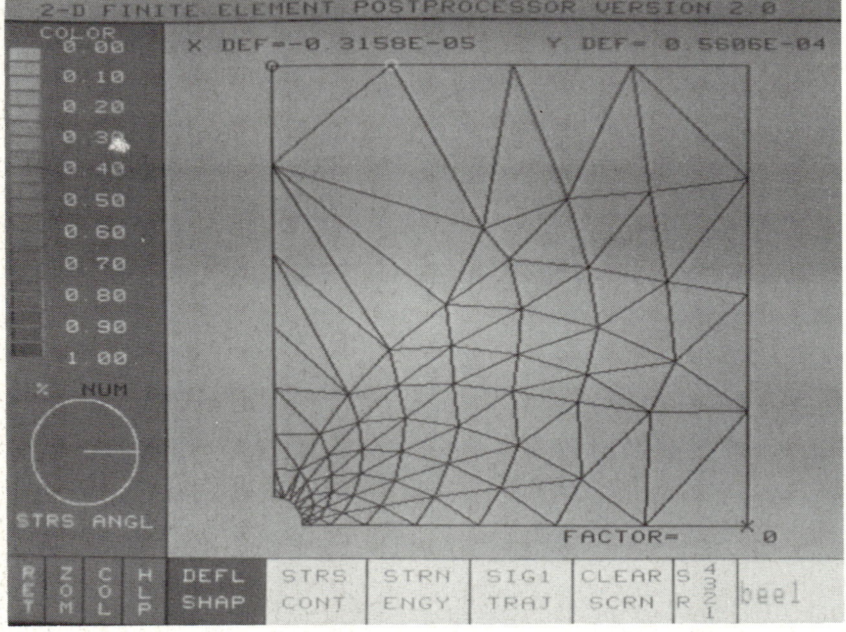

Figure 9. Interactive boundary element postprocessing for the
           displacements of the problem in Figures 7 and 8 using a two-
           dimensional finite postprocessor [14].

The first of these is parallel analysis in which an alternative analysis type is invoked to diagnose system behavior. Examples include buckling and free-vibration tests on nonlinear structures at an arbitrary stage of loading. The second is sequential analysis where different analyses are performed to infer the effect of the order of application of influences. For example, a static preloading may precede a nonlinear structural dynamic analysis.

It is apparent that interactive-adaptive analysis approaches hold strong promise for the exploration, development, and implementation of unified or combined numerical methods. For instance, the same capability for adaptive substructuring illustrated above could be used to determine interactively the partitioning of a geometric entity into "married" finite element and boundary element zones. One of the most exciting uses of interactive-adaptive analysis procedures at Cornell to date has, in fact, been as a "test bed" for research on numerical methods [1].

CONCLUSIONS

The intention of this paper has been to demonstrate how interactive graphics is affecting the environment in which analysis is performed and thereby is influencing possible unification or hybridization of numerical methods. Although this impact is only beginning to be felt in the computational mechanics community, the rapid growth in availability of iteractive graphical capabilities will definitely affect such developments in the near future. It takes only a small leap of imagination to see how interactive techniques already in use for specific numerical methods such

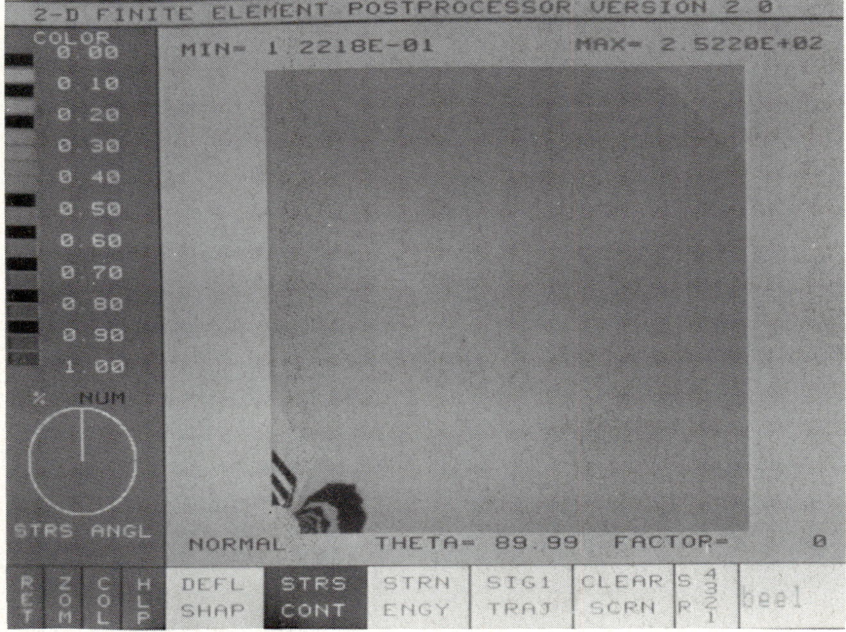

Figure 10.   Full-field stress visualization from a boundary element analysis using the same finite element postprocessor as in Figure 9 [14] [15].

as finite elements can be generalized for a variety of analytical
approaches.

Future developments of interactive graphical software systems for
engineering analysis and design should be designed to account for possible
diversification of analytical procedures. One obvious recommendation is
that strict separation be maintained between system models (geometry and
attributes) and analytical models (meshes or other discretizations).
Under such circumstances, developers of interactive software should be
urged to "go the extra mile" to provide the analysis flexibility that may
be needed for effective engineering design.

ACKNOWLEDGMENTS

Portions of the work used as examples in this paper have been sponsored by
the National Science Foundation, the National Aeronautics and Space
Administration, and by various private companies; the writers are grateful
for this support. However, the views expressed here are the writers' and
are not those of any sponsor. The writers would also like to express
their appreciation to their present and former colleagues at the Program
of Computer Graphics at Cornell University, particularly Donald P.
Greenberg, Director of the Program, and William McGuire, a faculty
co-investigator for much of the research cited.

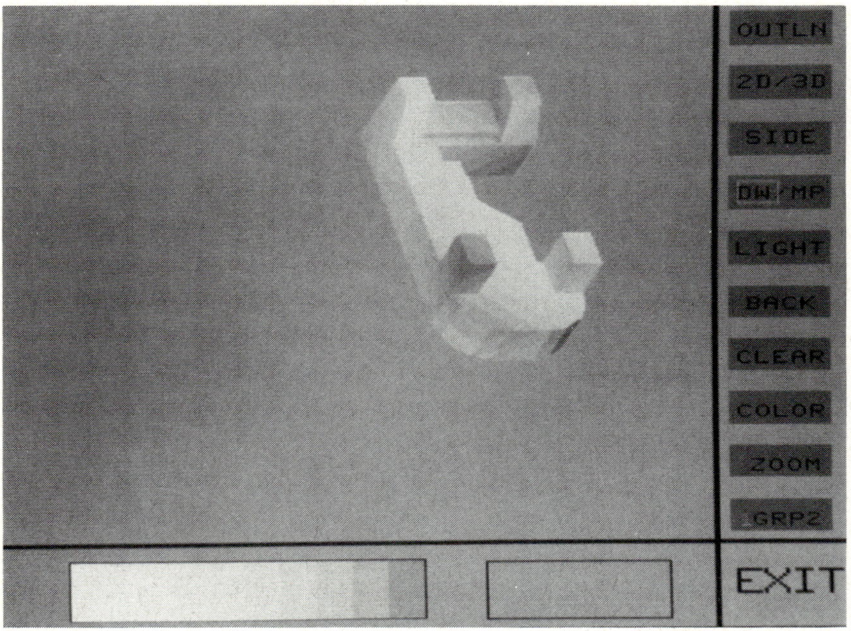

Figure 11.   Postprocessing using color contours on the surface of a solid
             in conjunction with a diffuse reflection model to convey
             three-dimensionality [15].

REFERENCES

[1] Gattass, M., and Abel, J. F., Interactive-adaptive large displacement
     analysis with real-time computer graphics, Computers and Structures
     16 (1983) 141-152.

[2] Kamel, H. A., Design and implementation of interactive engineering
     software, in: Abel, J. F., et al. (eds.), Interdisciplinary Finite
     Element Analysis (College of Engineering, Cornell University, Ithaca,
     NY, 1981) 773-803.

[3] Pesquera, C. I., McGuire, W., and Abel, J. F., Interactive graphical
     preprocessing of three-dimensional framed structures, Computers and
     Structures 17 (1983) 97-104.

[4] Abel, J. F., Interactive computer graphics in applied mechanics, in:
     Proc. 9th U. S. Nat. Cong. App. Mechs., (ASME, New York, 1983)
     97-104.

[5] Yerry, M. A., and Shephard, M. S., Authomatic three-dimensional mesh
     generation by the modified-octree technique, Intl. J. Num. Meth.
     Engrg. (to appear).

[6] Shephard, M. S., and Abel, J. F., The integration of finite element
     methods into CAD/CAM, in: Kardestuncer, H. (ed.), Finite Element
     Handbook (McGraw Hill, New York, to appear 1984).

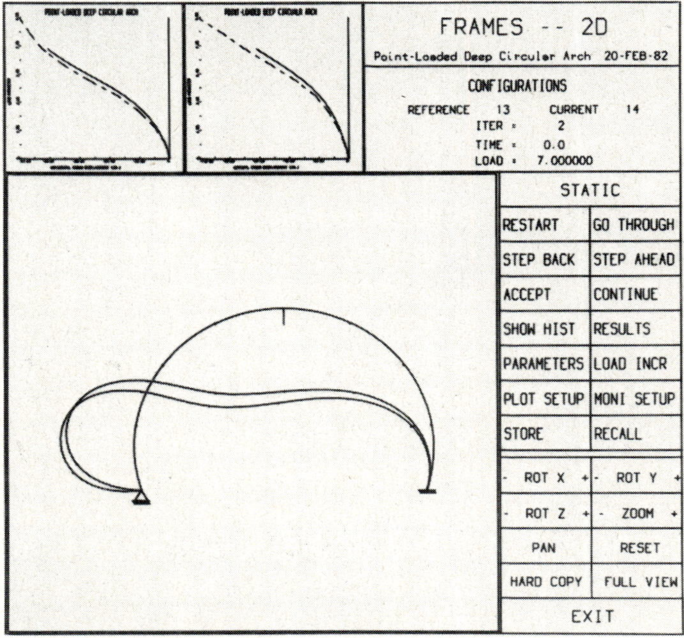

Figure 12. Use of multiple display windows to monitor nonlinear
            structural behavior as it is being calculated during
            interactive-adaptive analysis [1].

[7] Perucchio, R., Ingraffea, A. R., and Abel, J. F., Interactive
     computer graphic preprocessing for three-dimensional finite element
     analysis, Int. J. Num. Meths. Engrg. 18 (1982) 909-926.

[8] Han, T. Y., Adaptive substructuring and interactive graphics for
     three-dimensional finite element analysis, Ph.D. Thesis, Department
     of Structural Engineering, Cornell University (1984).

[9] Smith, B. M., et al., Initial graphics exchange specification (IGES),
     Version 2.0 (NBSIR 82-2531 (AF), National Bureau of Standards, 1982).

[10] Chang, S. C., An integrated finite element nonlinear shell analysis
     system with interactive computer graphics, Ph.D. Thesis, Department
     of Structural Engineering, Cornell University (1981).

[11] Perucchio, R., and Ingraffea, A. R., Interactive computer graphics
     preprocessing for three-dimensional boundary integral element
     analysis, Computers and Structures 16 (1983) 153-166.

[12] Perucchio, R., An integrated boundary element analysis system with
     interactive computer graphics for three-dimensional linear-elastic
     fracture mechanics, Ph.D. Thesis, Department of Structural
     Engineering, Cornell University (1984).

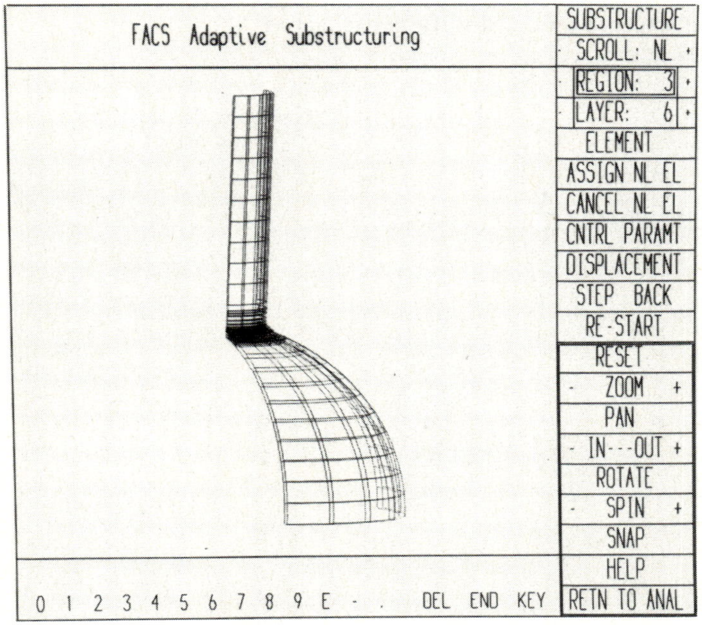

Figure 13.   Interactive-adaptive substructuring during elastoplastic
             analysis [8].   The analyst has selected those elements (shown
             darker) which should constitute a nonlinear substructure
             during the next stage of analysis.

[13] Han, T. Y., A general two-dimensional, interactive graphical
     finite/boundary element proprocessor for a virtual stage environment,
     M.S. Thesis, Department of Structural Engineering, Cornell University
     (1981).

[14] Han, T. Y., unpublished student project (Cornell University, 1982).

[15] Schulman, M. A., The interactive display of parameters on two- and
     three- dimensional surfaces, M.S. Thesis, Department of Architecture,
     Cornell University (1981).

Unification of Finite Element Methods
H. Kardestuncer (Editor)
© Elsevier Science Publishers B.V. (North-Holland), 1984

CHAPTER 3

HYBRID METHODS OF ANALYSIS

S.N. Atluri & T. Nishioka

Three different types of hybrid methods of analysis are discussed: (i) a hybrid-analytical-numerical method, (ii) a hybrid-experimental-numerical method, and (iii) a hybrid-numerical method, for treating complex engineering problems wherein either an analytical (closed form solution) method, or an experiment, or a single-type of numerical method, by itself, is insufficient; and thus a marriage of convenience between the methods is mandatory.

INTRODUCTION

In his invitation to us to contribute to this volume dedicated to Professor J. H. Argyris, one of the greatest mechanicians of this century, the chairman of this conference, Professor Kardestuncer, suggested that we dwell on the theme of "hybrid methods of analysis". It often happens that one's name gets associated with a certain line or school of thought. And so it is that the first author wondered if Professor Kardestuncer suggested the title of this paper because of his (the first author's) happy association with another distinguished academician, Professor Theodore Pian, whose name has become more or less synomymous with "hybrid finite element methods", wherein the word 'hybrid' connotes "inter-finite-element constraints", "Lagrange multipliers", and so forth. Further reflection convinced us that Professor Kardestuncer had something quite different in mind for us to write on - "hybrid methods" - and we were led to ponder about "compatibility" between "methods" and "inter-method" constraints, etc. Then we discovered, much to our pleasant surprise, that in fact we had been using "hybrid methods of analyses" in our research. Thus, the contents of the paper, which we hope would do justice to its predetermined title, began to unfold.

We present in the following, three such "hybrid methods of analyses". The first deals with a "hybrid-analytical-numerical method" of solving for the intensity of stress-state near the front of semi-elliptical or quarter-elliptical surface cracks in commonplace engineering structures. (We use the word 'analytical' in the context of closed form solutions.) The second deals with a "hybrid-experimental-numerical method" used in studies of dynamic fast crack propagation. The last deals with a "hybrid-numerical method" which employs two or more types of discretization of a continuum (space as well as time) problem. A brief discussion of each of these topics ensues.

I.   HYBRID-ANALYTICAL-NUMERICAL METHOD FOR SURFACE FLAWS

One of the most challenging problems of applied fracture mechanics of the past decade has been that of accurate and cost-effective evaluation of "stress-intensity factors" (or the strengths of the singularities of the stress and strain states) along the borders of embedded or surface flaws in complex structural geometries such as aircraft attachment lugs, nuclear reactor pressure-vessel-nozzle junctions, etc. The shapes of these flaws are often assumed, to a first approximation, as elliptical or part-elliptical.

The most common approach has been to numerically model the crack-tip region with either "finite-elements" or their variants (such as "boundary-elements", etc). A variety of such techniques, with a majority of them being classifiable as singular finite element methods (wherein the basis functions for elements near the crack-front are augmented by those that would induce the known "inverse-square-root" singularity in strains near the crack-front), have been reported [1,2] in literature. Even though excellent results for the stress-intensity factors were obtained through these methods, since the crack-front region was explicitly modeled <u>numerically</u> (by some form of discrete elements), the number of "elements" and hence the number of algebraic equations were excessive, thus rendering these methods prohibitively expensive, if not altogether impractical, for routine application purposes.

To circumvent the above problem, the authors have recently developed a methodology based on the Schwarz-Neuman alternating method. In the alternating technique, as applied to the problem of cracks in finite solids, two solutions are needed, generally. One of these solutions is for stresses in the uncracked finite body at the location of the considered crack, and the other solution is for the problem of an infinite body with a crack whose faces are subject to arbitrary normal as well as shear tractions. For cracks in complex-shaped finite bodies subject to complex external loads, the first solution previously mentioned would, in general, lead to a rather complex stress-field at the location of the considered crack. The analytical results for the second solution, mentioned above, have so far been limited to only crack-face tractions that can be represented by polynomials of the cubic order or lower [3,4]. Recently Vijayakumar and Atluri [5] have overcome this limitation and derived a general solution for an embedded elliptical crack, subject to <u>arbitrary-order</u> polynomial type crack-face tractions (both normal as well as shear). This general solution was recast in a form more suitable for application, in a finite-element alternating technique, by Nishioka and Atluri [6]. The methodology developed in [5,6] is applicable, in general, for combined Modes I, II, and III problems of arbitrary-shaped flawed structural components subject to arbitrary external loading. For typical three-dimensional surface-flaw problems, it was shown [6,7,8,9] that the "hybrid-analytical-numerical" procedure developed in [6] is <u>at</u> <u>least</u> <u>one</u> <u>order</u> <u>of</u> <u>magnitude</u> less expensive than the singular-finite-element procedures [1,2] developed in the mid 1970's.

We provide, in the following, some technical details of this "hybrid-analytical-numerical" method.

I.2 <u>Analytical Solution for an Elliptical Crack in an Infinite Solid with Arbitrary Crack-Face Tractions</u>

In this section, only the Mode I problem is considered. The complete, general solution including the Modes II and III is given in Refs. [5,6]. Suppose that x_1 and x_2 are cartesian coordinates in the plane of the elliptical crack and x_3 is normal to the crack plane such that

$$(x_1/a_1)^2 + (x_2/a_2)^2 = 1 , \quad a_1 > a_2 \tag{I.1}$$

describes the border of the elliptical crack of aspect ratio (a_1/a_2). The ellipsoidal coordinates ξ_α (α=1,2,3) are defined as the roots of the cubic equation:

$$\omega(s) = 1 - \left(\frac{x_1^2}{a_1^2+s}\right) - \left(\frac{x_2^2}{a_2^2+s}\right) - \left(\frac{x_3^2}{a_3^2+s}\right) = 0 \tag{I.2}$$

Let the normal traction along the crack surface be expressed in the form:

$$\sigma_{33}^{(0)} = \sum_{i=0}^{1} \sum_{j=0}^{1} \sum_{m=0}^{M} \sum_{n=0}^{m} A_{3,m-n,n}^{(i,j)} \, x_1^{2m-2n+i} \, x_2^{2n+j} \tag{I.3}$$

where A's are undetermined coefficients and the parameters i and j specify the symmetries of the load with respect to the axes of the ellipse, x_1 and x_2.

Using the well-known Trefftz's formulation as discussed in detail in [5], the problem is reduced to that of finding appropriate potential functions. As shown in [5], the solution corresponding to the load expressed by (I.3) can be found in terms of the potential function:

$$f_3 = \sum_{i=0}^{1} \sum_{j=0}^{1} \sum_{k=0}^{M} \sum_{\ell=0}^{k} C_{3,k-\ell,\ell}^{(i,j)} \, F_{2k-2\ell+i,2\ell+j} \tag{I.4}$$

where

$$F_{2k-2\ell+i,2\ell+j} = \frac{\partial^{2k+i+j}}{\partial x_1^{2k-2\ell+i} \partial x_2^{2\ell+j}} \int_{\xi_3}^{\infty} [\omega(s)]^{2k+i+j+1} \frac{ds}{\sqrt{Q(s)}} \tag{I.5}$$

and

$$Q(s) = s(s+a_1^2)(s+a_2^2) \tag{I.6}$$

and C's are also undetermined coefficients. The components of displacement u_i and stress σ_{ij} in terms of the derivatives of the potential functions (a comma followed by an index such as j implies partial differentiation w.r.t. x_j) are given by:

$$u_1 = (1-2\nu)f_{3,1} + x_3 f_{3,31} \tag{I.7a}$$

$$u_2 = (1-2\nu)f_{3,2} + x_3 f_{3,32} \tag{I.7b}$$

$$u_3 = -2(1-\nu)f_{3,3} + x_3 f_{3,33} \tag{I.7c}$$

and

$$\sigma_{11} = 2\mu(f_{3,11} + 2\nu f_{3,22} + x_3 f_{3,311}) \tag{I.8a}$$

$$\sigma_{22} = 2\mu(f_{3,22} + 2\nu f_{3,11} + x_3 f_{3,322}) \tag{I.8b}$$

$$\sigma_{12} = 2\mu(f_{3,12} - 2\nu f_{3,12} + x_3 f_{3,312}) \tag{I.8c}$$

$$\sigma_{33} = 2\mu(-f_{3,33} + x_3 f_{3,333}) \tag{I.8d}$$

$$\sigma_{31} = 2\mu x_3 f_{3,331} \tag{I.8e}$$

$$\sigma_{32} = 2\mu x_3 f_{3,332} \tag{I.8f}$$

where μ and ν are the shear modulus and Poisson's ratio. For later convenience, the stresses given by Eq. (I.8) through Eqs. (I.4-6) are expressed in a matrix form:

$$\{\sigma\} = [P] \ \{C\} \tag{I.9}$$
$$6 \times 1 \quad 6 \times N \quad N \times 1$$

where $[P]$ is the function of the coordinates (x_1, x_2, x_3) and N is the total number of coefficients A or C.

Satisfying the boundary condition on the crack surface, the relation between the coefficients A and C can be summarized in a matrix form:

$$\{A\} = [B] \ \{C\} \tag{I.10}$$
$$N \times 1 \quad N \times N \quad N \times 1$$

The detailed complete expression of components of $[B]$ is given in Ref. [6].

Expressions for the first, second, and third derivatives of the potential functions (I.4,.5) are given in [5,6]. It is shown [5,6] that the evaluation of far-field displacements and stresses in the "infinite body" containing an elliptical crack hinges on the evaluation of a generic elliptical integral of the type:

$$\int_{\xi_3}^{\infty} \partial_1^{k_1} \partial_2^{\ell_1} \partial_3^{m_1} \omega^{k+\ell+1} \frac{ds}{\sqrt{Q(s)}} \tag{I.11}$$

wherein, ∂_α^j implies the jth order derivative with respect to x_α. To evaluate (I.11), we expand $\omega^{k+\ell+1}$ in terms of x_α^2 and carry out term by term differentiations. Then we get:

$$\int_{\xi_3}^{\infty} \partial_1^{k_1} \partial_2^{\ell_1} \partial_3^{m_1} \omega^{k+\ell+1} \frac{ds}{\sqrt{Q(s)}} = (k+\ell+1)! \sum_{p=0}^{k+\ell+1} \sum_{q=0}^{p} \sum_{r=0}^{q} \frac{(-1)^p}{(k+\ell+1-p)!}$$

$$\cdot \frac{(2p-2q)! \ (2q-2r)! \ (2r)!}{(p-q)! \ (q-r)! \ (r)!} \frac{x_1^{2p-2q-k_1} \ x_2^{2q-2r-\ell_1} \ x_3^{2r-m_1}}{(2p-2q-k_1)! \ (2q-2r-\ell_1)! \ (2r-m_1)!}$$

$$\cdot J_{p-q,q-r,r}(\xi_3) \tag{I.12}$$

70   *S.N. Atluri & T. Nishioka*

where ($\cdot$) denotes a multiplication, and

$$J_{p-q,q-r,r}(\xi_3) = \int_{\xi_3}^{\infty} \frac{ds}{(s+a_1^2)^{p-q}(s+a_2^2)^{q-r}(s+a_3^2)^r\sqrt{Q(s)}} \qquad (I.13)$$

Thus, one of the key algebraic steps in the successful application of the analytical results originally derived in [5], in conjunction with an alternating method, is the evaluation of the elliptical integrals of the generic type given in (I.13). An efficient and accurate algorithm for accomplishing this key step has been presented in [6].

I.3   The "Hybrid-Analytical-Numerical" or "Finite-Element Alternating" Method

As noted earlier, the present "hybrid-analytical-numerical" technique is based on two solutions:

Solution 1:

Analytical solution for the problem of an elliptical crack subjected to arbitrary normal and shear loading on the surface of the crack, in an infinite solid, as presently discussed.

Solution 2:

A general numerical technique such as the finite element method or boundary element method for the solution of stresses in the uncracked finite body with geometry and loading being identical to the cracked structure in question.

Solution 1 is discussed earlier, while the finite element method is used in the present paper to generate Solution 2. The present finite element alternating method requires the following steps.

(1) Solve the uncracked body under the given external loads by using finite element method. The uncracked body has the same geometry with the given problem except the crack. To save computation time in solving the finite element equations, a special solution technique is implemented. This wil be explained later.

(2) Using the finite element solution, we compute the stresses at the location of the original crack.

(3) Compare the residual stresses calculated in Step (2) with a permissible stress magnitude. In the present study one percent of the maximum external applied stress is used for the permissible stress magnitude.

(4) To satisfy the stress boundary condition, reverse the residual stresses. Then determine coefficients A of Eq. (I.3) by using the following least-squares method:[3]

$$I_\alpha = \int_{S_c} \left(\sigma^R_{3\alpha} - \sigma^{(0)}_{3\alpha} \right)^2 dS \qquad (\alpha = 1,2,3) \text{ is a minimum} \qquad (I.14)$$

where $\sigma^R_{3\alpha}$ is the reversed residual stresses calculated by the finite element solution, $\sigma^{(0)}_{3\alpha}$ is defined by equation of the type (I.3), S_c is the region of the crack, and I_α are the functionals to be minimized.

Rewriting Eq. (I.3) in a matrix form:

$$\sigma^{(0)}_{3\alpha} = \{L\}^T \{A_\alpha\} \qquad (I.15)$$

and substituting Eq. (I.15) into Eq. (I.14), we obtain the relation between the coefficients A and the residual stresses:

$$\{A_\alpha\} = [H]^{-1} \{R_\alpha\} \qquad (I.16)$$

where

$$[H] = \int_{S_c} \{L\} \{L\}^T \, dS \qquad (I.17)$$

$$\{R_\alpha\} = \int_{S_c} \{L\} \sigma^R_{3\alpha} \, dS \qquad (1.18)$$

(5) Determine the coefficients C of Eq. (I.4) in the potential functions by solving Eq. (I.10) ($\{C\} = [B]^{-1}\{A\}$).

(6) Calculate the stress intensity factors for the current iteration by substituting coefficients C in the following analytical expressions [5,6] specialized here for the "Mode I" problem:

$$K_I = 8\mu \left(\frac{\pi}{a_1 a_2} \right)^{\frac{1}{2}} A^{1/4} \sum_{i=0}^{1} \sum_{j=0}^{1} \sum_{k=0}^{M} \sum_{\ell=0}^{k} (-2)^{2k+i+j} (2k+i+j+1)!$$

$$\frac{1}{a_1 a_2} \left(\frac{\cos\theta}{a_1} \right)^{2k-2\ell+i} \left(\frac{\sin\theta}{a_2} \right)^{2\ell+j} C^{(i,j)}_{3,k-\ell,\ell} \qquad (I.19)$$

where θ is the elliptic angle measured from x_1 axis, and

$$A = a_1^2 \sin^2\theta + a_2^2 \cos^2\theta \qquad (I.20)$$

(7) Calculate the residual stresses on external surfaces of the

body due to the loads in Step (4). To satisfy the stress boundary conditions, reverse the residual stresses and calculate equivalent nodal forces. These nodal forces {Q} can be expressed in terms of coefficients C:

$$\{Q\}_m = - [G]_m \{C\} \qquad (I.21)$$

and

$$[G]_m = \int_{S_m} [N]^T [n][P]dS \qquad (I.22)$$

where m denotes the number for a finite element, [N] is the matrix of isoparametric element shape functions, [n] is the matrix of the normal direction cosines and [P] is the basis function matrix for stresses and can be derived from Eq. (I.8). Although the matrix [P] has the singularity of order $1/\sqrt{r}$ at the crack-front, the functions in [P] decay very rapidly with the distance from the crack-front. Thus, the matrices $[G_m]$ are calculated only at the surface elements which satisfy the condition $r_{min} < 5a_1$, where r_{min} is the distance of the closest nodal point of each surface element from the center of the ellipse and a_1 is the semi-major axis of the ellipse.

(8) Consider the nodal forces in Step (7) as external applied loads acting on the uncracked body. Repeat all steps in the iteration process until the residual stresses on the crack surface become negligible (Step 3). To obtain the final solution, add the stress intensity factors of all iterations. As seen from above, for the finite element alternating method, we need to solve the following type of finite element equations:

$$[K][\underline{q}^0, \underline{q}^1, \ldots, \underline{q}^n] = [\underline{Q}^0, \underline{Q}^1, \ldots, \underline{Q}^n] \qquad (I.23)$$

and

$$\underline{Q}^i = \underline{Q}^i(\underline{q}^{i-1}): \quad i = 1, 2, \ldots, n \qquad (I.24)$$

in which the superscript denotes the cycle of iteration, [K] is the global (assembled) stiffness matrix of the uncracked body and remains the same during the iteration process, and $\underline{q}^i$ is the nodal displacement vector for ith iteration. $\underline{Q}^i$ is the nodal force vector for ith iteration and depends on the solution for the previous iteration $\underline{q}^{i-1}$ as expressed by Eq. (I.24).

An efficient equation solver OPTBLOK developed by Mondkar and Powell [10] is used to save computational time in solving Eq. (I.23).

The solution algorithm is divided into three parts, i.e. (i) reduction of stiffness matrix, (ii) reduction of load vector, and (iii) back substitution. In OPTBLOK the reduction of stiffness matrix is done only once, although the reduction of load vector and back substitution may be repeated for any number of load cases. Thus, denoting CPU time for each part by T_1, T_2, and T_3, respectively. The total CPU time T in solving Eq. (I.23) using OPTBLOK can be expressed by

$$T = T_1 + (n + 1)(T_2 + T_3) = (T_1 + T_2 + T_3) + n(T_2 + T_3) \qquad (I.25)$$

where n is the total number of iterations. Since T_1 is much larger than $(T_2 + T_3)$, a great amount of computational time can be expected to be saved by comparing with the case in which Eq. (I.23) is solved for each iteration $(T^* = (n + 1)(T_1 + T_2 + T_3))$. To illustrate this situation, we consider the example of a set of linear equations with the number of equations of 1960, and half band width of 200, wherein the CPU time for reduction of load vector and back substitution was about 5.6% of the total CPU time $(T_2 + T_3 \simeq 0.056T)$. Since for a typical problem, the present alternating method needs three iterations (n = 3), the additional cost in this case is only about 16.8%, which is considerably smaller than 300% in the case when Eq. (I.25) is solved for each iteration.

Now, some comments concerning the solution of <u>surface flaw</u> problems in finite bodies, through the present "hybrid-analytical-numerical" procedure are in order. Since the analytical solution for an elliptical crack in an infinite solid is implemented as Solution 1, it is necessary to define the residual stresses over the entire crack plane including the fictitious portion of the crack which lies outside of the finite body. Moreover, it is well known that the accuracy of the least squares fitting inside of the fitted region can be increased with the number of polynomial terms; however, the fitting curve may change drastically in the region outside of the fitting. For these reasons, in Ref. [6] numerical experimentation was carried out for arriving at an optimum pressure distribution on the crack surface extended into the fictitious region. For a semi-elliptical crack which lies in the region of $-a_1 \leq x_1 \leq a_1$ and $0 \leq x_2 \leq a_2$, it was concluded that the fictitious pressure which, for the region of $-a_2 \leq x_2 \leq 0$, remains constant in the x_2 direction but varies in the x_1 direction, gives the best result among the several numerical experiments performed in Ref. [6], even though the results for other types of assumed pressure in the fictitious region differed only slightly ($\pm 2\%$). The procedure of the fictitious pressure distribution for a semi-elliptical surface crack was successfully used on the analyses of surface cracks, in finite thickness plates subject to remote tension as well as remote bending [6], and in pressure vessels [7,9]. In the present paper, taking account of the conclusion drawn in Ref. [6], the fictitious pressure distribution given below is employed for the analysis of a quarter-elliptical corner crack. For the first quadrant $(x_1, x_2 \geq 0)$ (namely the actual surface crack), the residual stress can be calculated by the finite element method and is a function of the coordinates x_1 and x_2. For the other quadrants, the fictitious residual stress is defined as

$$\sigma_{33}^R = \begin{cases} \sigma_{33}^R(0, x_2) & \text{for the second quadrant} \quad (x_1 \leq 0, x_2 \geq 0) \\[2mm] \sigma_{33}^R(0,0) & \text{for the third quadrant} \quad (x_1 \ , x_2 \leq 0) \\[2mm] \sigma_{33}^R(x_1,0) & \text{for the fourth quadrant} \quad (x_1 \geq 0, x_2 \leq 0) \end{cases} \qquad (I.26)$$

I.4 Results for Quarter-Elliptical Corner Cracks in Aircraft Attachment Lugs

The geometry of the lug with two symmetric quarter-elliptical corner cracks is shown in Fig. 1. The lug material is 7075-76 Aluminum with Young's modulus E=71.71 GPa $(10.4 \times 10^6$ psi$)$ and Poisson's ratio $\nu = 0.33$. To simulate pin loading, the cosine bearing pressure defined by $\bar{\sigma}_{RR} = -\dfrac{2P}{(\pi R_i t)} \cos \psi$ acting on only a half of the boundary $-\pi/2 \leq \psi \leq \pi/2$, as shown in Fig. 1 is considered. The analysis was performed for nine crack geometries as follows

$$a_p/a_h = 0.5, 1.2, \text{ and } 2.0$$

$$a_h/t = 0.2, 0.5, \text{ and } 0.8$$

where a_p and a_h denote crack lengths at the surfaces of the plate and hole, respectively. Thus, $a_p = a_2$ and $a_h = a_1$ for $a_p/a_h = 0.5$, and $a_p = a_1$ and $a_h = a_2$ for $a_p/a_h = 1.2$ and 2.0.

The typical finite element model used for the uncracked lug is shown in Fig. 2. This model consists of 140 twenty-noded isoparametric elements with 2250 degrees of freedom (before imposition of boundary condition). Due to the symmetry, a half of the lug was used in the analysis. The displacements were imposed as $u_3 = 0$ at $x_3 = -L$, and $u_1 = 0$ at $x_1 = -R_i$. The matrices $[G]_m$ were calculated on the surface of $x_2 = 0$, t, $R = R_o$ $(x_3 \geq 0)$, and $x_1 = R_o - R_i$ $(x_3 < 0)$ satisfying the earlier stated condition, $r_{min} < 5a_1$.

First, only stress analyses of the uncracked lug shown in Fig. 2 were performed to examine the effect of the lug length, changing $L = 5R_i$ to $6R_i$. The average value of normal stress σ_{33} at the original crack location differs only 0.02% as the lug length changes from $5R_i$ to $6R_i$. Thus, the following analyses were done with $L = 5R_i$. In addition, the magnitude of shear stresses which produce the Mode II and Mode III stress intensity factors was also examined. The average value of the shear stresses σ_{31} and σ_{32} were, respectively, 0.5 and 0.1% of the normal stress σ_{33}. Thus, the Mode I stress intensity factor is dominant and other modes are negligible in this case.

To quantify the effects of a finite body, crack aspect ratio, etc., a magnification factor (normalized stress intensity factor) F_i, defined by the following equation, is used

$$F_i(\theta) = \frac{K_I(\theta)}{\left\{ \dfrac{\sigma_i}{E(k)} \sqrt{\dfrac{\pi a_2}{a_1}}\, A^{1/4} \right\}} \tag{I.27}$$

where σ_i is a reference stress magnitude, $E(k)$ is the complete elliptic integral of second kind, $k^2 = (a_1^2 - a_2^2)/a_1^2$ and A is defined by Eq. (I.20). The denominator of the right-hand side of Eq. (I.27) corresponds to the exact stress intensity factor for the elliptical crack subject to the constant pressure σ_i on the crack surface, in an infinite solid. The reference stress σ_i is chosen to be: $\sigma_i = \sigma_p = (P/2R_i t)$. The complete elliptic integral of second kind, $E(k)$, in Eq. (I.27) is given by 1.2111 for $a_p/a_h = 0.5$ and 2.0, and 1.4429 for $a_p/a_h = 1.2$. Figures 3a–c show the magnification factors as a function of the elliptical angle for the aspect ratios of $a_p/a_h = 0.5$, 1.2, and 2.0, respectively. The elliptical angle is always measured from the hole surface in these cases. Figures 3a–c show the magnification factors as a function of the crack length at the plate surface, a_p, for the crack depth of $a_h/t = 0.2$, 0.5, and 0.8. Magnification factors increase as the crack length a_p decreases, due to the fact that the stress concentration exists around the pin hole.

The CPU time for all the above analyses was approximately 1800 seconds using the CYBER 74.

II. HYBRID-EXPERIMENTAL-NUMERICAL (H-E-N) METHODS OF ANALYSIS

In this instance, a combined use of the experimental as well as numerical techniques is made to ascertain information that is, in general, not possible from the use of either an experiment or a numerical analysis alone. Such hybrid techniques are rather commonplace [11].

Here we discuss one hybrid-experimental-numerical technique in particular, involving dynamic crack propagation and arrest. In dynamic fracture mechanics, under either quasi-static or dynamic loading conditions, crack propagation is postulated to commence when the dynamic

S.N. Atluri & T. Nishioka

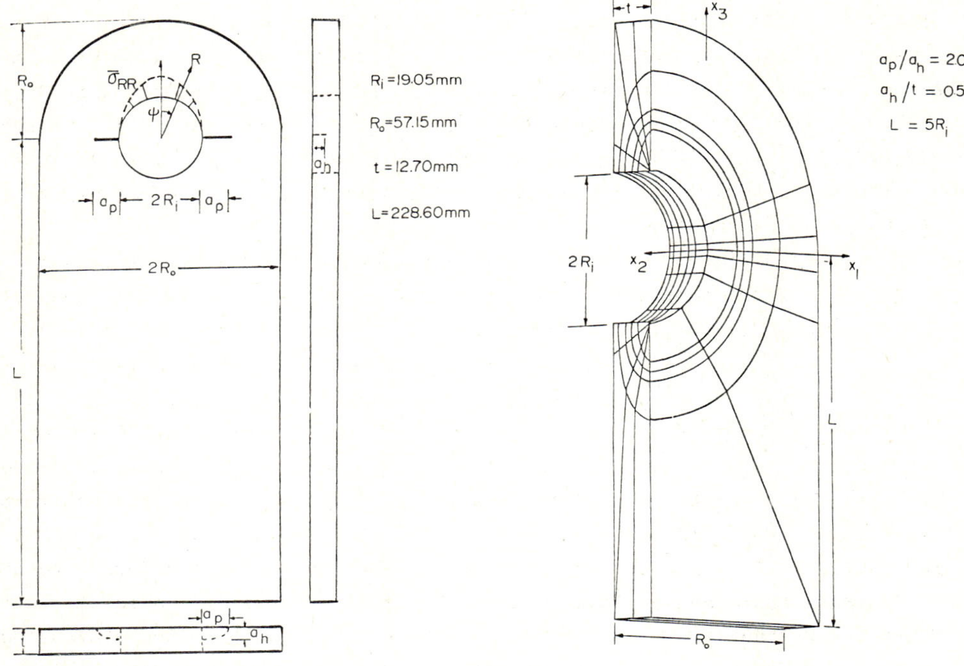

Fig. 1. Pin-Loaded, Corner-Cracked Attachment Lug

Fig. 2. Finite Element Model of the Uncracked Lug

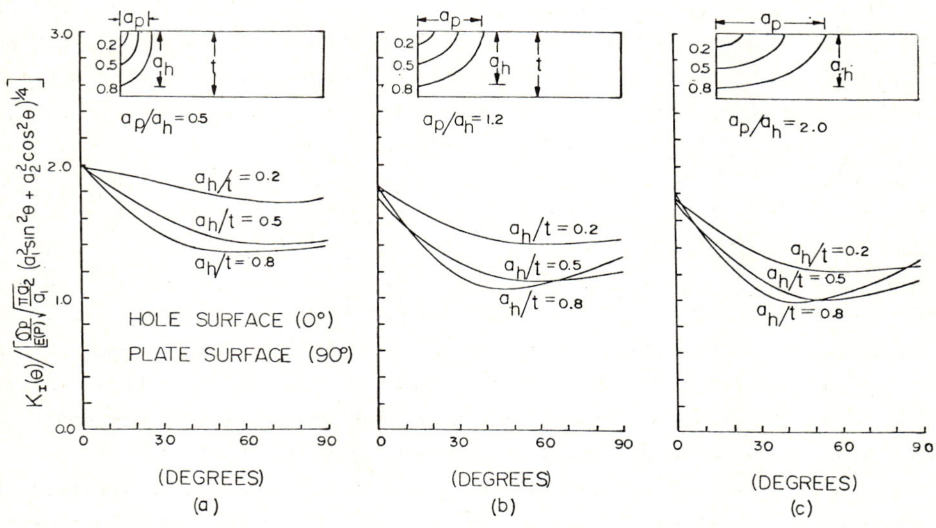

Fig. 3. Stress-Intensity Factor Solutions for 9 Different Crack Geometries in an Attachment Lug

stress-intensity factor near the crack-tip exceeds a given material property, K_{Id}, the so-called initiation toughness. Thereafter, the crack-propagation is governed by the criterion, $K_{It} = K_{ID}(v)$, where K_{It} is the time-dependent-stress-intensity factor near the tip of the propagating crack in the given problem, and K_{ID} is the material's dynamic fracture toughness which depends on the velocity, v, of the propagating crack-tip.

One of the primary aims of experimentation in dynamic fracture is to determine the material's fracture toughness as a function of crack-velocity, $K_{ID}(v)$. This involves measuring directly, the crack-tip stress data. To this end, dynamic photoelastic methods, or methods of caustics, have been developed [12]. However, these techniques for optically measuring the stress-state near the crack-tip are limited to <u>transparent</u> materials, such as photoelastic polymers and resins, and to plane-stress situations which cause thickness changes in the specimen near the crack-tip. However, such direct measurements are not possible for structural steels or other metallic materials under loading conditions that deviate from that of plane-stress. In such situations, a H-E-N technique (also referred to as a "generation calculation") is useful. Thus, for instance, in the dynamic crack-propagation experiment on a specimen, one may carefully measure the initial conditions on the specimen as well as the crack-length versus time history. These experimental data can then be used as input to a well-formulated numerical analysis program to determine the dynamic stress-intensity factor at the tip of the propagating crack. When such analyses are repeated on a number of specimens, one can accurately determine the dynamic fracture toughness property of the material as a function of crack-velocity. It may be worthwhile to note that for dynamic crack propagation in finite elastic bodies, such as the above-mentioned test specimen, the interaction with the crack-tip of stress waves reflected from the boundaries and/or emanated from other moving crack-tips plays an important role in determining the intensity of the dynamic singular stress-field at the considered crack-tip. Because of the analytical intractability of such elasto-dynamic crack problems, computational techniques are mandatory. This, coupled with the above described problems in experimentation, make the H-E-N methods indispensible in studies of dynamic crack propagation and arrest.

Once the material property $K_{ID}(v)$ is determined, it may be used, in conjunction with specified initial conditions and loading conditions on the cracked structure, to <u>predict</u> the history of crack propagation and arrest,if any, in the structure under question. Such a calculation is often called a "prediction calculation".

In the following, we present briefly the details of both the "generation" and "prediction" calculations.

II.1 A Synopsis of the Formulation of the "Generation" Problem

Consider two instants of time t_1 and $t_2 = t_1 + \Delta t$. Assuming, without loss of generality, that the crack propagation is in pure Mode I, let the crack lengths at t_1 and t_2 be Σ_1 and $\Sigma_2 = \Sigma_1 + \Delta\Sigma$, respectively. Let the displacements, strains, and stresses at t_1 and t_2 be, respectively, $(u_i^1, \ \varepsilon_{ij}^1, \ \text{and} \ \sigma_{ij}^1)$ and $(u_i^2, \ \varepsilon_{ij}^2, \ \text{and} \ \sigma_{ij}^2)$. The variables at time t_1 are presumed known. It has been shown [13] that the variational principle governing the dynamic crack propagation between t_1 and t_2 can be written as:

$$\int_{V_2} \dot{\imath}(\sigma_{ij}^2 + \sigma_{ij}^1)\delta\varepsilon_{ij}^2 + \rho(\ddot{u}_i^2 + \ddot{u}_i^1)\delta u_i^2\} \ dV$$

$$= \int_{S_{\sigma_2}} (\bar{T}_i^2 + \bar{T}_i^1)\delta u_i^2 ds + \int_{\Sigma_1^+} (\bar{T}_i^2 + \bar{T}_i^1)^+(\delta u_i^2)^+ \ ds$$

$$+ \int_{\Delta\Sigma^+}(\bar{T}_i^2 + \sigma_{ij}^1 \nu_j^1)^+(\delta u_i^2)^+ \ ds \qquad\qquad (II.1)$$

In the above, V_2 is the domain, and $s_{\sigma 2}$ the external boundary where time-dependent tractions are prescribed, at time t_2; $\bar{T}_i^1$ are the prescribed tractions at t_1 at $s_{\sigma 1}$ $(\simeq s_{\sigma 2})$ as well as at $\Delta\Sigma^+$; $(\ )^+$ indicates the upper half of the crack-face, which only is considered in the present Mode I problem. It is seen that the integrand $(\sigma_{ij}^1 \nu_j^1) \ (\delta u_i^2)$ in the last term of the rhs of Eq. (II.1) corresponds to the term of energy-release rate due to dynamic crack propagation. The Eq. (II.1) may thus be viewed as a virtual energy-balance relation for dynamic crack-propagation, and hence the present numerical method based on Eq. (II.1) is inherently energy-consistent.

In Eq. (II.1), $(\ddot{u}_i^1, \ \sigma_{ij}^1)$ are known, while $(\sigma_{ij}^2, \ \varepsilon_{ij}^2, \ \text{and} \ u_i^2)$ are the variables. Now, Eq. (II.1) is used to develop a finite element approximation at time t_2. Thus, the domain V_2 is discretized into a finite number of elements, with a domain V_s immediately surrounding the crack-tip being treated as the so-called "singular element", and the domain $V_2 - V_s$ being mapped by the well-known, 8-noded, isoparametric elements. In the singular-element V_s, the basis functions for assumed displacements are the crack-velocity dependent eigen-function solutions to the elasto-dynamic problem of crack-propagation in an infinite domain, as discussed in this paper.

Note that at time t_2, in the present Mode I problem, the crack-tip is located at $x_1 = \Sigma_1 + \Delta\Sigma$ and hence the singular-element is centered at $x_1 = \Sigma_1 + \Delta\Sigma$. In developing the equations for the finite element mesh at t_2, it is seen from Eq. (II.1) that the variation of σ_{ij}^1 and u_i^1 must be known in the finite element mesh at t_2. However, σ_{ij}^1, u_j^1, and $\ddot{u}_j^1$ were solved for in the finite element mesh at t_1. In the mesh at t_1, the crack-tip was located at $x_1 = \Sigma_1$, and hence the crack element was centered at Σ_1. Thus, between t_1 and $t_2(t_1 + \Delta t)$ the crack element is translated by an amount $\Delta\Sigma$. While the crack-element is translated, only the elements surrounding the moving crack-tip are distorted. Thus, the finite element meshes at times t_1 and t_2 differ only in the location of the crack-tip (and hence the crack-element) and the shapes of the immediately surrounding isoparametric elements. Thus, the known data at σ_{ij}^1 and u_j in the mesh at t_1 is interpolated easily into corresponding data in the mesh at t_2. Further details of the above translating singularity-element method of simulating dynamic crack propagation in arbitrary shaped finite bodies can be found in [13,14].

We now remark briefly on the basis functions for assumed displacements used in the singular element. Let $x_\alpha(\alpha = 1,2)$ be fixed rectangular coordinates in the plane of the present two-dimensional elastic body, with the crack-tip moving along the x_1 axis and x_2 is normal to the crack-axis. We introduce a coordinate system (ξ, x_2) which remains fixed w.r.t. the propagating crack-tip, such that $\xi = x_1 - vt$, where v is, without loss of generality, the constant speed of crack-propagation. It can be shown [13,14] that the elastodynamic equations, governing this problem, for the wave potentials ϕ (dilatational) and ψ (shear) are:

$$[1 - (v/c_d)^2](\partial^2\phi/\partial\xi^2) + (\partial^2\phi/\partial x_2^2) = -(2v/c_d^2)(\partial^2\phi/\partial t\partial\xi) + (1/c_d^2)(\partial^2\phi/\partial t^2) \quad (\text{II.2})$$

and a similar equation for ψ, except that c_d in Eq. (II.2) is to be replaced by c_s, where c_d and c_s are the dilatational and shear wave speeds respectively. The "steady-state" eigen-function solution to the homogeneous part of Eq. (II.2), namely, the solution which appears time-invariant to an observer moving with the crack-tip, and satisfies the prescribed traction conditions on the crack-face ($\xi < 0$, $x_2 = \pm 0$) can be derived easily, as indicated in [13] and elsewhere. We use these eigen function solutions for an infinite body, as basis functions for assumed displacements within the "crack-tip-singularity-element". However, to satisfy the full Eq. (II.2), the undertermined coefficients, β_j below, in the eigen function expansion are taken to be functions of time. Thus, within the singular element

$$u_\alpha(\xi,x_2,t) = u_{\alpha j}(\xi,x_2,v)\beta_j(t) \quad [\alpha = 1,2; \; j = 1,2...N] \tag{II.3}$$

where $u_{\alpha j}$ are the above described eigen-functions, and β_j are undetermined parameters, which are to be determined from the finite element equations for the cracked body.

As seen from Eq. (II.3), the eigen functions $u_{\alpha j}$ depend on the crack-tip velocity. In the present numerical approach, the crack-tip velocity is assumed to be constant within each time-increment Δt, say v_1 between t_1 and $t_1 + \Delta t$, and v_2 between t_2 and $t_2 + \Delta t$, etc. Thus, between t_1 and $t_1 + \Delta t$, the eigen-functions embedded in the singularity-element correspond to velocity v_1 and those between t_2 and $t_2 + \Delta t$ correspond to velocity v_2. Thus, the present finite element method is capable of handling non-uniform-velocity crack propagation.

The total velocities and accelerations of a material particle in the singular element, within each time step, corresponding to Eq. (II.3), can be written as:

$$\dot{u}_\alpha = u_{\alpha j}\dot{\beta}_j - vu_{\alpha j,\xi}\beta_j \tag{II.4}$$

and

$$\ddot{u}_\alpha = u_{\alpha j}\ddot{\beta}_j - 2vu_{\alpha j,\xi}\dot{\beta}_j + v^2 u_{\alpha j,\xi\xi}\beta_j \tag{II.5}$$

where $(\ ),_\xi = \partial(\ )/\partial\xi$, and $(\cdot)$ implies a time derivative.

The salient features, pertinent to the studies reported in this paper, of the present method, the mathematical details of which are reported elsewhere [13,14], are as follows:

(i) The eigen functions $u_{\alpha 1}$ ($\alpha = 1,2$) lead to the familiar ($1/\sqrt{(r)}$) singularities in strains and stresses. Thus, the coefficient $\beta_1(t)$ is directly related (to within a scalar constant) to the dynamic stress intensity factor, $K_I(t)$.

(ii) The compatibility of displacements, velocities, and accelerations of material particles at the boundary of surrounding elements with those of the surrounding (usual) isoparametric elements is satisfied   through   a   continuous   least-squares   approach.   If the displacements, velocities, and accelerations of the nodes at the boundary of the singular-element, V_s, are q, $\dot{q}$, and $\ddot{q}$ respectively, the above least-squares technique leads to linear algebraic relations between the sets (q, $\dot{q}$, $\ddot{q}$) and (β, $\dot{\beta}$, and $\ddot{\beta}$) where are undetermined parameters in the eigen-function expansion, Eq. (II.3), in the singular-element. From these equations and the final finite element equations governing the nodal   displacements,   velocities,   and   accelerations   of   the   cracked structure, the variables β, $\dot{\beta}$, $\ddot{\beta}$ can be computed directly. Thus, the dynamic   stress-intensity   factor,   as   well   as   its   first   two   time derivatives, are <u>computed directly</u> in the present procedure.

(iii) The "transient" finite element equations are integrated in time, using the well-known Newmark's β-method [13,14].

(iv) Because of the use of the eigen functions in a moving coordinate system, as in Eq. (II.3), in the singular-element, there is the presence of an "apparent" damping matrix for the singular element. Further, for the same reason, this damping matrix as well as the stiffness matrix of the singular-element, are unsymmetric. However, the stiffness and mass matrices of the surrounding isoparametric elements are, of course, symmetric. Thus, the final finite element equation system will have a "small" degree of unsymmetry. This equation system is solved, in the present studies, using a simple iterative scheme.

In the "generation" calculation, the crack velocity history, v(t) in (II.4,.5), is provided from the experiment. This history, as well as the initial conditions, are used as input to the finite element program to determine $K_I(t)$.

II.2 Details of the "Prediction" Calculation

The problem here is to predict the time histories of crack-length [$\Sigma(t)$], crack-velocity [$\dot{\Sigma}(t) \equiv v(t)$], and possible crack-arrest, for a specified relationship of dynamic fracture toughness [K_{ID}] versus crack-velocity [v].

Let the prediction problem be considered to have been solved up to time t_1. In order to find the solution at $t_2(\equiv t_1 + \Delta t)$, the crack-velocity at t_2, namely, $v_2 \equiv \Sigma(t_2)$ must be found. To this end, it is first noted that the dynamic stress-intensity factor can be written as:

$$K_I = K_I(t,v) \qquad\qquad (II.6)$$

Since, in the present procedure, the velocity of crack-propagation is assumed to be constant within each time-step, an approximate procedure to predict the velocity at $[t_1 + (\Delta t)/2]$ will be sought. Using double Taylor series expansion, it is seen from Eq. (II.6) that:

$$K_{Ip}\left(t_1 + \frac{\Delta t}{2};\ v_1 + \Delta v\right) = \sum_{n=0}^{\infty} \frac{1}{n!} \left\{\left(\frac{\Delta t}{2}\right)\frac{\partial}{\partial t} + \Delta v \frac{\partial}{\partial v}\right\}^n K_I(t,v)\Big|_{t_1,v_1} \qquad (II.7)$$

where K_{Ip} is the predicted value of K_I at $t_1 + (\Delta t/2)$. One can, upon expanding terms, write Eq. (II.7) as:

$$K_{Ip} = K_I(t_1,v_1) + \left(\frac{\Delta t}{2}\right)\dot{K}_I(t_1,v_1) + \tfrac{1}{2}\left(\frac{\Delta t}{2}\right)^2 \ddot{K}_I(t_1,v_1) + R \qquad (II.8)$$

$$\equiv \beta_1(t_1) + \left(\frac{\Delta t}{2}\right)\dot{\beta}_1(t_1) + \tfrac{1}{2}\left(\frac{\Delta t}{2}\right)^2 \ddot{\beta}_1(t_1) + R \equiv K_{Ip}^* + R \qquad (II.8)$$

where $(\dot{\ }) = \partial(\ )/\partial t$, and R is "residue" of the Taylor expansion indicated in Eq. (II.8). Note that use is made of the salient feature of the present analysis procedure, that $\beta_1(t) \equiv K_I(t)$, β_1 being the coefficient of the first eigen-functions as in Eq. (II.3).

Since during dynamic crack propagation, $K_I = K_{ID}$, using the predicted K_{Ip} or Eq. (II.8) and the specified K_{ID} vs $\dot{\Sigma}(t)$ relation, the crack velocity $v(\equiv \Sigma)$ at the time $[t_1 + (\Delta t/2)]$ can be predicted. If the arrest dynamic-toughness is K_{ID}^{arr}, crack-arrest is predicted if $K_{Ip} < K_{ID}^{arr}$. Thus, in the present procedure, crack arrest is predicted as a terminal event, if any, in the propagation analysis.

Using the above predicted crack-velocity value, the finite element system of equations at time t_2, based on Eq. (II.1), are constructed; and, from these, the actual dynamic stres-intensity factor $K_I(t_2)[\equiv \beta_1(t_2)]$ is computed. Thus, the actual K_I at $t_1 + (\Delta t/2)$ is computed, as,

$$K_I\left(t_1 + \frac{\Delta t}{2}\right) \simeq (1/2)\left[K_I(t_1) + K_I(t_2)\right] \qquad (II.9)$$

The correlation between the predicted K_{Ip} of Eq. (II.8) and the actual K_I of Eq. (II.9) can be seen to depend on the residue, "R", of Eq. (II.8). To ensure this correlation, a further approximation is introduced in the present work that the residue R at $t_1 + (\Delta t/2)$ can be approximated by its known value at $[t_1 - (\Delta t/2)]$, in the generic sense. Thus, in the present procedure, the generic algorithm used to find K_{Ip} at $t_1 + (\Delta t/2)$ can be written as:

$$K_{Ip}\left(t + \frac{\Delta t}{2}\right) = \beta_1(t_1) + \left(\frac{\Delta t}{2}\right)^2 \dot{\beta}(t_1) + \tfrac{1}{2}\left(\frac{\Delta t}{2}\right)^2 \ddot{\beta}_1(t_1)$$

$$- \left[K_I\left(t_1 - \frac{\Delta t}{2}\right) - K_{Ip}^*\left(t_1 - \frac{\Delta t}{2}\right)\right] \tag{II.10}$$

In all the presently reported computations, when Eq. (II.8) with $R \equiv 0$ was used, a maximum error of the order of 3% between K_{Ip} and K_I was noted. However, when Eq. (II.10) was used, this maximum error reduced to the order of 0.5%.

II.3 Example of a H-E-N or "Generation" Type Analysis

To demonstrate the "generation" type calculations, we first treat a wedge-loaded rectangular double cantilever beam specimen (WL-RDCB), the crack-propagation histories and dynamic stress-intensity factor histories in which were directly measured by Kalthoff et al. [12]. The relevant geometric data of the WL-RDCB specimen are indicated in Fig. 4 which also shows the finite element model wherein the moving-singularity-element is shown hatched, at the beginning of crack propagation. The material constants used in the present analysis are: $E = 3380$ MN/m^2 and Poisson's ratio, $\nu = 0.33$. In the experiments of [12], several test specimens, wherein cracks were initiated from blunted notches with crack-propagation initiation stress-intensity factors K_{Iq} larger than the fracture toughness K_{Ic}, were studied.

Note that the actual loading mechanism in the experiment is closer, in numerical simulation, to loading the finite element model at point A in Fig. 4, with the material to the left-hand side of line BA in Fig. 4 also considered to be participating in the motion. In the first attempt at the analysis, however, the loading was modeled to act a point B in Fig. 4 instead, and the material to the left of line AB was not modeled. In the remainder of the paper, the numerical model wherein load was applied at point A of Fig. 4 and the material to the left of line AB (Fig. 4) was also modeled, is often referred to as the "actual loading condition" and the other one as the "simplified loading condition", respectively.

In their report, Kalthoff et al. [12] identify the RDCB specimen with K_{Iq} value of 2.32 as specimen 4. For convenience, the same identification is used in the presently reported numerical simulation.

As noted earlier, the "generation" calculation uses as input the experimentally measured crack length (and hence crack-velocity) history. The output of the calculation is the directly computed dynamic stress-intensity factor at the tip of the propagating crack for various time instants.

Figure 5 shows the considered crack velocity and length history for RDCB specimen 4 as reported in [12]. Figure 5 also shows the presently computed dynamic K_I as a function of time, along with conparison experimental results of [12]. The present calculation for K_I was performed in three alternate ways: (i) direct computation, since K_I is the same as the undetermined parameter β_1 in the element basis functions as mentioned earlier, (ii) from a crack-tip integral which gives directly the crack-opening energy, and using the crack-velocity dependent relation between K_I and the energy-release rate, and (iii) calculating fracture energy from a global energy balance relation. It is seen that all the three values agree excellently, thus pointing to the inherent consistency of the present numerical procedure. It should be pointed out that the results in Fig. 5 were based on using the forementioned "simplified loading condition". As seen from Fig. 5, the present numerical results exhibit a pronounced peak as compared to the experimental results.

Figure 6 shows variation of different energy quantities: input energy (W); kinetic energy (T); strain energy (U); and fracture energy (F), for RDCB specimen 4, when the "simplified loading condition" is used. It should be noted that in the present procedure, each of the quantities W, T, U, and F is calculated separately and directly. Thus, the fact that U + T + F is equal to W at all times (no other energy dissipation mechanisms are accounted for here) is an inherent check on the accuracy of the calculation. That this is so can be seen from Fig. 6.

Figure 7 demonstrates the effects of the alternate loading-conditions employed in the finite element model of RDCB specimen. In both the cases, the model is loaded so that $K_{Iq} = 2.32$ MN/m$^{3/2}$. For this value of K_{Iq}, the deformation profiles of the crack face when the load is applied at points A and B, respectively, are shown in Fig. 7. It is seen from Fig. 7 that for the same value of K_{Iq}, load (and displacements) at points A and B, respectively, are: 970.7N (and 0.615 mm) and 972.8N (and 0.74 mm). Thus, when the load is modeled to act at B (the so-called "simplified loading case") there is more apparent input of energy to the specimen than when the load is modeled to act at A (the so-called "actual loading case"). When an identical crack-length history as in Fig. 5 is used, but with the "actual loading condition", the

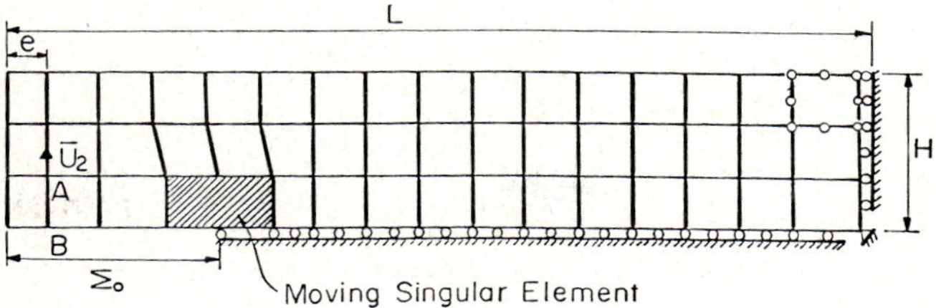

e = 16 mm
h = 10 mm
H = 63.5
L = 321 mm
Σ₀ = 67.8 + 16 mm

Fig. 4.   Finite Element Model of Double Cantilever
Beam Specimen [Propagating Singular Element
Shown Hatched at time t = 0]

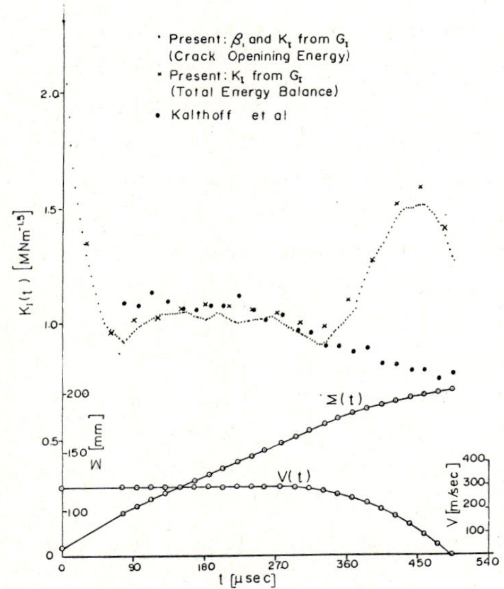

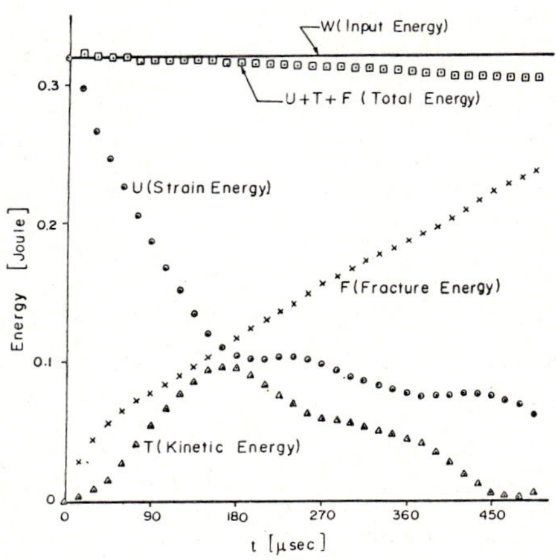

Fig. 5.   Stress-Intensity Factor
Variational with Time for
a Propagating Crack
[Simplified Loading Condition]

Fig. 6.   Energy Variations During
Crack Propagation [Simpli-
fied Loading Condition]

computed dynamic k-factors are shown in Fig. 8. Comparing Figs. 5 and 8, it is seen that an apparently small modification in the load-condition modeling contributes to a substantial difference in the K-factor variation. It is seen that the results in Fig. 8, for the "actual loading case", agree remarkably well with the experimental results (considering the possible rate-sensitive behaviour of Araldite B as opposed to the present linear elastic modeling). The variation in energies W, U, T, F for the "actual loading case" is shown in Fig. 9. Comparing Figs. 6 and 9, it is seen that W in the "simplified loading case" is higher than in the "actual", T is higher in the "simplified" than in the "actual", and that the variations of U and F are qualitatively similar in both the "loading cases".

These results indicate clearly the need to numerically simulate the experimental boundary/initial conditions as closely as possible in order to obtain meaningful results from a H-E-N analysis of the type presented here. In closing it may also be pointed out that even though the above analysis is for a transparent specimen (in which a direct measurement of the K-factor, shown in the above comparison with the numerical results from the present H-E-N analysis, was possible), similar H-E-N analysis on steel specimen were peformed by the authors [15].

III. HYBRID-NUMERICAL METHODS

In the following, we present procedures whereby two (or more) seemingly different numerical methods may be employed simultaneously. Even though the two methods in question may be arbitrary, for convenience we consider the two currently popular ones: the finite element method (FEM) and the boundary element method (BEM). Such coupling methods were discussed elaborately in [16]. Here we present a synopsis.

Let the region, V, occupied by the material be decomposed into two regions V_1 and V_2, where V_1 is the BEM modeled region and V_2 is the FEM modeled region. Let ρ be the interface between V_1 and V_2, as indicated in Fig. 10. The matching conditions are

$$\text{(a)} \quad u_i^+ = u_i^- \text{ at } \rho$$

$$\text{(b)} \quad t_i^+ = -t_i^- \text{ at } \rho \tag{III.1}$$

$$\text{where } u_i^+ = \lim_{\underline{x} \to \underline{x}_\rho} u_i(\underline{x}), \qquad u_i^- = \lim_{\underline{x} \to \underline{x}_\rho} u_i(\underline{x})$$

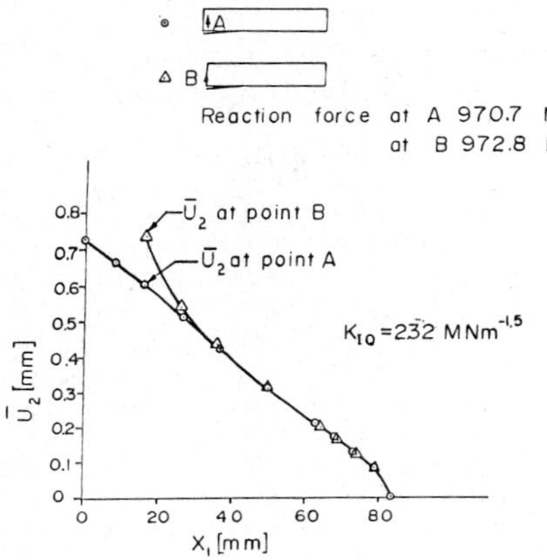

Fig. 7.　Crack Surface Deformation Profiles: A Comparison for 2 Different Loading Conditions

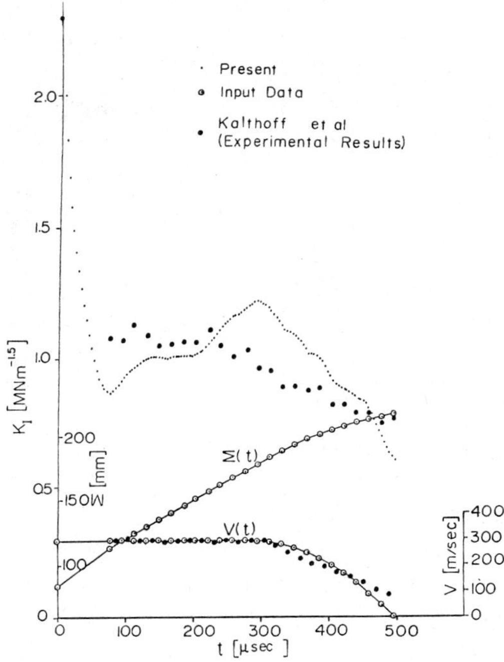

Fig. 8.　Stress-Intensity Factor Variation with Time for a Propagating Crack [Actual Loading Condition]

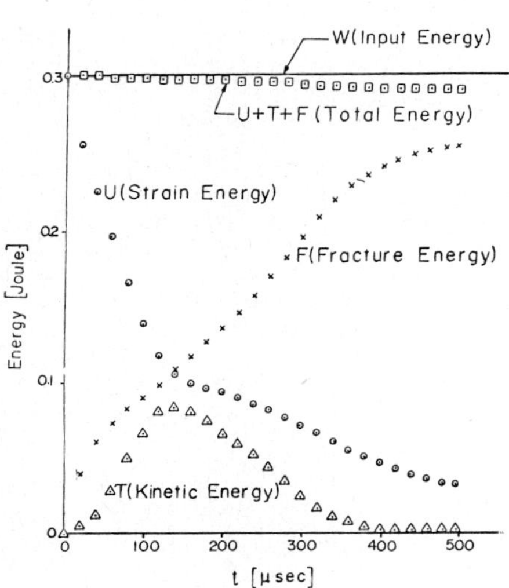

Fig. 9.　Energy Variations During Crack Propagation [Actual Loading Condition]

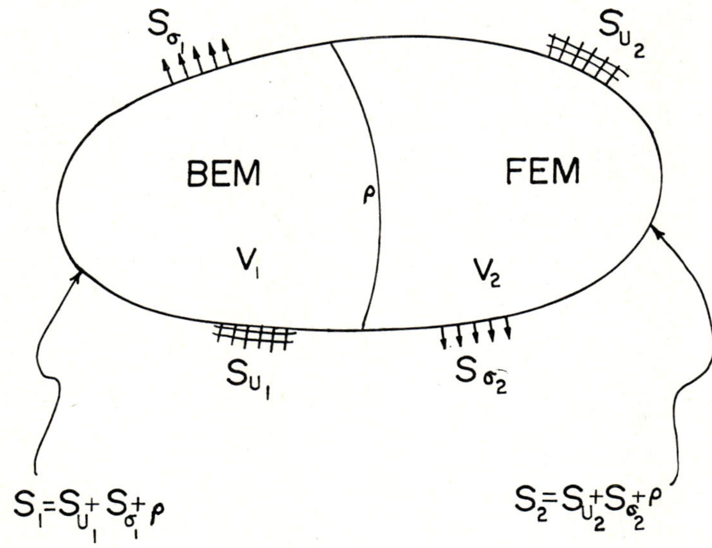

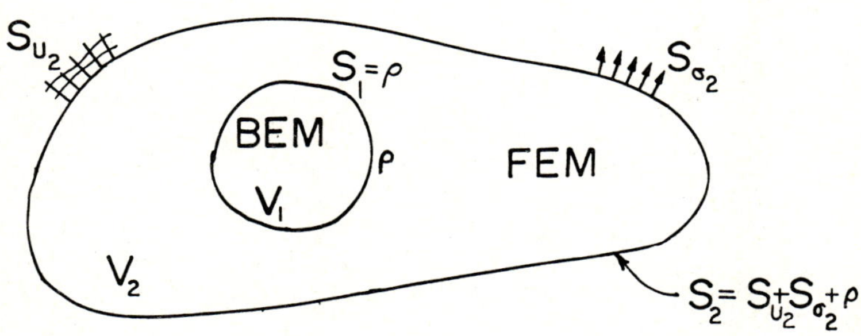

Fig. 10. Schematic Representation of Domains
         Modeled by Different Discretization
         Methods

$$\underline{x} \in V_1 \qquad\qquad \underline{x} \in V_2$$

$$\underline{x}_\rho \in \rho \qquad\qquad \underline{x}_\rho \in \rho$$

with similar definitions for the tractions t_i^+ and t_i^-. Three cases of coupling are considered.

III.1 Coupling of FEM with Direct BIE Method

The notation is given in Fig. 10. Application of the Direct BIE technique yields the equations for V_1:

$$\underline{A}\,\underline{q}^* = \underline{B}\,\underline{Q}^* \tag{III.2}$$

Application of the FEM to V_1 yields the equation

$$\underline{K}\,\underline{q} = \underline{Q} \tag{III.3}$$

III.1.A Direct Coupling.

Equations (II.2,.3) may be writen as follows:

$$\underline{A}\left\{ \begin{array}{c} \underline{q}_a^* \\ \hline \underline{q}_b^* \end{array} \right\} = \underline{B}\left\{ \begin{array}{c} \underline{Q}_a^* \\ \hline \underline{Q}_b^* \end{array} \right\}$$

$$\underline{K}\left\{ \begin{array}{c} \underline{q}_a \\ \hline \underline{q}_b \end{array} \right\} = \left\{ \begin{array}{c} \underline{Q}_a \\ \hline \underline{Q}_b \end{array} \right\} \tag{III.4}$$

where $\underline{q}_a^*$ is the vector of nodal displacements at ρ for the BIE modeled region and where $\underline{q}_a$ is the vector nodal displacements at ρ for the FEM modeled region. The vector $\underline{Q}_a^*$ is that of <u>nodal</u> <u>tractions</u> at ρ for the BIE region, and $\underline{Q}_a$ is the vector of <u>equivalent</u> <u>nodal</u> <u>forces</u> for the FEM region. Two possibilities arise:

(i) lump the tractions at BIE nodes,

(ii) distribute the forces for the FEM region.

By satisfying the equations $\underline{q}_a = \underline{q}_a^*$ and the condition $t_i^+ = -t_i^-$ at ρ, using either (i) or (ii), <u>direct</u> <u>coupling</u> is achieved as in a substracturing procedure. However, the <u>assembled</u> <u>equations</u> for $V_1 \cup V_2$ are <u>unsymmetric.</u> Thus, in the direct-coupling procedure, a significant advantage of the FEM (viz: symmetric banded matrices) is lost without appearing to gain much.

III.1.B Coupling Through the Variational Method.

The function $u_i(p)$, where p is a point in V_1, generated from the solution of (III.2), using the Direct BIE method, satisfies the Navier equations exactly. Let the FEM interpolation for the displacement field at ρ be $\underline{u}_{F\rho}$. Let this be written in the form

$$\underline{u}_{F\rho} = M_{F\rho}[\rho(\underline{x})]\underline{q}_{F\rho} \tag{III.5}$$

where $\underline{q}_{F\rho}$ is the vector of FEM nodal displacements at ρ. Let the known boundary values of u_i and t_i on S_{u_i} and S_{σ_i} be not yet substituted into (III.2). Then the solution $u_i(p)$ in V_1, that satisfies the inter-region continuity condition:

$$\underline{u}^{+} = \underline{u}^{-}$$

at all points of ρ, and the relevant boundary conditions at S_{u_1} and S_{σ_1} can be obtained from the stationary condition of the functional[4]

$$[\pi_p(\underline{u})]_{BIE} = \int_{S_{\sigma_1}} (\tfrac{1}{2}t_i - \bar{t}_i)u_i\,ds - \int_{S_{u_1}} (\tfrac{1}{2}u_i - \bar{u}_i)t_i\,ds - \int_{\rho}(\tfrac{1}{2}u_i - u_{iF_\rho})t_i\,ds \tag{III.6}$$

The displacements and tractions may be interpolated over $S_1 = S_{u_1} \cup S_{\sigma_1} \cup \rho$, as:

$$\underline{u}(Q) = \underline{N}^*(Q)\underline{q}^* \tag{III.7a}$$

$$\underline{t}(Q) = \underline{M}^*(Q)\underline{Q}^* \tag{III.7b}$$

where $\underline{M}^*$ and $\underline{N}^*$ are functions defined appropriately over S_1, Q is a point on S_1, and $\underline{q}^*$ and $\underline{Q}^*$ are the master-vectors of nodal displacements and tractions over S_1. From (III.2) it follows that

$$\underline{Q}^* = \underline{B}^{-1} \underline{A}\, \underline{q}^*$$

(requiring the inversion of an unsymmetric, densely populated matrix). Equation (III.7b) then yields

$$\underline{t}(Q) = \underline{M}^*\underline{B}^{-1}\underline{A}\, \underline{q}^* \tag{III.8}$$

Substitution of (III.8) into (III.6) gives:

$$[\pi_p]_{BIE} = \tfrac{1}{2}\int_{S_{\sigma_1}} \underline{q}^{*T}[\underline{M}^*\underline{B}^{-1}\underline{A}]^T\underline{N}^*\underline{q}^*\,ds$$

$$- \tfrac{1}{2}\int_{S_{u_1}} \underline{q}^{*T}[\underline{M}^*\underline{B}^{-1}\underline{A}]^T\underline{N}^*\underline{q}^*\,ds$$

$$- \int_{S_{\sigma_1}} \underline{t}^T \underline{N}^* \underline{q}^* \, ds + \int_{S_{u_1}} \underline{u}^T \underline{M}^* \underline{B}^{-1} \underline{A} \, \underline{q}^* \, ds$$

$$- \int_{\rho} [\tfrac{1}{2} \underline{q}^{*T} \underline{N}^{*T} - (\underline{M}_{F\rho} \underline{q}_{F\rho})^T] \underline{M}^* \underline{B}^{-1} \underline{A} \, \underline{q}^* \, ds \qquad (III.9)$$

It should be noted that

$$[\pi_p]_{BIE} = \pi_p(\underline{q}^* , \underline{q}_{F\rho})$$

where $\underline{q}_{F\rho}$ are, as yet, unknown. The stationary condition

$$\delta\pi_p \ (\delta\underline{q}^*) = 0$$

leads to algebraic equations for $\underline{q}^*$ in terms of $\underline{q}_{F\rho}$. On expressing $\underline{q}^*$ thus in terms of $\underline{q}_{F\rho}$ the functional π_p can be expressed in the form

$$[\pi_p]_{BIE} = \tfrac{1}{2} \, \underline{q}_{F\rho}^T \, [\underline{K}]_{BIE} \, \underline{q}_{F\rho} + \underline{Q}_{BIE}^T \, \underline{q}_{F\rho} \qquad (III.10)$$

where $\underline{K}_{BIE}$ is now symmetric. Thus, a symmetric equivalent stiffness matrix has been obtained for the BIE modeled region, which can be added to that of the FEM modeled region. Thus, a symmetric system matrix is obtained, at the expense, however, of inverting the unsymmetric matrix $\underline{B}$.

The procedure given in Eq. (III.9) is general and can be used to link several BIE and FEM modeled regions.

A simplification occurs if the condition

$$\underline{u}_B = \underline{u}_{F\rho} \text{ on } \rho \qquad (III.11)$$

where $\underline{u}_B$ is the BIE interpolation for $\underline{u}$. The integral over ρ in (III.6) served the purpose of enforcing this condition. This integral now reduces to

$$\int_{\rho} \tfrac{1}{2} \underline{u}_{F\rho}^T \, \underline{t} \, ds$$

if (III.11) is satisfied a priori. Thus, for the BIE region

$$\underline{u}_B(Q) = \underline{N}^*(Q) \, \underline{q}^* \quad \text{on } S_1 - \rho$$

$$\underline{u}_B(Q) = \underline{u}_{F\rho} = \underline{M}_{F\rho} \, \underline{q}_{F\rho} \quad \text{on } \rho \qquad (III.12)$$

This results in some simplification to the algebra leading to the equivalent stiffness matrix $[K]_{BIE}$ defined in (III.10). However, the inversion of B still remains.

Further simplifications arise if the BIE region is completely surrounded by FEM regions. In this case $S_1 = \rho$ and $S_{u_1} = S_{\sigma_1} = 0$. If the assumed displacement field at ρ for the BIE is identical to that for the FEM assumed displacement field at ρ, then an equivalent symmetric stiffness matrix for the BIE region can, a priori, be obtained as:

$$[K]_{BIE} = \tfrac{1}{2} \int_{\rho} [(M^*B^{-1}A)^T N^* + N^{*T}(M^*B^{-1}A)] \, d\rho \qquad (III.13)$$

The inversion of B still remains.

III.2 Coupling of FEM with Indirect Boundary Solution Method

Consider the mixed boundary value problem for the BEM region V_1:

$$u_i = \bar{u}_i \quad \text{at} \quad S_{\sigma_1} \qquad (III.14a)$$

$$t_i = \bar{t}_i \quad \text{at} \quad S_{\sigma_1} \qquad (III.14b)$$

$$t_i^+ = -t_i^- \quad \text{at} \quad \rho \qquad (III.14c)$$

$$S_1 = S_{u_1} \cup S_{\sigma_1} \cup \rho \qquad (III.14d)$$

where S_1 is the boundary of V_1. The solution may be represented by a single-layer potential:

$$u_j(p) = \int_{S_1} S_i(Q)U_{ji}(p,Q) \, dS_Q \qquad (III.15)$$

where p is a point in V_1, Q is a point on S_1, $S_i(Q)$ the unknown single layer potential, and U_{ji} is the known Kelvin solution [16]. The stress field corresponding to (III.15) may be written as:

$$t_j(P) = -\tfrac{1}{2} S_j(P) + \int_{S_1} S_i(Q)T_{ji}(P,Q) \, dS_Q \qquad (III.16)$$

where P is also on S_1. In vector form, (III.15,.16) may be written as:

$$\underline{u}(p) = \int_{S_1} U(p,Q)\underline{S}(Q) \, dS_Q \qquad (III.17)$$

$$\underline{t}(P) = -\tfrac{1}{2}\,\underline{S}(P) + \int_{S_1} \underline{\underline{T}}(P,Q)\underline{S}(Q)\ dS_Q \qquad\qquad \text{(III.18)}$$

since $\underline{\underline{U}}$ is continuous at the boundary [16], it follows that:

$$\underline{u}(P) = \int_{S_1} \underline{\underline{U}}(P,Q)\underline{S}(Q)\ dS_Q \qquad\qquad \text{(III.19)}$$

Now, $S(Q)$ may be interpolated over S_1 as:

$$\underline{S}(Q) = \underline{\underline{M}}(Q)\underline{\alpha} \qquad\qquad \text{(III.20)}$$

where $\underline{\alpha}$ is a vector of unknown parameters; or the boundary S_1 may be partitioned into elements S_1, S_2, ...S_M; and the potenial $\underline{S}$ may be locally interpolated over __each__ __boundary__ __element.__ The resulting interpolation could still be written in the form of (III.20) where, now, $\underline{\alpha}$ is a vector of nodal values of $\underline{S}$. On substituting (III.20) into (III.17,.18,and .19), we obtain:

$$\underline{u}(p) = \underline{\overline{\underline{M}}}(p)\underline{\alpha} \qquad\qquad \text{(III.21)}$$

$$\underline{u}(p) = \underline{\overline{\underline{M}}}(P)\underline{\alpha} \qquad\qquad \text{(III.22)}$$

$$\underline{t}(P) = \underline{\overline{\underline{N}}}(P)\underline{\alpha} \qquad\qquad \text{(III.23)}$$

Since $\underline{u}(p)$ in (III.17) satisfies the Navier equations of elasticity identically, the one that satisfies the boundary conditions (III.14a,b,c) may be determined from the simplified potential energy functional [16], as the condition:

$$(\ddot{\pi}_p)_{\text{IBS}} = \int_{S_{\sigma_1}} [\tfrac{1}{2}\underline{t}(Q) - \overline{\underline{t}}]^T\!\cdot\underline{u}\ ds - \int_{S_{u_1}} [\tfrac{1}{2}\underline{u} - \overline{\underline{u}}]^T\!\cdot\underline{t}\ ds$$

$$- \int_{\rho} [\tfrac{1}{2}\underline{u} - \underline{u}_{F\rho}]^T\!\cdot\underline{t}\ ds \qquad \underline{\text{is minimum}} \qquad \text{(III.24)}$$

and

$$\underline{u}_{F\rho} = \underline{\underline{M}}_{F\rho}\underline{q}_{F\rho} \qquad\qquad \text{(III.25)}$$

where $\underline{q}_{F\rho}$ are nodal displacements at ρ, and $\underline{\underline{M}}_{F\rho}$ are the interpolation functions. On substituting (III.21,.22,.23, and .25) into (III.24) and varying $(\pi_p)_{\text{IBS}}$ with respect to both $\underline{\alpha}$ and $\underline{q}_{F\rho}$, one obtains:

$$\tfrac{1}{2}[\underset{\sim}{P}{}^* + \underset{\sim}{P}{}^{*T}] \left\{ \frac{\alpha}{q_{F\rho}} \right\} = \underset{\sim}{Q}{}^T \tag{III.26}$$

where $(\underset{\sim}{P}{}^* + \underset{\sim}{P}{}^{*T})/2$ is the symmetric stiffness matrix of the region V_1 modeled by indirect boundary solution (IBS) method. Eqs. (III.26) may now be added to other symmetric equations of the FEM modeled region V_2. Thus, unlike the direct boundary integral method, no unsymmetric-matrix inversions arrive in the case of coupling of IBS method with FEM. However, a close examination of Eqs. (II.17,.18,.19, and .24) reveals that surface integrations must be performed twice.

ACKNOWLEDGEMENTS

The results presented herein were obtained during the course of investigations supported by the U.S. AFOSR under grant 81-0057C to Georgia Institute of Technology. The authors gratefully acknowledge this support as well as the encouragement of Dr. A. Amos. It is a pleasure to sincerely thank Ms. J. Webb for her assistance in the preparation of this manuscript.

FOOTNOTES

1. Regents' Professor of Mechanics
2. Visiting Assistant Professor
3. Note, however, that in the presently considered symmetric 'mode I' problem only σ_{33}^R and $\sigma_{33}^{(0)}$ are <u>nonzero.</u>
4. This can be derived [16] from the usual potential energy functional, when the displacement field, in addition to satisfying the compatibility condition, also satisfies the equilibrium equations.

REFERENCES

[1]   Atluri, S. N., "Higher-Order, Special, and Singular Finite Elements", Chapter 4 in: <u>State-of-the-Art Surveys on Finite Element Technology</u> (Eds.: A. K. Noor and W. D. Pilkey), ASME, New York, NY, (1983), pp. 37-126.

[2]   Atluri, S. N. and Kathiresan, K., "3-D Analyses of Surface Flaws in Thick-Walled Reactor Pressure Vessels Using a Displacement-Hybrid Finite Element Method", <u>Nuclear Engineering and Design</u>, Vol. 51, No. 2, (1979), pp. 163-176.

[3]   Kobayashi, A. S., Enetanya, A. N., and Shah, R. C., "Stress-Intensity Factors for Elliptical Cracks" in: <u>Prospects of Fracture Mechanics</u> (Eds.: G. C. Sih, H. C. van Elst, and D. Brock), Noordhoff Int. Pub., (1975), pp. 525-544.

[4]   Sorensen, D. R. and Smith, F. W., "Semielliptical Surface Cracks Subjected to Shear Loading" in: <u>Pressure Vessel Technology, Part II (Materials and Fabrication) Proceedings</u>, Vol. 3, ICPVT, Tokyo,

ASME, NY, (1977), pp. 545-551.

[5]   Vijayakumar, K. and Atluri, S. N., "An Embedded Elliptical Crack, in an Infinite Solid, Subject to Arbitrary Crack-Face Tractions", Journal of Applied Mechanics, Vol. 48, (March 1981), pp. 88-96.

[6]   Nishioka, T. and Atluri, S. N., "Analytical Solution for Embedded Elliptical Cracks, and Finite Element Alternating Method for Elliptical Surface Cracks, Subjected to Arbitrary Loadings", Engineering Fracture Mechanics, Vol. 17, No. 3, (1983), pp. 247-268.

[7]   Nishioka, T. and Atluri, S. N., "Analysis of Surface Flaws in Pressure Vessels by a New 3-Dimensional Alternating Method" in: Aspects of Fracture Mechanics in Pressure Vessels and Piping, ASME PVP, Vol. 58, (1982), pp. 17-35.

[8]   Nishioka, T. and Atluri, S. N., "Integrity Analyses of Surface-Flawed Aircraft Attachment Lugs: A New, Inexpensive, 3-D Alternating Method," AIAA Paper No. 82-0742, 23rd SDM Conference, AIAA/ASCE/ASME/AHS, (10-12 May 1982), New Orleans, pp. 287-300.

[9]   O'Donoghue, P., Nishioka, T., and Atluri, S. N., "Multiple Surface Cracks in Pressure Vessels", Engineering Fracture Mechanics (In Press), Georgia Tech Report (1983).

[10]   Mondkar, D. P. and Powell, G. H., "Large Capacity Eqn. Solver for Structural Analysis", Computers and Structures, Vol. 4, (1974), pp. 699-728.

[11]   Kobayashi, A. S., "Hybrid Experimental-Numerical Stress Analysis", Experimental Mechanics, Vol. 23, No. 3, (1983), pp. 338-347.

[12]   Kalthoff, J. F., Beinert, J., and Winkler, S., "Measurements of Dynamic Stress Intensity Factors for Fast Running and Arresting Cracks in Double Cantilever Beam Specimens" in Fast Fracture and Crack Arrest (Eds.: G. T. Hahn, and M. F. Kanninen), ASTM STP 627, (1977), pp. 161-176.

[13]   Atluri, s. N., Nishioka, T., and Nakagaki, M., "Numerical Modeling of Dynamic and Nonlinear Crack Propagation in Finite Bodies by Moving Singular Elements" in Nonlinear and Dynamic Fracture Mechanics (Eds.: N. Perrone and S. N. Atluri), ASME-AMD Vol. 35, (1979), pp. 37-67.

[14]   Nishioka, T. and Atluri, S. N., "Numerical Modeling of Dynamic Crack Propagation in Finite Bodies, by Moving Singular Elements, Part 1 - Formulation, Part II-Results", Journal of Applied Mechanics, Vo. 47, (1980), pp. 570-583.

[15]   Nishioka, T. and Atluri, S. N., "Finite Element Simulation of Fast Fracture in Steel DCB Specimen", Engineering Fracture Mechanics, Vol. 16, No. 2, (1982), pp. 157-175.

[16]   Atluri, S. N. and Grannell, J. J., Boundary Element Methods (BEM) and Combination BEM-FEM, Report GIT-ESM-SA-78-16, Georgia Institute of Technology, (1978), 84 pp.

Unification of Finite Element Methods
H. Kardestuncer (Editor)
© Elsevier Science Publishers B.V. (North-Holland), 1984

CHAPTER 4

THE POSTPROCESSING TECHNIQUE IN THE FINITE ELEMENT METHOD. THE THEORY & EXPERIENCE

I. Babuška, K. Izadpanah, & B. Szabo

The paper addresses the h, p, and h-p versions of the finite element method in connection with a postprocessing technique for extracting the values of a functional. This technique combines the finite element method with the analytical ideas of the theory of partial differential equations of elliptic type.

1. INTRODUCTION

Finite element computations in structural mechanics usually have two purposes: (1) to determine the stress and displacement fields and (2) to determine the values of certain functionals defined on displacement fields as, for example, the stress intensity factors, stresses at specific points, reactions, etc. Computations of these values involve the finite element solution. For example, the stress components are often computed at the Gauss points of the elements and the stresses at any other points are then computed by the interpolation technique, the stress intensity factors is determined by the J-integral or curve fitting technique, etc. We shall refer to these operations as **postprocessing**.

Usually the values of these functionals are needed to be known with higher accuracy and reliability than the displacement or stress field itself.

---

[1]Partially supported by the Office of Naval Research under grant number N00014-77-C-0623.

[2]Partially supported by the Office of Naval Research under grant number N00014-81-K-0625.

Assuming that we have the finite element solution and wish to determine certain functional values the following questions arise:

1) What should the relationship be between the computational effort spent on the finite element solution and the effort spent on postprocessing: Is it better to use a very simple and inexpensive postprocessing technique as for example direct evaluation of the stresses from the finite element solution in the desired points or should one select a more expensive technique. Of course we have to relate the answer to the achieved accuracy and to the reliability and robustness of the postprocessing procedures under consideration.

2) Given a finite element solution, what is the largest accuracy of the functional values one can achieve by the postprocessing technique. In other words, what is the maximal information contained in the finite element solution which could be used for the extraction of the desired value.

3) How do the various versions of the finite element method, i.e., the h-version, the p-version and the h-p version bear on the importance of proper selection of the postprocessing techniques.

These questions are discussed in some details.

2. THE EXTENSION OPERATORS. THE h, p AND h-p VERSIONS OF THE FINITE ELEMENT METHOD

There are three versions of the finite element methods based on the common variational (energy) principle. They are characterized by the systematic selection (extension) of the finite element spaces leading to the convergence of the finite element solutions to the exact one.

The **h-version** is the classical and most commonly used method of extension: the polynomial degree of elements p is fixed and mesh refinement is used for controlling the error of approximation (h refers to the size of the element). Typically the polynomial degree of elements is low, usually $p = 1$ or 2. Proper selection of the mesh and its refinement strongly influences the error and its dependence on the computational effort.

In the **p-version** the mesh is fixed and the polynomial degree of elements is increased either uniformly or selectively over the mesh.

The **h-p version** combines the h and p-versions, i.e., error reduction is achieved by a proper mesh refinement and concurrent changes in the distribution of the polynomial degree of elements.

The performance of the various extensions operators can be compared from various points of view, the most important of which are human and computer resources requirement in relation

to the desired level of precision. Such relationships are
difficult to quantify and are subject due to various factors,
therefore the performance of the extension operators is usually
related to the number of degrees of freedom N. Of course
evaluation of an extension operator would not be meaningful
without considering the goals of computation. For example, if
only stress intensity factors are desired, then the accuracy of
the computed displacements, reactions or stresses are not of
importance. In many cases the computation has multiple goals.

3. THE MODEL PROBLEM

In order to illustrate the essential properties of finite
element solution and extraction techniques, we
have selected a model problem which represents
some of key features of a large class of engineering
problems. Specifically let us consider the plane
strain problem of two-dimensional elasticity
(homogeneous isotropic material) with E and ν
representing the modulus of elasticity and Poisson
ratio respectively (E > 0, 0 < ν < .5). The domain
D, a square panel with a crack is shown in Fig. 1.

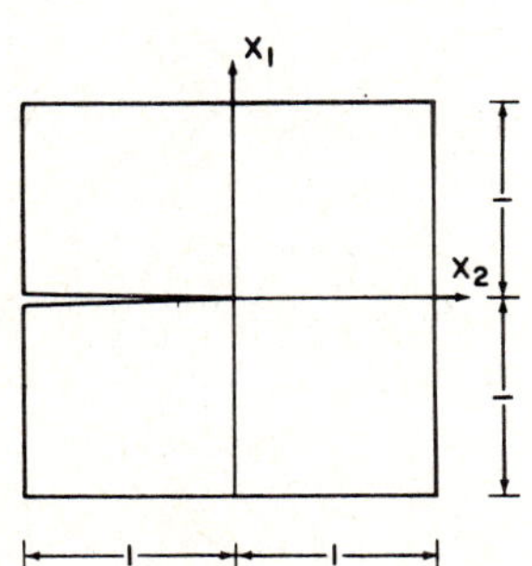

Figure 1
The model problem

We shall be concerned here with problems in which only
tractions are prescribed at the boundary (i.e., first boundary
value problem of elasticity).

We denote the displacement vector function by $\underline{u} = \{u_1, u_2\}$
and the corresponding stress tensor by

$$T = \begin{bmatrix} \tau_{11} & \tau_{12} \\ \tau_{21} & \tau_{22} \end{bmatrix}, \qquad \tau_{12} = \tau_{21}.$$

The strain energy functional is

$$W(u) = \frac{E}{2(1-2\nu)(1-\nu)} \int_D \left[(1-\nu)\left(\frac{\partial u_1}{\partial x_1}\right)^2 + 2\nu \frac{\partial u_1}{\partial x_1}\frac{\partial u_2}{\partial x_2} \right.$$

$$\left. + (1-\nu)\left(\frac{\partial u_2}{\partial x_2}\right)^2 + \frac{1-2\nu}{2}\left(\frac{\partial u_1}{\partial x_1} + \frac{\partial u_2}{\partial x_2}\right)^2 \right] dx_1\, dx_2 . \qquad (3.1)$$

The solution u satisfies the Navier-Lamé equations. It is possible to express the solution through two holomorphic functions $\phi(z)$, $\psi(z)$ using the theory of Muskhelishvili [1].

$$2\mu(u_1 + iu_2) = \kappa\phi(z) - z\overline{\phi'(z)} - \overline{\psi(z)} \tag{3.2}$$

where

$$z = x_1 + ix_2, \quad \mu = \frac{E}{2(1+\nu)}, \quad \kappa = 3 - 4\nu \tag{3.3}$$

and $\bar{z} = x_1 - ix_2$, resp. $\overline{\phi'(z)}$ mean conjugate values to z and $\phi'(z)$.

The components of the stress tensor are expressed by Kolosov-Muskhelishvili formulae

$$\tau_{11} + \tau_{22} = 2(\phi'(z) + \overline{\phi'(z)}) = 4\,\mathrm{Re}\,\phi'(z)$$

$$= 2(\Phi(z) + \overline{\Phi(z)}) \tag{3.4}$$

$$\tau_{11} - \tau_{12} + 2i\tau_{12} = 2[\bar{z}\phi''(z) + \psi'(z)] = 2[\bar{z}\Phi'(z) + \Psi(z)] \tag{3.5}$$

where

$$\Phi(z) = \phi'(z), \qquad \Psi(z) = \psi'(z) \tag{3.6}$$

and $\mathrm{Re}\,\phi'(z)$ is the real part of $\phi'(z)$.

The correspondence between the displacements (and the stress) field and the functions ϕ and ψ is one to one up to the constants γ and γ' in ϕ and ψ, respectively, satisfying the relation $\gamma - \overline{\gamma}' = 0$.

In our model problem we consider the following (exact) solution

$$\Phi(z) = (1+i)z^{-1/2} \tag{3.7}$$

$$\Omega(z) = \Phi(z) \tag{3.8}$$

$$\Omega(z) = \Phi(z) + z\overline{\Phi}'(z) + \overline{\Psi}(z) \tag{3.9}$$

where $\overline{\Phi}(z) = \overline{\Phi(\overline{z})}$, $\overline{\Phi'}(z) = \overline{\Phi'(\overline{z})}$, $\overline{\Psi}(z) = \overline{\Psi(\overline{z})}$.

$\Omega(z)$ is a holomorphic function on D. Function $z^{-1/2}$ is to be understood as the principal branch of $z^{-1/2}$ on D. Function $\Psi(z)$ is uniquely defined by (3.9) and (3.7) (3.8). The tractions on the boundary of D are defined by (3.4) (3.5). It can be readily verified that the two edges of the crack are traction free.

We will now discuss the finite element solution and the postprocessing technique if the tractions are prescribed on the boundary of D so that the exact solution to the problem is given by (3.7)-(3.9). Specificaly we now consider the case E = 1, $\nu = 3$. The strain energy of the exact solution is: W = 42.16491240.

4.   THE FINITE ELEMENT SOLUTION

We have solved the model problem by the h and p-versions of the finite element method. The p-version of the finite

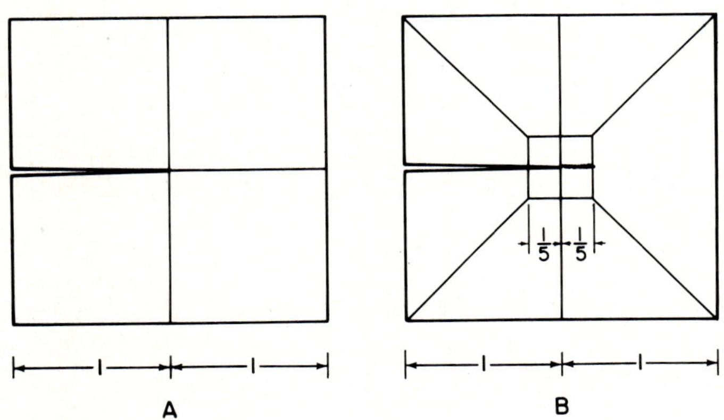

Figure 2
The meshes for the p-version, A: Mesh 1, B: Mesh 2

element method was implemented in the experimental computer program COMET-X developed at the Center for Computational Mechanics of Washington University in St. Louis [2]. The two meshes shown in Fig. 2A,B were used. The polynomial degrees were the same for all elements and ranged from 1 to 8. The shape functions on trapezoidal elements of mesh 2 were constructed by blending function technique.

The h-version solution was obtained by means of the computer program FEARS developed at the University of Maryland [3, 4].

FEARS uses quadrilateral elements of degree one. The program
is adaptive and produces a sequence of nearly optimal meshes.
See [3] [4] [5] [6] [7]. The mesh from this sequence with 319
elements and number of degrees of freedom N = 617 is shown in
Fig. 3.

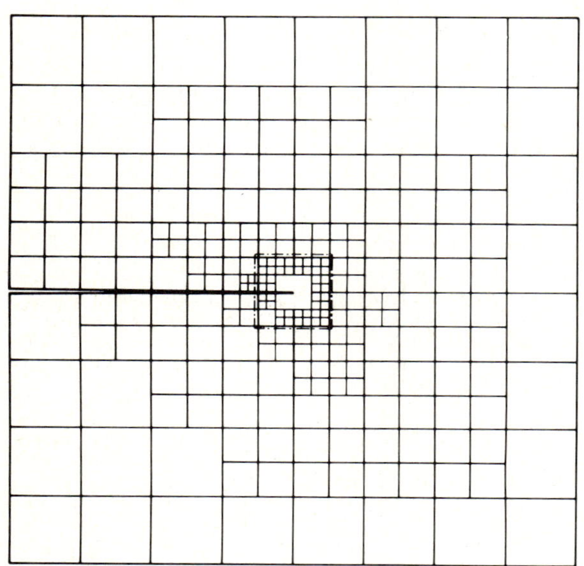
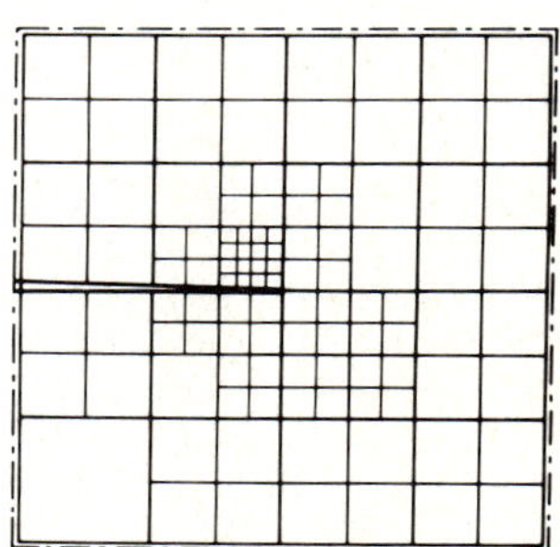

Figure 3
The mesh constructed by the adaptive program FEARS

5. ERROR OF THE FINITE ELEMENT SOLUTION MEASURED IN ENERGY NORM

We denote the exact solution by $\underline{u}_0$ and the finite element
solution by $\underline{u}_{FE}$. The error of the finite element solution is
denoted by $\underline{e}$,

$$\underline{e} \; = \; \underline{u}_0 - \underline{u}_{FE}.$$

We measure the magnitude of the error by the **energy norm** $\| \cdot \|_E$,

$$\|\underline{e}\|_E \; = \; \left[W(\underline{e}) \right]^{1/2} . \tag{5.1}$$

This measure is equivalent to measuring the error in the stress components by integrals of its squares (the L_2 norm). In our case when tractions are specified at the boundary

$$W(\underline{u}_{FE}) \ < \ W(\underline{u}_0) \tag{5.2}$$

and

$$\|e\|_E \ = \ \left[W(\underline{u}_0) - W(\underline{u}_{FE})\right]^{1/2} . \tag{5.3}$$

The extensions operators under consideration monotonically increase the finite element spaces either by increasing the degree of elements or refining the mesh. Therefore the energy norm of the error monotonically decreases. We can write

$$\|e\|_E \ \leq \ C(N)N^{-\mu} \tag{5.4}$$

and expect that for properly chosen μ the function $C(N)$ is nearly constant especially for larger N. The number $\mu > 0$ is the rate of convergence of the error measured in the energy norm.

It is possible to estimate the value of μ. In our case the rate μ is governed by the strength of the singularity of the solution. It can be shown that for the p-version [8],[9]

$$\|e\|_E \ < \ C(\epsilon)N^{-(1/2 - \epsilon)} \tag{5.5}$$

with $\epsilon > 0$ arbitrarily small and C independent of N. The h-version using the uniform mesh yields the estimate

$$\|e\|_E \ < \ CN^{-1/4} \tag{5.6}$$

with the rate independent of the degree of elements. The optimal refinement of the mesh leads to the estimate

$$\|e\|_E \ < \ CN^{-p/2} \tag{5.7}$$

(FEARS uses $p = 1$) where the rate is independent of the strength of the singularity.

The h-p version with optimal mesh and p-distribution leads to the estimate

$$\|\underline{e}\|_E \ < \ Ce^{-\gamma N^\theta}$$

where $\theta = 1/3$ independently of the strength of the singularity and $\gamma > 0$.

The relative error in the energy norm defined as

$$\|\underline{e}\|_{E,R} = \frac{\|\underline{e}\|_E}{\|\underline{u}_0\|_E} \tag{5.8}$$

has been plotted in Fig. 4 on log-log scale for the p-version (mesh 1,2), for the h-version with

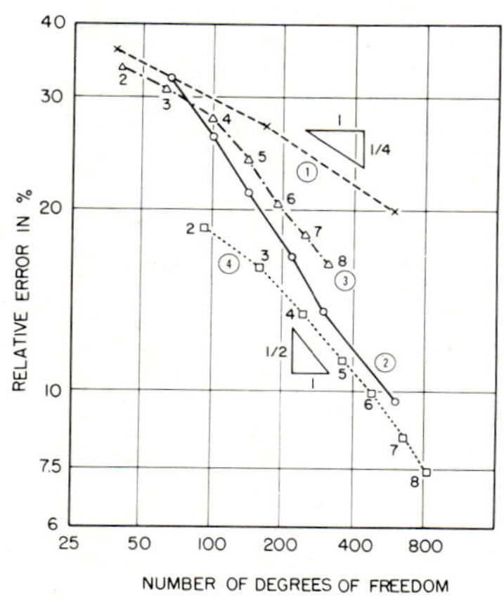

Figure 4
Relative error in the energy norm vs degrees of freedom
(1) h-version, uniform mesh, (2) h-version, adaptively constructed mesh, (3) p-version Mesh 1, (4) p-version Mesh 2

adaptively constructed mesh and for the h-version with uniform mesh. The polynomial degree of elements is also shown in the figure. The shown slopes are the theoretical slopes of the rate of convergence [$\mu = 1/2$ and $1/4$]. It is seen that the observed rate of convergence closely agrees with (5.5)(5.7). From (5.4) we can compute $C(N)$ for the p-version. The results are given in Table 1.

Tables 2 and 3 show analogous results for the h-version. The comparison between Tables 1-3 shows that for 5% accuracy we need $N = 1770$ when using p-version Mesh 2, $N = 2290$ for h-version with adaptively refined mesh and $N = 146000$ for h-version with uniform mesh.

TABLE 1

Relationship between $\|e\|_{E,R}$ and N for the p-version, Mesh 2 $[\mu = \frac{1}{2}]$

p	N	$\|e\|_{E,R}$	$C(N)/\|u_0\|_E$
1	35	32.61%	2.010
2	95	18.35%	1.816
3	135	15.89%	1.997
4	239	13.24%	2.059
5	347	11.06%	2.061
6	479	9.47%	2.079
7	635	8.27%	2.088
8	815	7.37%	2.099

TABLE 2

Relationship between $\|e\|_{E,R}$ and N for the h-version with adaptively constructed mesh $[\mu = \frac{1}{2}]$

N	$\|e\|_{E,R}$	$C(N)/\|u_0\|_E$
67	32.91%	2.035
101	26.38%	2.665
143	21.35%	2.562
221	16.79%	2.501
301	13.61%	2.366
617	9.63%	2.394

TABLE 3

Relationship between $\|e\|_{E,R}$ and N for the h-version with uniform mesh $[\mu = \frac{1}{4}]$

N	$\|e\|_{E,R}$	$C(N)/\|u_0\|_E$
51	36.02%	.967
167	27.07%	.974
591	19.81%	.977

5. COMPUTATION OF THE STRESSES

The finite element method provides the solution u_{FE} which converges to the exact solution in the energy norm. We have seen that the error measured in this norm decreases monotonically and very orderly. We now examine the pointwise error an stresses for the h and p-version. We denote the error in the stress components as

I. Babuška et al.

$$e_{ij}(x_1,x_2) = \tau_{ij}^{[0]}(x_1,x_2) - \tau_{ij}^{[FE]}(x_1,x_2) \qquad (6.1)$$

and the relative error by

$$e_{ij}^{R}(x_1,x_2) = \frac{|e_{ij}(x_1,x_2)|}{|\tau_{ij}^{[0]}(x_1,x_2)|}$$

where $\tau_{ij}^{[0]}$ and $\tau_{ij}^{[FE]}$ are respectively the stress components corresponding to the exact and finite element solution.
We will compute the stresses directly from the derivatives of
$\bar{u}_{FE}$ and stress-strain law. Fig. 5 shows the relative error
e_{ij}^{R} in τ_{ij} at the point $(.0,.1)$ computed by the p-version.

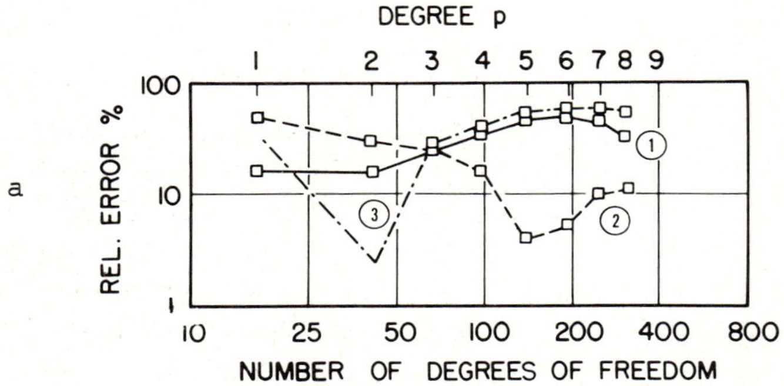

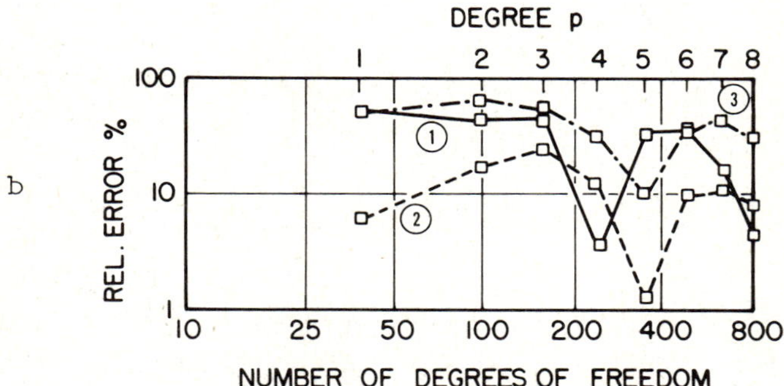

Figure 5

The relative error of e_{ij}^{R} computed by the p-version
a) Mesh 1, b) Mesh 2. (1) e_{11}^{R}, (2) e_{22}^{R}, (3) e_{12}^{R}

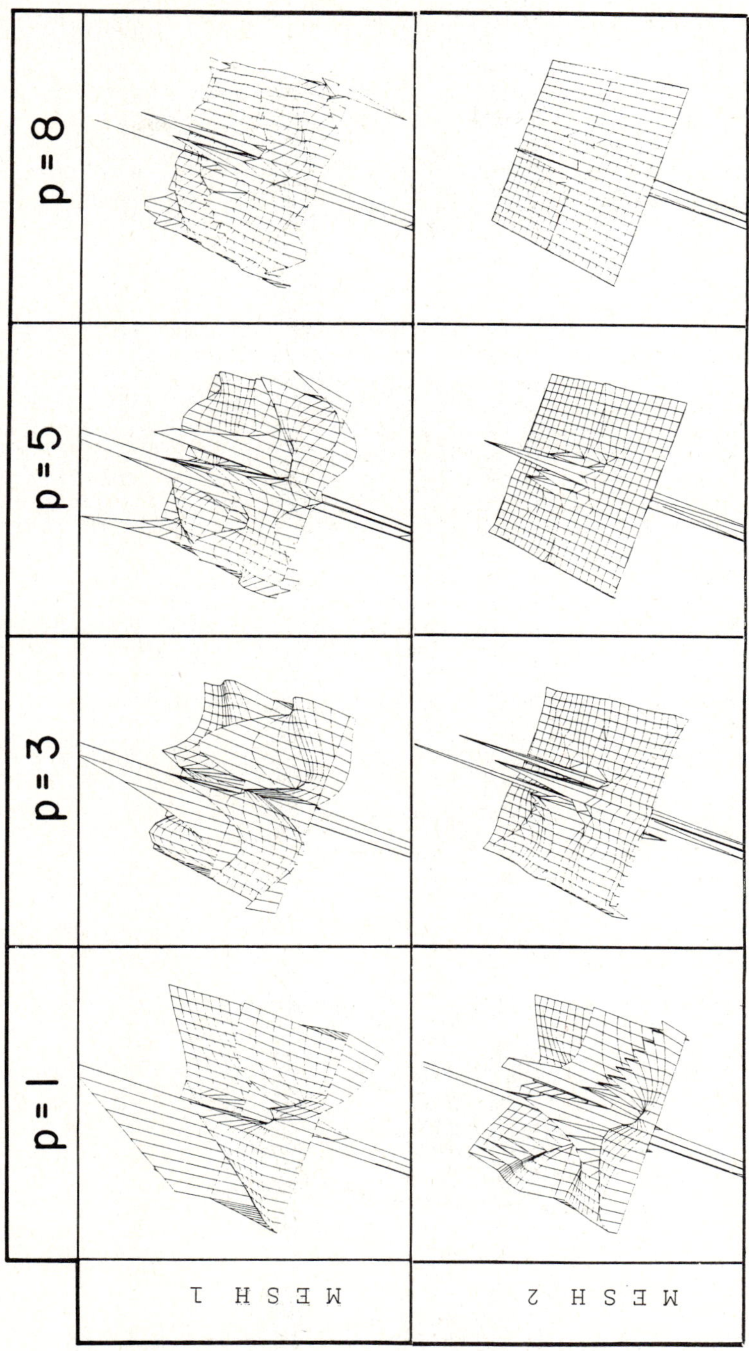

Figure 6

The behavior of the error e_{22} of τ_{22} computed by the p-version.

Fig. 6 shows isometric drawings of the error in τ_{22} for various p values for meshes 1 and 2. The error values were computed on a uniform grid with the grid points (ih,jh) h = .1, i,j = − 10, 10. At points other than the grid points, the values were computed by linear interpolation.

In the case of the h-version, the error is discontinuous at the boundary of every element. Therefore we compute the stresses in the center of every element where increased accuracy can be expected.

In Fig. 7 we show the level-lines of the error in τ_{22} (using the mesh shown in Fig. 3) in the upper right quarter of the domain D. The local maxima and minima are shown also in the figure. The error is large in the neighborhood of the tip of the crack. The level-lines and the local maxima and minima depend on the used interpolation technique. We see in contrast to the p-version that the oscillating behaviour of the error is not so strong here; nevertheless, it has to be underlined that if the stresses will be computed everywhere directly from displacements strong oscillatory behaviour will appear in every element.

The center of the elements are changing with the mesh. To show the convergence of the stresses, we selected for the Table 4 the center points which are closest to the tip of the crack (in the first quarter of D). The table shows the error in % and the magnitude of the exact values of the stress.

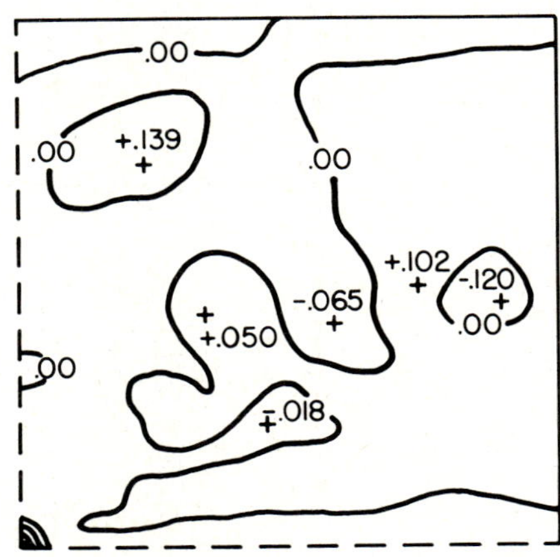

Figure 7

The level lines of the error e_{22} of τ_{22} in the upper

quarter of D computed by the h-version

TABLE 4

The relative error of the stresses in the neighborhood of the origin.

No. of elements	N	Coordinates		$\left|e_{11}^{R}\right|$ $\left|\tau_{11}^{[0]}\right|$	$\left|e_{22}^{R}\right|$ $\left|\tau_{22}^{[0]}\right|$	$\left|e_{12}^{E}\right|$ $\left|\tau_{12}^{[0]}\right|$	
		x_1	x_2				
16	51	.25	.25	35.54% 5.037	19.90% 3.751	20.70% 1.553	Adaptive Mesh
43	143	.125	.125	33.95% 7.124	10.84% 5.304	15.14% 2.197	
106	221	.3125(−1)	.3125(−1)	31.01% 14.251	6.09% 10.609	12.35% 4.394	
319	617	.7125(−2)	.7125(−2)	29.77% 28.501	2.40% 21.218	17.65% 8.789	
16	51	.25	.25	35.54% 5.037	19.90% 3.751	20.70% 1.553	Uniform Mesh
64	167	.125	.125	36.34% 7.124	20.07% 5.304	14.52% 2.197	
256	591	.625(−2)	.625(−1)	34.46% 10.076	17.16% 7.502	14.09% 3.107	

To depict the behaviour in a fixed point (.25, .25) we select
the center points closest to it. Table 5 shows the results.
If we desire to compute the stress components in the nodal
point (.25, .25) we have 4 values for disposition and also
their average. Table 6 we shows the relative errors. The
value in the lines 1, 2, 3, 4 are computed from the elements
ordered counterclockwise starting with the upper-right one.
The line A shows relative error of the average of the stress
values computed in the four elements.

In contrast to the monotonic and orderly behaviour of the error
measured in the energy norm, the accuracy in the stresses is
poor and nonmonotonic, although the stresses are converging in
integral sense (in the energy norm) monotonically. In
addition, the quality of the computed stress components is very
different.

TABLE 5

The relative error of the stresses in the neighborhood of (.25,.25).

No. of elements	N	Coordinates		$\dfrac{|e_{11}^R|}{|\tau_{11}^{[0]}|}$	$\dfrac{|e_{22}^R|}{|\tau_{22}^{[0]}|}$	$\dfrac{|e_{12}^R|}{|e_{12}^R|}$	
		x_1	x_2				
16	51	.25	.25	35.54% 5.037	19.90% 3.751	20.70% 1.553	Adaptive Mesh
43	143	.375	.375	1.09% 4.113	2.10% 3.062	43.51% 1.268	
106	221	.1875	.1875	3.89% 5.817	8.26% 4.331	40.28% 1.794	
319	617	.21875	.21875	.464% 5.386	.812% 4.010	12.12% 1.661	
16	51	.25	.25	35.54% 5.037	19.90% 3.751	20.70% 1.553	Uniform Mesh
64	167	.373	.375	8.01% 4.113	7.79% 3.062	16.06% 1.268	
256	591	.1875	.1875	10.20% 5.817	10.78% 4.331	6.45% 1.794	

7. POSTPROCESSING

We have seen that stresses computed directly from finite element solutions are not accurate. Nevertheless, often the values of the stresses is the main aim of the computation.

We will show now that by utilizing the analytical structure of the Navier-Lamé equations it is possible to compute stresses with the accuracy comparable to the accuracy of the energy of the finite element solution (which is the square of the error measured in the energy norm). We will outline the main idea. For more, see [8], [9], [10].

Let $x_0 = (x_{0,1},x_{0,2}) \in D$ and denote by $S(x_0,\rho)$ the disc of radius ρ centered in $\underline{x}_0$. Further, let $D(x_0,\rho) = D -S(x_0,\rho)$. See Fig. 8. The boundary of $D(x_0,\rho)$ is denoted by $\partial D(x_0,\rho) = \partial\Omega \cup \Gamma$ where Γ is the boundary of the disk $S(x_0,\rho)$. We now define the **extraction** (displacement) function $\underline{w}(x_0,x) \equiv (w_1,w_2)$ which corresponds to the functions $\hat{\phi}$, $\hat{\psi}$ in the sense of (3.2) (3.3) and are defined as follows

$$\hat{\phi}(z) \;=\; A(z - z_0)^{-1} + \hat{\psi}*(z) \qquad\qquad (7.1)$$

TABLE 6

The relative error of the stresses in (.25.25)

No. of elements	N		e_{11}^{R} %	e_{22}^{R} %	e_{12}^{R} %	
43	143	A 1 2 3 4	.034 .042 2.41 .026 3.09	1.60 4.97 6.60 1.76 3.14	33.95 4.35 3.38 72.22 64.41	Adaptive Mesh
106	221	A 1 2 3 4	10.99 4.79 12.21 17.27 9.71	7.57 2.31 2.01 17.42 13.01	11.16 35.86 .093 106.43 71.49	
319	617	A 1 2 3 4	4.09 1.46 4.41 4.41 6.07	5.47 2.68 .099 13.43 11.74	13.42 22.96 55.69 3.88 28.85	
64	167	A 1 2 3 4	12.12 19.37 6.75 5.87 17.51	10.65 15.35 8.68 5.94 12.63	17.36 13.36 5.47 68.41 60.22	Uniform Mesh
256	591	A 1 2 3 4	8.11 8.10 5.38 8.12 10.84	10.13 6.62 5.05 13.64 15.84	12.66 13.38 8.15 38.71 33.48	

$$\hat{\xi}(z) \;=\; B(z-z_0)^{-1} + \hat{\xi}*(z) \qquad\qquad (7.2)$$

$$\hat{\psi}(z) \;=\; \hat{\xi}(z) - z_0\hat{\phi}'(z) \qquad\qquad (7.3)$$

where $\hat{\phi}*(z)$ and $\hat{\xi}*(z)$ are arbitrary holomorphic functions on D (not only on $D(x_0,\rho)$). Note that $\hat{\phi}$, $\hat{\psi}$ are holomorphic on $\Omega(x_0,\rho)$ for any $0 < \rho$.

Although the domain $D(x_0,\rho)$ is doubly connected, the

displacement function w defined by (7.1) through (7.3) by
(3.2) is a single valued function, and it is an admissible
displacement function.

Denote by $T^{[u]}$, $T^{[w]}$ the stress tensors associated with the

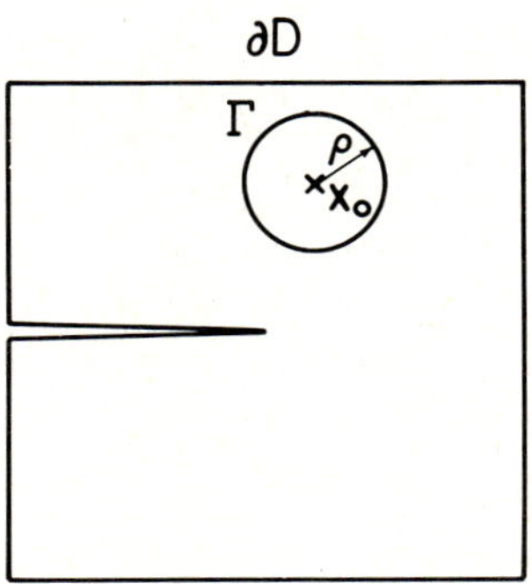

Fig. 8. The domain $D(x_0,\rho)$.

displacement functions u and w. Denote the outward normal to
$\partial\Omega(x_0,\rho)$ by n. Then Betti's law can be written in the form

$$\int_{\partial D(x_0,\rho)} (\underline{u},T^{[w]}\cdot n)\,ds \;=\; \int_{\partial D(x_0,\rho)} (\underline{w},T^{[u]}\cdot n)\,ds \;. \qquad (7.4)$$

This equation can be rewritten

$$\int_{\partial D} [(\underline{u},T^{[w]}\cdot n) - (\underline{w},T^{[u]}\cdot n)]\,ds$$

$$(7.5)$$

$$= \int_{\Gamma} [-(\underline{u},T^{[w]}\cdot n) + (\underline{w},T^{[u]}\cdot n)]\,ds \;.$$

The functions ϕ, ψ associated to the solution $\underline{u}$ can be
written in the neighborhood of z_0:

$$\phi(z) \;=\; a_0 + a_1(z-z_0) + O((z-z_0)^2) \qquad (7.6)$$

$$\xi(z) = b_0 + b_1(z-z_0) + 0((z-z_0)^2) \tag{7.7}$$

$$\psi(z) = \xi(z) - \bar{z}_0\phi'(z). \tag{7.8}$$

Using (7.1)-(7.3) and (7.6)-(7.8) in (7.5) and letting $\rho \to 0$ we get

$$\int_{\partial D} (u,T^{[w]}\cdot n)ds - \int_{\partial D} (w,T^{[u]}\cdot n)ds$$

$$= \frac{\Pi}{2\mu}[b_1(\kappa A + A) + \bar{b}_1(\kappa\bar{A} + \bar{A}) + a_1(1+\kappa)B + \bar{a}_1(1+\kappa)\bar{B}]. \tag{7.9}$$

By application of (3.4)-(3.6) we get

$$\tau_{11}(\underline{x}_0) = 2Re(a_1 + \bar{a}_1 - b_1) \tag{7.10}$$

$$\tau_{22}(\underline{x}_0) = 2Re(a_1 + \bar{a}_1 + b_1) \tag{7.11}$$

$$\tau_{12}(\underline{x}_0) = Im\, b_1 . \tag{7.12}$$

By proper selection of A, B we can obtain that the right hand side of (7.9) be $\tau_{i,j}$. Note that any choice of $\hat{\phi}_*$ and $\hat{\xi}_*$ in (7.1) and (7.2) does not change the right hand side of (7.9).

In our problem when the tractions are prescribed at ∂D, the function $g(x) = T^{[u]}\cdot n$ is given. (7.9) can therefore be written in the form

$$F = \int_{\partial D} (\underline{u}_0,T^{[w]}\cdot n)ds - \int_{\partial D} (\underline{w},g)ds \tag{7.13}$$

where F is (for proper choice of A, B) the exact value of the stress component at $x = x_0$. Of course u_0 is not known but u_{FE} is. Therefore we define

$$F_{FE} = \int_{\partial D} (\underline{u}_{FE},T^{[w]}\cdot n)ds - \int_{\partial D} (\underline{w},\underline{g})ds \tag{7.14}$$

By subtracting (7.13) (7.14), the error in the extracted functional F_{FE} (provided that integrals are evaluated exactly) is

$$F - F_{FE} = \int_{\partial D} (\underline{u}_0 - \underline{u}_{FE},T^{[w]}\cdot n)ds. \tag{7.15}$$

Let us analyze now (7.15). To this end, let $\underline{v} = (v_1, v_2)$ be
the (exact) solution of the problem when tractions $T^{[w]} \cdot n$ are
prescribed at ∂D. $\underline{v} \neq \underline{w}$ because $\underline{w}$ is singular at $x = x_0$,
but $\underline{v}$ is not. Existence of v is guaranteed because $T^{[w]} \cdot n$
satisfy the equilibrium condition. We can write

$$\int_{\partial D} (\underline{u}_0 - \underline{u}_{FE}), T^w \cdot n)\,ds \;=\; 2W(\underline{u}_0 - \underline{u}_{FE}, v) \qquad (7.16)$$

where $W(u,v)$ is the usual energy scalar product associated
with $W(u)$ defined in (3.1). Using one of the basic property
of the finite element method, namely

$$W(\underline{u} - \underline{u}_{FE}, \underline{v}_{FE}) \;=\; 0, \qquad (7.17)$$

we obtain from (7.15) (7.16)

$$F - F_{FE} \;=\; 2W(\underline{u}_0 - \underline{u}_{FE}, \; \underline{v} - \underline{v}_{FE})$$

and hence

$$|F - F_{FE}| \;<\; 2\|\underline{u}_0 - \underline{u}_{FE}\|_E \|\underline{v} - \underline{v}_{FE}\|_E \cdot \qquad (7.18)$$

So far we did not discuss the choice of $\hat{\phi}_*(z)$ and $\hat{\xi}_*(z)$.
(7.18) shows that $\hat{\phi}_*$ and $\hat{\xi}_*$ should be selected so that
$\|\underline{v} - \underline{v}_{FE}\|$ is at least of the order of $\|u_0 - u_{FE}\|$.

If $\|\underline{v} - \underline{v}_{FE}\|_E \approx C\|\underline{u} - \underline{u}_{FE}\|_E$ we get $|F - F_{FE}| <$
$C\|u_0 - u_{FE}\|_E^2 < C(W(u_0) - W(u_{FE}))$ and the rate of convergence
is twice that of the rate of the error measured in the energy
norm. Note that inequality (7.18) is upper bound which
neglects possible cancellation in the energy integral.

8. SELECTION OF THE EXTRACTION FUNCTION

When x_0 is not close to the boundary of D, then we can
select $\hat{\phi}_* = \hat{\xi}_* = 0$. When x_0 is close to ∂D, then
$\hat{\phi}_*$ and $\hat{\xi}_*$ should be selected so that $T^{[w]} \cdot n = 0$ on that
part of the boundary which is close to x_0. Otherwise, we
would not achieve that $\|\underline{v} - \underline{v}_{FE}\|_E$ will be small.

In the following we outline briefly the procedure for
constructing $\hat{\phi}_*$ and $\hat{\xi}_*$ so that $T^{[w]} \cdot n = 0$ on the crack
surfaces. To simplify the notation we will write ϕ instead
of $\hat{\phi}$, etc.

Define an auxiliary function $\Omega(z)$ on D

$$\Omega(z) = \bar{\Phi}(z) + z\bar{\Phi}'(z) + \bar{\Psi}(z) \tag{8.1}$$

Using (3.4) and (3.5) the tractions on the crack surface can be written as follows

$$\tau_{22}(z_+) - i\tau_{12}(z_+) = \Phi(z_+) + \Omega(z_-) \tag{8.2a}$$

$$\tau_{22}(z_-) - i\tau_{12}(z_-) = \Phi(z_-) + \Omega(z_+) \tag{8.2b}$$

where z_+ and z_- respectively denote the upper and lower surface of the crack. Using (7.1)-(7.3) we get

$$\Omega(z) = -\bar{A}(z-\bar{z}_0)^{-2} + 2\bar{A}(z-z_0)(z-\bar{z}_0)^{-3} - \bar{B}(z-\bar{z}_0)^{-2} + \Omega_*(z) \tag{8.3}$$

where

$$\Omega_*(z) = \bar{\Phi}_*(z) + z\bar{\Phi}'_*(z) + \bar{\Psi}_*(z). \tag{8.4}$$

Setting

$$\Omega_*(z) = \Phi_*(z) \tag{8.5}$$

we obtain

$$\tau_{22}(z_+) - i\tau_{12}(z_+) = Q(z_+) + \Phi_*(z_+) + \Phi_*(z_-) \tag{8.6a}$$

where

$$Q(z) = -A(z-z_0)^{-2} + \bar{A}(z-\bar{z}_0)^{-2}$$
$$\tag{8.7}$$
$$+ 2\bar{A}(z - z_0)(z - \bar{z}_0)^{-3} - \bar{B}(z - \bar{z}_0)^{-2} .$$

Note that $Q(z_+) = Q(z_-)$. Similarly

$$\tau_{22}(z_-) - i\tau_{12}(z_-) = Q(z_+) + \Phi_*(z_-) + \Phi_*(z_+). \tag{8.6b}$$

Now we select Φ_* so that

$$\Phi_*(z_+) + \Phi_*(z_-) = -Q(z_+). \tag{8.8}$$

(8.5) and (8.8) define now ϕ_* and ψ_*. By this selection we achieve that $\tau_{22}(z_+) = \tau_{12}(z_+) = \tau_{22}(z_-) = \tau_{12}(z_+) = 0$. The relation (8.8) can be easily achieved. For example, for

$$\Phi = -\frac{z^{-1/2} z_0^{-1/2}}{4(z^{1/2} + z_0^{1/2})^2}$$

we get

$$\Phi(z_+) + \Phi(z_-) = \frac{1}{(-z_+ + z_0)^2}$$

which is one term in (8.7). Consequently we get the other terms and combining them (8.8) is achieved.

9. NUMERICAL PERFORMANCE OF THE EXTRACTION TECHNIQUE

We now present the results of computational experiments based on our model problem and the extraction function described in Section 8 (using $\hat{\phi}*$, $\hat{\xi}*$).

Fig. 9. shows the results analogous to those shown in Fig. 5 but stress components τ_{ij} was computed by the extraction technique. The slope shown in the figure shows the rate $\mu = 1$ (i.e., the rate of the convergence of the energy and not the energy norm). For comparison the error e_{12}^R for mesh 2 computed directly (see Fig. 5b) is shown also in Fig. 9. Fig. 10 shows the isometric drawings (in the same scale as in

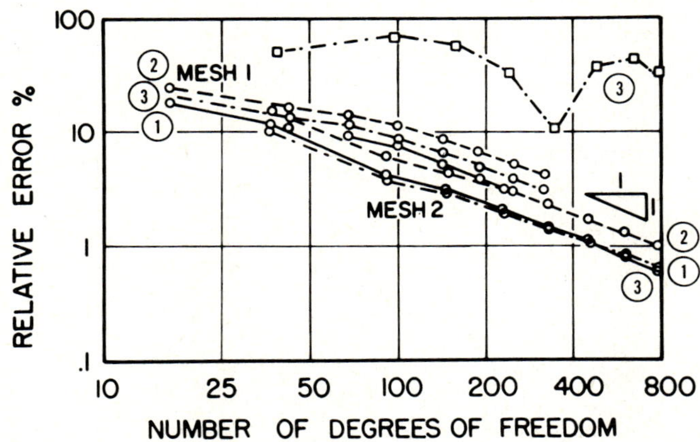

Figure 9

The relative error of τ_{ij} computed by postprocessing of the p-version for Mesh 1 and Mesh 2. (1) e_{11}^R, (2) e_{22}^R, (3) e_{12}^R

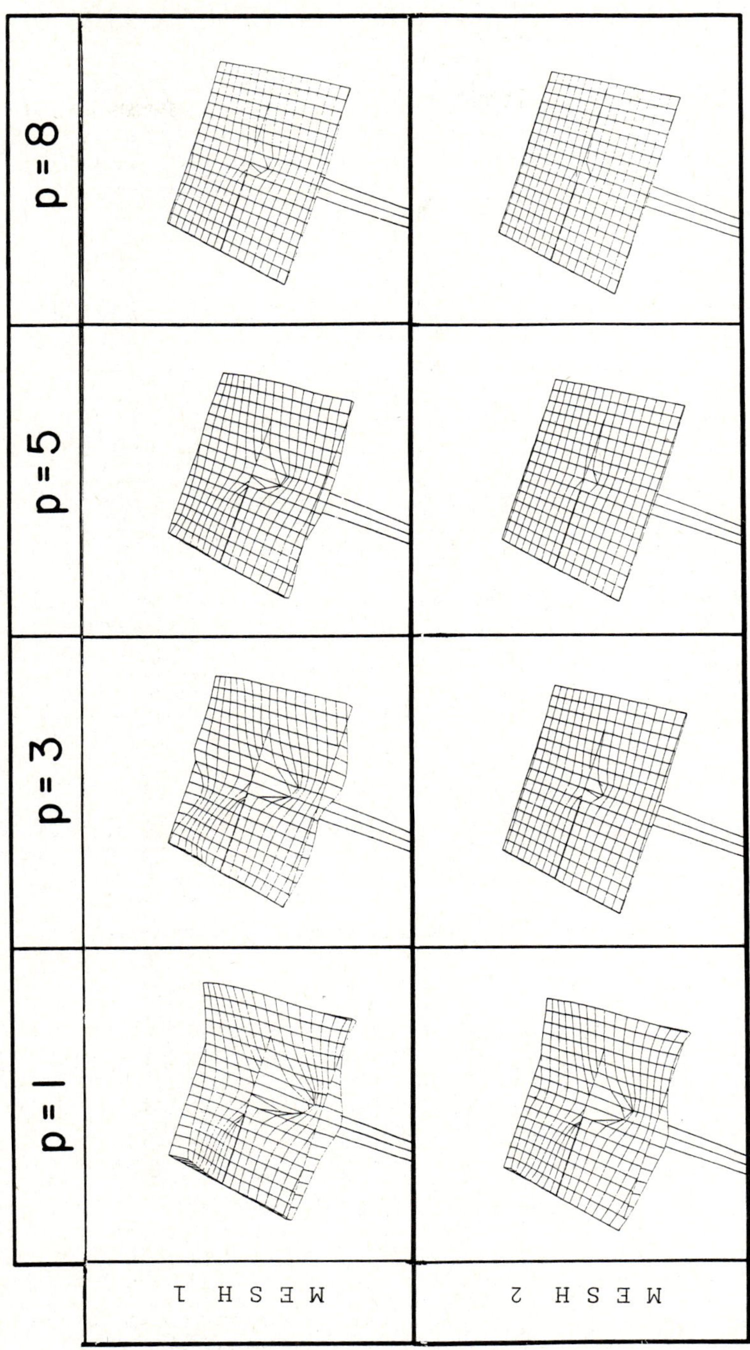

Figure 10

The behavior of the error e_{22} of τ_{22} computed by postprocessing of the p-version.

Fig. 6) of the error in τ_{22} computed by the postprocessing technique.

Table 7 shows the relative error e^R_{ij} in the stresses τ_{ij} at the point (.25,.25) computed by the postprocessing technique taking $\hat{\phi}* = \hat{\xi}* = 0$ (because the point is not close to the boundary). This data should be compared with the results of Table 5 and 6.

TABLE 7

The relative error e^R_{ij} in the stresses τ_{ij} at the point (.25,.25) computed by postprocessing.

No. of elements	N	e^R_{11}	e^R_{22}	e^R_{12}	
16	51	20.91%	22.51%	12.70%	Adaptive Mesh
43	143	11.40%	13.07%	8.29%	
106	221	4.60%	5.92%	3.74%	
319	617	1.47%	2.01%	1.31%	
16	51	20.91%	22.51%	12.70%	Uniform Mesh
64	167	12.51%	14.88%	9.86%	
256	591	6.87%	8.90%	6.28%	

We see that for adaptive meshes the error is of order N^{-1} and for uniform meshes of order $N^{-1/2}$. Similarly, as in the case of the p-version we see an orderly convergence with the rate as the square of the error measured in the energy norm (as theoretically expected).

10. CONCLUSIONS

The shown computations are characteristic in the following way. The convergence in the energy norm is monotonic and very orderly. For the smooth solution the p-version is especially effective. For unsmooth solutions the refinement of the meshes in the h-version is very essential.

The convergence of stresses in a fixed point is very "chaotic," the accuracy in various components can be very different. The rate of convergence of the postprocessed values are as the square of the error measured in the energy norm. In the case of the h-version, uniform (or piecewise uniform) meshes and smooth solution the **superconvergence** occurs in the center of

the elements. The rate is $h^2 \log h$, i.e., effectively as the square of the error in energy norm (h). Therefore, the gain

for the elements of degree 1 is not in the rate of
convergence of the postprocessed value but is in the
magnitude. (For p > 1 the gain of the postprocessing appears
also in the rate.)

The postprocessing is especially important for the p-version,
although it is also essential for the h-version especially for
unsmooth solutions and for general meshes.

11. EFFECTIVITY OF THE POSTPROCESSING TECHNIQUE

In the introduction we raised a number of questions concerning
the postprocessing. We now briefly address these question in
the light of our results. Detailed analysis will be made in a
forthcoming paper.

1) It is cost effective not to save computational effort
on a postprocessing procedure especially when not an excessive
number of extractions is made. The cost of obtaining reliable
and accurate values by postprocessing is much smaller than to
obtain comparable accuracy by increasing p in the p-version
or refine the meshes in the h-version. The postprocessing
usually removes very reliably the "chaotic" behaviour of the
errors in stresses. The effectivity of the postprocessing is
characterized by higher rate of convergence than in the energy
norm.

2) The rate of convergence as the square of the rate of
the error in the energy norm is theoretically the maximal one
which can be directly extracted. The postprocessing technique
we outlined leads to this rate.

3) Developoment and implementation of the postprocessing
techniques in finite element programs is practically not a very
simple task. We mention some aspects:

a) A number of extraction functions must be
developed. Although many analytical solutions of special
problems are very helpful for such development, the
general approach especially for nonhomogeneous material
still needs further research.

b) Special care must be excercised in the numerical
evaluation of integrals because the extraction function
can have singular character.

c) The postprocessing technique for nonlinear
problems could be especially important but additional
research is necessary.

REFERENCES

[1]   Muskhelishvili, N. I., Some basic problems of the
       mathematical theory of elasticity (P. Noordhoff,
       Groningen, Netherlands, 1963).

[2]   Basu, P. K., M. P. Rossow, B. S. Szabo, Theoretical

manual and user's guide for COMET-X (Center for
Computational Mehanics, Washington University, St.
Louis).

[3]   Mesztenyi, C., W.Szymczak, FEARS user's manual for
UNIVAC 1100 (University of Maryland, Institute for
Physical Science and Technology Tech. Note BN-991,
October 1982).

[4]   Gignac, D. A., I. Babuška, C. Mesztenyi, An introduction
to the FEARS program, David W. Taylor Naval Ship
Research and Development Center Report DTNSRDC/CMLD-
83/04, February 1983.

[5]   Babuška, I., A. Miller, M. Vogelius, Adaptive methods
and error estimation for elliptic problems of structural
mechanics, University of Maryland, Institute for
Physical Science and Technology, Tech. Note BN-1009,
June 1983, to appear in the Proceedings of ARO Workshop
on Adaptive Methods for Partial Differential Equations,
SIAM, 1984.

[6]   Babuška, I., M. Vogelius, Feedback and adaptive finite
element solution in one-dimensional boundary value
problems, University of Maryland, Institute for Physical
Science and Technology Tech. Note 1006, October 1983.

[7]   Babuška, I., W. C. Rheinboldt, Reliable error estimation
and mesh adaptation for finite element method; in
Computational methods in nonlinear mechanics (J. T.
Oden, ed., North-Holland Publ. Co., Amsterdam, 1980, pp.
67-109).

[8]   Babuška, I., B. A. Szabo,I. N. Katz, The p-version of
the finite element method, SIAM, J. Numer. Anal 18
(1981) 515-545.

[9]   Babuška, I., B. A. Szabo, On the rates of convergence of
the finite element method, Internal J. Numer. Methods
Engrg. 18 (1982) 323-341.

[10]   Babuška, I., A. Miller, The post-processing in the
finite element method, Part 1, Calculation of
displacements stresses and other higher derivatives of
displacements, University of Maryland, Institute for
Physical Science and Technology, Tech. Note BN-992,
December 1982. To appear in Internal J. Numer. Methods
Engrg., 1984.

[11]   Babuška, I., A. Miller, The post-processing approach in
the finite element method, Part 2, The calculation of
stress intensity factors, University of Maryland,
Institute for Physical Science and Technology, Tech.
Note BN 993, December 1982, to appear in Internal J.
Numer. Methods Engrg., 1984.

[12]	Babuška, I., A. Miller, The post-processing approach in the finite element method, Part 3, A posteriori error estimates and adaptive mesh selection, University of Maryland, Institute for Physical Science and Technology, Tech. Note BN 1007, June 1983, to appear in Internal J. Numer. Methods Engrg., 1984.

CHAPTER 5

ON FINITE ELEMENT ANALYSIS OF
LARGE DEFORMATION FRICTIONAL CONTACT PROBLEMS

K.-J. Bathe & A. Chaudhary

We consider the solution of contact problems involving
large deformations of the contacting bodies and sticking
or sliding on the contacting surfaces. A finite element
solution procedure is described and the results of some
numerical studies are presented. The objective in this
presentation is to give further insight into the solution
procedure already presented in an earlier paper [8].

1. INTRODUCTION

Much progress has been made during the last decade in the development
of computational techniques for nonlinear analysis. These advancements have
come about through significant fundamental contributions in discretization
theories and numerical algorithms, but important has also been the cross-
fertilization that has taken place between the various approaches used for
the numerical solution of problems. For this cross-fertilization the conferences on the unification of numerical methods — at one of which this
paper is presented — have provided an excellent forum.

One area of research and development that has obtained much attention
by analysts using various possible numerical approaches is the analysis
of contact problems [1-8]. These problems can be most difficult to solve
and although much research effort has been expended on the solution of
contact problems using finite difference methods, finite element techniques,
surface integral methods, etc., there is still much room for more reliable
and effective algorithms to analyse general contact conditions.

The objective in this presentation is to discuss certain aspects of
a contact solution algorithm that we have developed and researched [8].
We consider two-dimensional plane stress, plane strain or axisymmetric conditions. The contacting bodies can be subjected to large deformations with
sticking, sliding and separation on the contacting surfaces.

In the next section we describe in detail the contact problem we consider, and in Section 3 we discuss our numerical solution procedure. The

---

$(†)$Professor of Mechanical Engineering
$(*)$Research Assistant

objective in this section is to give insight into our algorithm. The governing equations are derived in detail in ref. [8], which should be consulted for a more full account of our solution technique. In Section 4 we then present the results of various numerical experiments to illustrate our observations on the algorithm. These experiments show how the contact solution procedure works, what the different assumptions are, and how the method can be applied. In the presentation we consider static analysis, but the algorithmic steps used can also be employed in a dynamic solution.

2. STATEMENT OF CONTACT PROBLEMS CONSIDERED

We can use our algorithm for the analysis of a number of flexible bodies coming into contact with each other or with rigid bodies. However, for ease of presentation of the theory and our algorithm we now consider two bodies, both flexible, that partly come into contact. Figure 1 shows schematically the two bodies, which we call the contactor and the target.

The bodies can undergo very large deformations, and can come into contact, with or without sliding and can separate again. However, we only consider static analysis conditions (hence the motions of the bodies are "slow" so that the effects of inertia and damping forces can be neglected).

The following equations govern the problem we consider.

a) The linear momentum equation for the target and the contactor,

$$\tau_{ij,j} + f_i^B = 0 \tag{1}$$

where τ_{ij} is the (i,j)th component of the Cauchy stress tensor and f_i^B is the i'th component of the body force vector, both are referred to the current configurations of the bodies. Equation (1) must be satisfied throughout the motion of the bodies subject to the appropriate constitutive relations to evaluate the Cauchy stresses.

b) The boundary conditions correspond to prescribed displacements on S_d, prescribed surface forces on S_f and a priori unknown contact conditions on the possible area of contact S_c. We note that in Fig. 1 the surfaces S_d, S_f and S_c are distinct from each other, and their sum constitutes the complete surface of the target and the contactor.

Considering the surfaces S_d and S_f on the bodies, we have the boundary conditions,

$$u_i = u_i^S \quad \text{on } S_d \tag{2}$$

$$n_j \tau_{ji} = f_i^S \quad \text{on } S_f \tag{3}$$

where u_i^S and f_i^S denote the imposed displacements and applied surface forces, and the n_j are the direction cosines of the normal to the surface. We may note that we assume both bodies to be supported against rigid body motions, hence the prescribed displacements on S_d must be such as to make the target and the contactor, without contact between them, stable structures.

Considering the surfaces S_c, the displacements are free from constraints and no forces are developed on these surfaces as long as there is no contact. However, contact is reached as soon as material particles of the target and contactor surfaces touch each other, and then the following considerations are valid. Let t_n^C and t_n^T be the contact tractions (forces/ unit area) in the direction normal to the contact surfaces, with t_n^C acting upon the contactor and t_n^T acting upon the target, see Fig. 2. Contact is established as long as t_n^C is positive (acting into the body), and during contact we have

$$t_n^C = t_n^T \tag{4}$$

Also, denoting the differential displacement increments in the direction normal to the contact surface as du_n^C for the contactor and du_n^T for the target, then during contact we have

$$du_n^C = du_n^T \tag{5}$$

For the evaluation of the tangential tractions that act onto the target and the contactor we use Coulomb's law of friction. Let t_t^C and t_t^T represent the <u>developed</u> tangential tractions along the contact surfaces, and let du_t^{T-C} be the relative incremental displacement between the material particles of the contactor and the target. Then du_t^{T-C} is zero as long as $|t_t^C| < \mu\, t_n^C$ (and hence $|t_t^T| < \mu\, t_n^T$) where μ is the coefficient of friction. Further, the maximum tangential traction that can be reached is $|t_t^C| = \mu\, t_n^C$ (and also $|t_t^T| = \mu\, t_n^T$), and when this tangential traction is developed we have $|du_t^{T-C}| \geq 0$, hence relative tangential motion between the target and contactor particles is then possible. The "direction" of relative motion is such that the developed tangential tractions oppose the motion.

We should note that at the beginning of the analysis, the <u>actual</u> area of contact (being a part of the <u>possible</u> area of contact) is unknown, and so are the contact tractions $t_n{}^C$ and $t_t{}^C$ (and $t_n{}^T$, $t_t{}^T$). The determination of the actual area of contact and the corresponding contact forces, while the target and contactor are subjected to small or large deformations with linear or nonlinear constitutive behavior, is the key task of the solution procedure.

3. SUMMARY OF CONTACT SOLUTION PROCEDURE

The solution procedure we have developed solves for the motion of the contactor and target bodies using the basic considerations summarized in the previous section. Since the equilibrium, compatibility and constitutive relations must be satisfied throughout the — in general — highly nonlinear response history, an incremental solution is performed. In this section we aim to describe the solution procedure presented already in ref. [8] to render more insight into the detailed operations of the algorithm.

3.1 The incremental equations of motion

As long as there is no contact, the incremental solution is performed as described in ref. [9, chapter 6]. Namely, assuming that the solution is known for the configuration at time t, the iteration is performed to obtain the solution corresponding to time t+Δt. Since the effect of inertia forces is neglected, the governing equations in this iteration are, using the full Newton method and the notation of ref. [9],

$$\,^{t+\Delta t}\underline{K}^{(i-1)}\,\Delta\underline{U}^{(i)} = \,^{t+\Delta t}\underline{R} - \,^{t+\Delta t}\underline{F}^{(i-1)} \tag{6}$$

$$\,^{t+\Delta t}\underline{U}^{(i)} = \,^{t+\Delta t}\underline{U}^{(i-1)} + \Delta\underline{U}^{(i)} \tag{7}$$

where $\,^{t+\Delta t}\underline{K}^{(i-1)}$ is the tangent stiffness matrix[†] corresponding to time t+Δt and the end of iteration (i-1), $\,^{t+\Delta t}\underline{F}^{(i-1)}$ is a nodal point force vector corresponding to the internal element stresses, $\,^{t+\Delta t}\underline{R}$ is the vector of externally applied loads and $\Delta\underline{U}^{(i)}$ is the vector of incremental nodal point displacements. Note that at the beginning of the iteration, for i=1, we have the initial conditions

$$\,^{t+\Delta t}\underline{K}^{(0)} = \,^{t}\underline{K};\; \,^{t+\Delta t}\underline{F}^{(0)} = \,^{t}\underline{F};\; \,^{t+\Delta t}\underline{U}^{(0)} = \,^{t}\underline{U}$$

The contributions in $\,^{t+\Delta t}\underline{K}^{(i-1)}$ and $\,^{t+\Delta t}\underline{F}^{(i-1)}$ are those of the contactor and the target. Since there is no contact as yet, these contributions

____

[†]Note that the time superscript t+Δt signifies here the configuration (and load) at time t+Δt and does not imply a dynamic analysis.

are uncoupled, but the solution is possible because both the target and the contactor bodies are properly supported (an assumption we stated in Section 2).

As described in detail in ref. [8], during each iteration using Eqs. (6) and (7), the algorithm checks using the current configurations of the contactor and target bodies whether the contactor has "penetrated" the target. If a contactor surface node is within the target, contact has been established and during such conditions the governing incremental equilibrium equations are

$$\left\{ \begin{bmatrix} {}^{t+\Delta t}\underline{K}^{(i-1)} & \underline{0} \\ \underline{0} & \underline{0} \end{bmatrix} + \begin{bmatrix} {}^{t+\Delta t}\underline{K}_C^{(i-1)} \end{bmatrix} \right\} \begin{bmatrix} \Delta\underline{U}^{(i)} \\ \Delta\underline{\lambda}^{(i)} \end{bmatrix}$$

$$= \begin{bmatrix} {}^{t+\Delta t}\underline{R} \\ \underline{0} \end{bmatrix} - \begin{bmatrix} {}^{t+\Delta t}\underline{F}^{(i-1)} \\ \underline{0} \end{bmatrix} + \begin{bmatrix} {}^{t+\Delta t}\underline{R}_C^{(i-1)} \\ {}^{t+\Delta t}\underline{\Delta}_C^{(i-1)} \end{bmatrix}$$

(8)

where ${}^{t+\Delta t}\underline{K}_C^{(i-1)}$ is a contact stiffness matrix, ${}^{t+\Delta t}\underline{R}_C^{(i-1)}$ is a vector of contact forces and ${}^{t+\Delta t}\underline{\Delta}_C^{(i-1)}$ is a vector of geometric overlaps, i.e. penetrations of the contactor nodes into the target. Figure 3 illustrates schematically the meaning of the vector ${}^{t+\Delta t}\underline{\Delta}_C^{(i-1)}$. The Lagrange multipliers $\Delta\lambda^{(i)}$ can be interpreted as increments in the nodal point forces acting on the contact surfaces required to prevent the overlap ${}^{t+\Delta t}\underline{\Delta}_C^{(i-1)}$. However, since Eq. (8) has been derived by linearizing about the state at the end of iteration (i-1), these contact force increments can be very approximate (due to geometric and material nonlinearities and the frictional restraints) and are not directly used to calculate the contact forces. Instead we use Eq. (8) to impose the geometric constraints of no material overlap and evaluate the contact forces from the internal stresses of the contacting bodies.

Before discussing the evaluation of the contact forces ${}^{t+\Delta t}\underline{R}_C^{(i-1)}$, let us consider some further important details regarding Eq. (8) by considering the following three different cases.

Case of perfect sliding, $\mu = 0.0$

In the case of no friction, our solution procedure establishes only one additional equation corresponding to $\Delta\lambda^{(i)}$ for each contactor node that has penetrated the target body. This equation corresponds to the displacement constraints of no material overlap normal to the contact surface, whereas tangent to this surface the contactor and target bodies can slide freely on each other. Hence, the only contact tractions developed are $t_n{}^C$ and $t_n{}^T$.

Note that at convergence using Eq. (8) we must have

$$t+\Delta t \underline{\Delta}_C{}^{(i-1)} \doteq \underline{0} \tag{9}$$

and since no external forces are applied on the contact surface[+]

$$t+\Delta t \underline{R}_C{}^{(i-1)} = t+\Delta t \underline{F}^{(i-1)} \tag{10}$$

where the "approximately equal sign" is used because convergence is only obtained within specific convergence tolerances. Also, since only contact forces normal to the contact surface can be established, the components in the vector $t+\Delta t \underline{R}_C{}^{(i-1)}$ must correspond to (compressive) tractions acting normal to the contact surface.

Case of perfect sticking, $\mu = \infty$

In the case of sticking the solution procedure establishes two equations corresponding to $\Delta\underline{\lambda}^{(i)}$ as soon as overlap is detected, and eliminates the geometric overlap. Note that as for the other cases the contactor node corresponding to which the equations are established can come into contact anywhere along the contact surface of the target. At convergence, we have again that Eqs. (9) and (10) are satisfied.

Case of sticking or sliding, $\mu > 0$

We discussed in Section 2 that when the friction coefficient is nonzero and small, the normal and tangential tractions developed during contact determine whether sliding occurs. Consequently, the solution procedure assumes in the first iteration from no contact to contact sticking conditions and establishes corresponding to $\Delta\underline{\lambda}^{(i)}$ two equations. The solution in this iteration yields contact tractions, calculated assuming sticking conditions, that are used to establish whether there actually are sticking or sliding conditions. The updated conditions on sticking and sliding together with the corresponding contact forces are employed in the next iteration to establish two (for sticking) or only one (for sliding) constraint equations corresponding to each contactor node in contact. Note that during the solution and the iterations, a node may also change its state from sliding back to sticking.

A very important part of the solution procedure is hence the evaluation of the contact forces, with their normal and tangential components, and the decision of whether sliding or sticking conditions prevail.

3.2 Evaluation of contact forces and sliding or sticking conditions

Using Eqs. (7) and (8) we calculate at the beginning of the (i+1)st

---

[+] Externally applied forces on the contact surface could actually be included in Eq. (10)

iteration $^{t+\Delta t}\underline{U}^{(i)}$, from which we can directly evaluate the nodal point force vector $^{t+\Delta t}\underline{F}^{(i)}$ which is equivalent (in the virtual work sense) to the current stresses $^{t+\Delta t}\tau_{k\ell}^{(i)}$. Next, the following vector of forces is evaluated

$$\Delta\underline{R}^{(i)} = {}^{t+\Delta t}\underline{R} - {}^{t+\Delta t}\underline{F}^{(i)} \tag{11}$$

where we note that with no contact between the contactor and the target the elements in $\Delta\underline{R}^{(i)}$ must all be small at convergence, i.e., $\Delta\underline{R}^{(i)} \to \underline{0}$ as $i \to \infty$. However, if there is contact, then the elements in $\Delta\underline{R}^{(i)}$ corresponding to the target and contactor nodes on the contact surface S_C (that are affected by the contact) must at convergence in mesh discretization and iteration be equal to the contact forces. This means that these forces must satisfy the friction conditions summarized in Section 2. The components in $\Delta\underline{R}^{(i)}$ corresponding to the nodes that are not in contact must of course still be small (approach zero) at convergence.

The first step in evaluating whether the forces $\Delta\underline{R}^{(i)}$ satisfy the frictional conditions is to evaluate from $\Delta\underline{R}^{(i)}$ normal and tangential tractions that act onto the contactor. Since, considering the contact surface, the conditions on the contactor are statically equivalent to those on the target, we consider only the contactor. The distributed contact tractions are those normal and tangential forces (per unit area) that give the consistent nodal point forces stored in $\Delta\underline{R}^{(i)}$. Let t_n^k and t_t^k be the intensities of the normal and tangential tractions at the contactor node k evaluated using this approach. Further, let the total resultant normal and tangential forces on segment j, evaluated from the values t_n^k and t_t^k for all nodes k in contact, be T_n^j and T_t^j, see Fig. 4. The solution procedure uses the values T_n^j and T_t^j to evaluate (globally) whether the force contact conditions are satisfied and, if necessary, updates these conditions for the next iteration.

If $T_n^j < 0$, the segment is released for the next iteration because the segment cannot be in tension. To evaluate the contact force vector $^{t+\Delta t}\underline{R}_C^{(i)}$, the tractions t_t and t_n are set to zero over the segment.

If $|T_t^j| \leq \mu\, T_n^j$, the segment is assumed to stick in the next iteration. To evaluate the contact force vector $^{t+\Delta t}\underline{R}_C^{(i)}$, the tractions t_t and t_n from the previous iteration are employed.

If $|T_t^j| > \mu\, T_n^j$, the segment is assumed to slide in the next iteration. To evaluate the contact force vector $^{t+\Delta t}\underline{R}_C^{(i)}$, the normal traction t_n from the previous iteration is employed but the tangential traction

t_t is updated to a constant value of $\dfrac{T_t^{\,j}}{A_j} \left| \dfrac{\mu T_n^{\,j}}{T_t^{\,j}} \right|$, where A_j is the surface area of the segment. Using this value of t_t, Coulomb's law of friction is satisfied globally over the segment, but a constant value of tangential traction is assumed to act over the segment.

The tractions t_t and t_n over each contact segment thus obtained are employed to evaluate the nodal point consistent contact forces $^{t+\Delta t}\underline{R}_c^{(i)}$.

With the states of the segments updated to "release", "sticking", or "sliding" and the calculation of $^{t+\Delta t}\underline{R}_c^{(i)}$ completed, the solution procedure establishes the states of the contactor nodes as summarized in Table 1 and then (see Section 3.1) establishes in Eq. (8) two contact equations for each node in the sticking condition and one equation for each node in the sliding condition. We may note that by means of the above calculations the distributed force and frictional effects on the segments are concentrated to the nodes, consistent with usual finite element procedures.

3.3 Convergence of the iterative scheme

To study the convergence of the iterations it is convenient to consider the three different cases, $\mu = 0.0$, $\mu = \infty$ and $\mu > 0$ but of small value.

When $\mu = 0.0$ (case of perfect sliding) the equilibrium relations in Eq. (8) reduce to those without contact conditions (see ref. [9], chapter 6) supplemented with the constraint that the contactor nodes cannot penetrate the target but instead will slide without resistance over the target segments. Hence, at convergence, only contact forces that act normal to the contact surface are transmitted. Note that since the contactor nodes slide over the target segments, the target nodes can be within or outside the contactor. Hence, the finite element discretization for the contactor and the target should be such that the resulting material overlap is acceptable. These considerations are also important when $\mu > 0$.

The case of $\mu = \infty$, i.e., perfect sticking, is achieved by simply choosing μ large enough so that the tangential tractions on the contact surface are always less than μ times the normal tractions. The solution obtained in the incremental analysis is in this case path-dependent because the contactor nodes stick throughout the analysis to the material points of the target segments with which they first come into contact (unless tension release occurs). Hence, a different sequence of external load application with the same final load distribution may lead to significantly different results. However, convergence using Eqs. (8) to (10) means that at each load level the equilibrium, constitutive and compatibility conditions, within the assumptions of the finite element discretization, are satisfied.

The most difficult types of problems to solve are those for which μ is greater than zero but small, so that depending on the unknown normal and tangential tractions along the contact surface, in some areas sticking and in other areas sliding may occur. Considering the equilibrium relations in Eq. (8) and the procedure for evaluating the contact forces, we can make the following important observations regarding the convergence of the iterations:

° Consider that corresponding to the configuration at time t, the conditions for all contactor nodes are known.

° With the increase in the externally applied load from time t to time t+Δt some nodes reach the "sliding condition" during the iterations, which results into incremental displacements and a redistribution of the internal element stresses, until finally the contact tractions satisfy Coulomb's law of friction (globally, for each of the segments, see Section 3.2). This means that at convergence of the iterations for the load level t+Δt, the contactor nodes are largely in the "sticking condition" although during the iterations they may have slid. However, note that regarding the physical interpretation of the solution results, a contactor node has been in sliding from time t to time t+Δt, whenever the contactor node is not any more at the same target material particle as it was at time t. Hence, the final condition of sticking for a contactor node at time t+Δt does not alone tell whether the node has or has not been sliding from time t to time t+Δt.

Considering the evaluation of the tangential tractions in sliding we recall that these tractions are calculated by reducing the developed tangential tractions to the magnitude compatible with the normal tractions (using Coulomb's law of friction). Hence, the direction of relative tangential sliding along the contact surface does not directly enter into the determination of the direction of the tangential contact tractions. However, our experience is that the calculated tangential tractions do oppose the motion provided the finite element representation is fine enough and the incremental solution is performed in small enough steps. We are pursuing further theoretical and computational studies of this observation.

So far we considered only how convergence is reached. However, in the iterations, actual convergence criteria are necessary that measure when to accept the calculated solution. The convergence criteria we have used measure the incremental energy and the change in the contact forces. Namely, referring to Eq. (8), the solution is accepted once the following relation is satisfied,

$$\frac{\Delta\underline{U}^{(i)^T}\left(^{t+\Delta t}\underline{R} + {}^{t+\Delta t}\underline{R}_c^{(i-1)} - {}^{t+\Delta t}\underline{F}^{(i-1)}\right)}{\Delta\underline{U}^{(1)^T}\left(^{t+\Delta t}\underline{R} + {}^t\underline{R}_c - {}^t\underline{F}\right)} \leq \text{ETOL} \qquad (12)$$

where ETOL is the energy convergence tolerance, and once for all nodes in contact, referring to Eq. (11), the components in $\Delta\underline{R}^{(i)}$ corresponding to the contact forces satisfy the relation

$$\frac{||\Delta \underline{R}^{(i-1)} - \Delta \underline{R}^{(i-2)}||_2}{||\underline{R}^{(i-1)}||_2} \leq \text{RCTOL} \tag{13}$$

where RCTOL is the contact force convergence tolerance. Typical values
used are ETOL = 0.001 and RCTOL = 0.01.

We illustrate how the algorithm proceeds in solutions by means of
some numerical results given in the next section.

4. NUMERICAL EXPERIMENTS

Various analysis results obtained with our solution algorithm and comparisons with solutions previously reported have been presented in ref. [8].
The objective in this section is to supplement the analysis results of
ref. [8] by showing in more detail how the solution is obtained and presenting some results on the effects of mesh selection and load step size.
We consider two problems already discussed in ref. [8]; namely, the analysis
of a buried pipe and the solution of a rubber sheet moving in a rigid converging channel.

4.1 Analysis of a Buried Pipe

Figure 5 shows the buried pipe considered. The objective of the
analysis is to predict the tractions along the pipe-soil interface. Both,
the pipe and the surrounding soil are considered linear elastic media.

In ref. [8] we presented the solution to the problem using the finite
element idealization of Fig. 6, now called mesh B. In order to study the
effect of discretization on the solution results we now also give the
solution to the problem using the coarse mesh (mesh A) and the fine mesh
(mesh C) shown in Figs. 7 and 8. Note that mesh B is obtained by sub-
dividing each 8-node isoparametric element of mesh A into four 8-node isoparametric elements, and mesh C is obtained in the same manner from mesh B.

Figure 9 shows the computed tractions using the different meshes. The
solutions have been obtained in a four step solution, i.e. by applying the
total overburden pressure P_0 in four equal steps and using ETOL = 0.001 and
RCTOL = 0.01. For comparison also the solutions for zero friction
and infinite friction, obtained using mesh B, are shown.

Note that the tractions t_n and t_t plotted in Fig. 9 are the mean
tractions over a segment; hence, for typical points representing the
tractions on segment j we have $t_t^j = T_t^j/A_j$ and $t_n^j = T_n^j/A_j$ (see Section
3.2). Figure 9 shows that the differences in the contact tractions calculated using meshes B and C are reasonably small.

Figures 10 and 11 show the traction distributions for each iteration
using meshes B and C for a one step solution. The figures show the calculation of tangential tractions (iteratively updated) to satisfiy Coulomb's law

of friction globally over each segment. At convergence — see Figs. 10(e) and 11(g) — the mean updated tangential tractions over a segment are essentially equal to the mean tractions prior to updating.

In order to study the effect of using a different load incrementation, we show in Fig. 12 the traction distributions calculated when using the 4 equal load increments to reach the total overburden pressure. We note that the solution obtained this way is very close to the solution calculated when one load increment is used to apply the total overburden pressure — see Figs. 11(g) and 12(d).

4.2 Analysis of a Rubber Sheet Moving in a Rigid Converging Channel

Figure 13 shows the rubber sheet considered. The right face of the sheet is subjected to a displacement history pulling it into the channel and then pushing it back to its original location. The displacements are imposed slowly so that inertia forces can be neglected.

This problem was analyzed in ref. [8] using the mesh shown in Fig. 14, here called mesh B. We now also give the solution to the problem using the coarser mesh shown in Fig. 15, called mesh A. Figure 16 shows the predicted tangential and normal tractions calculated using meshes A and B. We note the close correspondence between the results obtained although mesh A represents quite a coarse idealization of the rubber sheet.

In these solutions the convergence tolerances ETOL = 0.001 and RCTOL = 0.01 were used. In the first step a rather large number of iter- ations was necessary (19 for mesh A and 26 for mesh B), but from the second step onwards an average of about 4 iterations per step for mesh A and 5 iterations per step for mesh B was used.

5. CONCLUDING REMARKS

The objective in this paper was to describe certain important aspects of our contact solution algorithm and thus supplement the description given in ref. [8]. In the paper we focussed on some physical and numerical key aspects of the solution procedure, and we illustrated our observations through the results — presented in detail — of some numerical solutions.

Considering this work, we summarized in the conclusions of ref. [8] a number of areas where further research would be very valuable.

ACKNOWLEDGEMENTS

We are grateful for the financial support provided by the U.S. Army and the ADINA users group for this work.

Table 1

State of Contactor node as Decided by States of Adjoining Segments

STATE OF ADJOINING SEGMENTS		STATE OF NODE
one adjoining segment	other adjoining segment	
sticking	sticking sliding tension release	sticking
sliding	sliding tension release	sliding
tension release	tension release	tension release

REFERENCES

[1] Argyris, J.H., Doltsinis, J., Pimenta, P.M. and Wustenberg, H.,
    "Thermomechanical Response of Solids at High Strains - Natural
    Approach", J. Computer Methods in Applied Mechanics and Engineering,
    Vol.32-34, 1982, pp. 3-57.

[2] de Pater, A.D., and Kalkar J.J., "The Mechanics of the Contact Between
    Deformable Bodies", Delft University Press, 1975.

[3] Hallquist, J.O., "A Numerical Treatment of Sliding Interfaces and
    Impact", Computational Techniques for Interface Problems, AMD-Vol. 30,
    American Society of Mechanical Engineers, 1978.

[4] Hughes, T.J.R, Taylor, R.L., and Kanoknukulchai, W., "A Finite Element
    Method for Large Displacement Contact and Impact Problems", in
    Formulations and Computational Algorithms in Finite Element Analysis,
    K.J. Bathe et al. eds., M.I.T. Press,1977.

[5] Campos, L.T., Oden, J.T., and Kikuchi, N.,"A Numerical Analysis of a
    Class of Contact Problems with Friction in Elastostatics", Comp. Meth.
    in Appl. Mech. and Eng., Vol. 34, pp. 821-845, 1982.

[6] Kalkar, J.J., Allaert, H.J.C., and de Mul, J., "The Numerical
    Calculation of Contact Problem in the Theory of Elasticity", in
    Nonlinear Finite Element Analysis in Structural Mechanics,
    W. Wunderlich et al, eds., Springer Verlag, 1981.

[7] Hartnett, M.J., "The Analysis of Contact Stresses in Rolling Element
    Bearings", J. Lubrication Technology, ASME, Vol. 101, pp. 105-109, 1979.

[8] Bathe, K.J., and Chaudhary, A.B., "A Solution Method for Planar and Axisymmetric Contact Problems", Int. J. Num. Meth. in Engg., in Press.

[9] Bathe, K.J., "Finite Element Procedures in Engineering Analysis", Prentice-Hall, 1982.

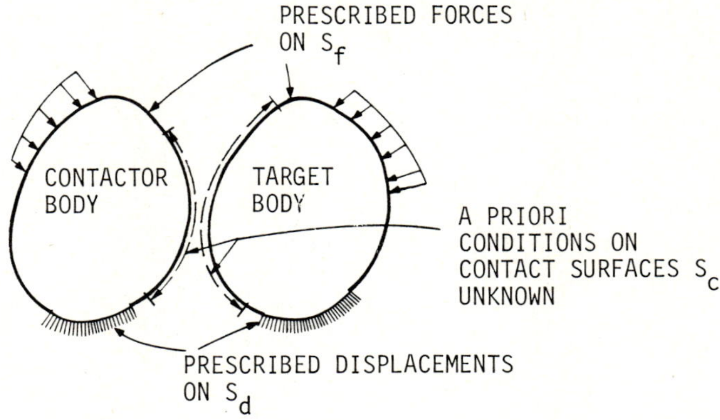

a)   Condition prior to contact

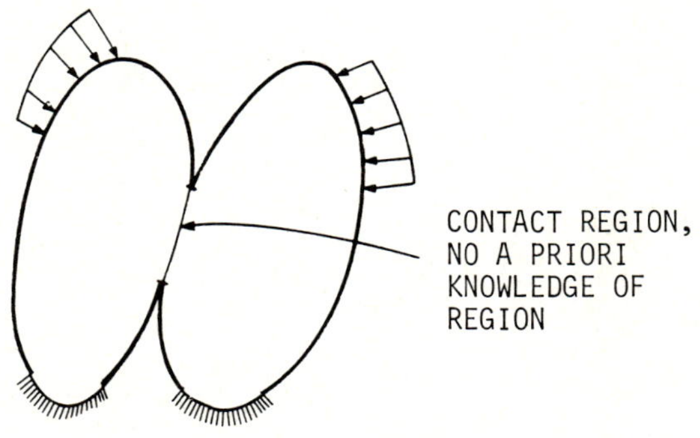

b)   Condition at contact

Figure 1   Schematic representation of two contacting bodies

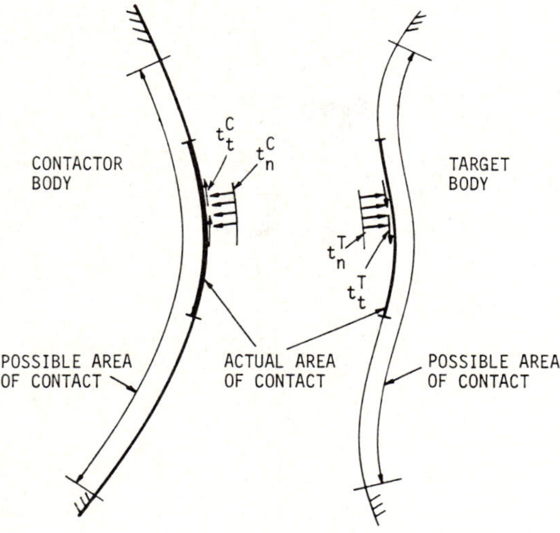

Figure 2   Contact tractions on actual area of contact

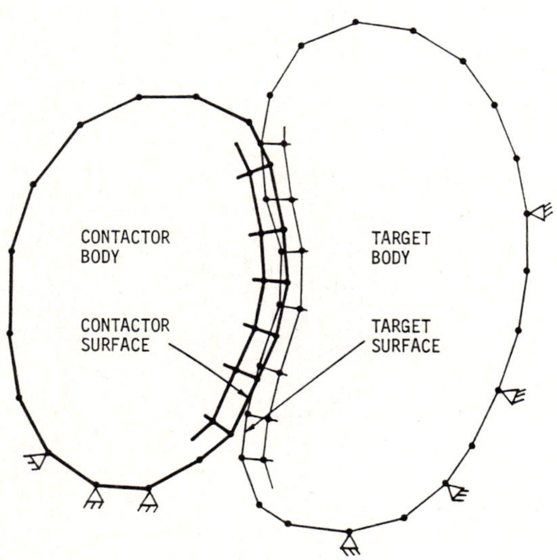

a)   Finite element discretization

Fig. 3

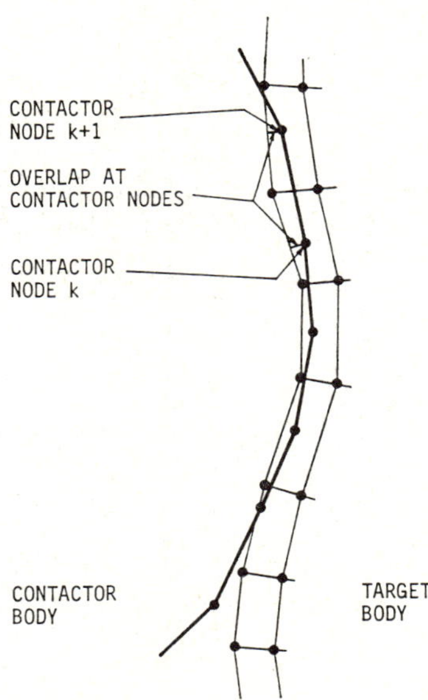

(b)   Overlap at contactor nodes

Figure 3   Schematic representation of overlap between
two contacting bodies

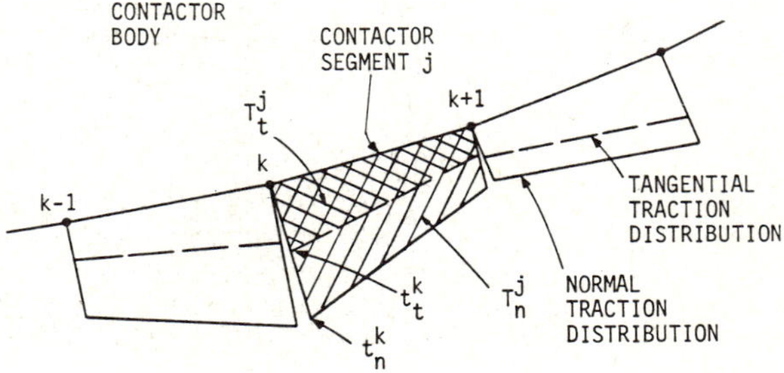

Figure 4   Normal and tangential tractions onto contractor
body. Normal traction is positive when acting
inward to the body, tangential traction is
positive when acting from node k to node (k+1)

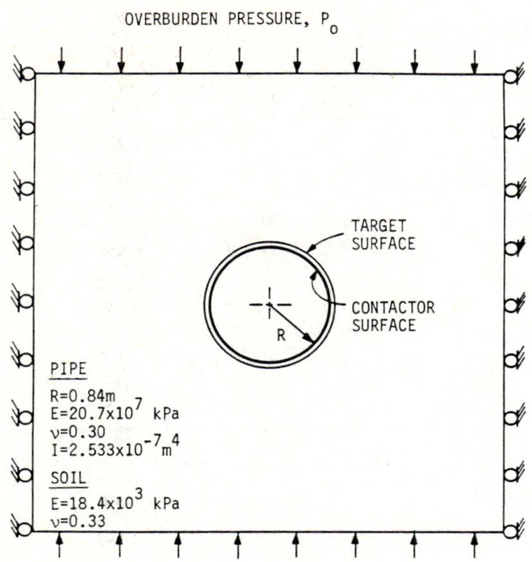

Figure 5 Pipe buried in soil subjected to total overburden pressure P_0 = 100 kPa

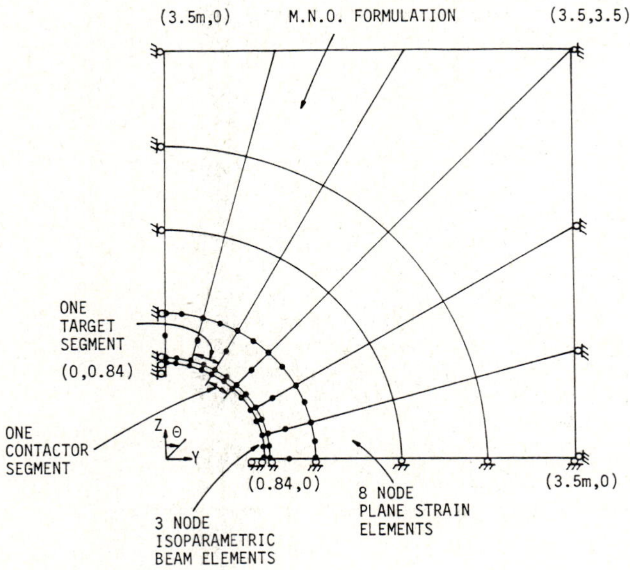

Figure 6 Finite element idealization of buried pipe in soil; mesh B

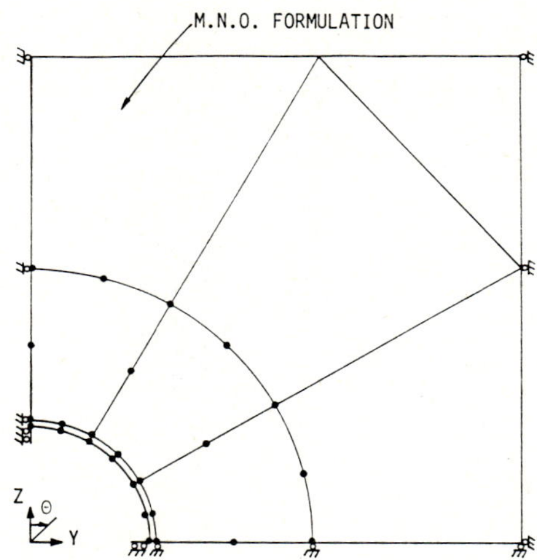

Figure 7   Coarse mesh finite element idealization
of buried pipe in soil; mesh A

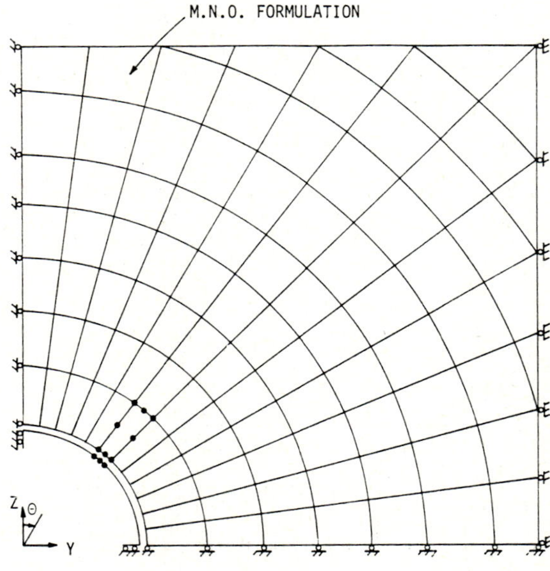

Figure 8   Fine mesh finite element idealization
of buried pipe in soil; mesh C

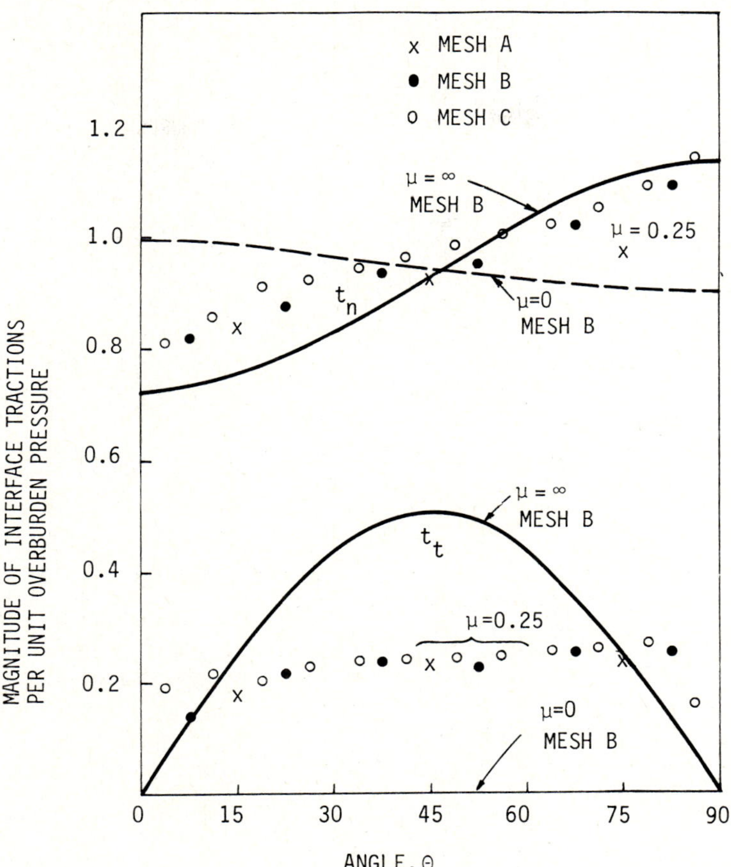

Figure 9 Computed tractions at total load along pipe/soil interface in analysis of buried pipe; solution obtained using four equal size increments to total load for each mesh.

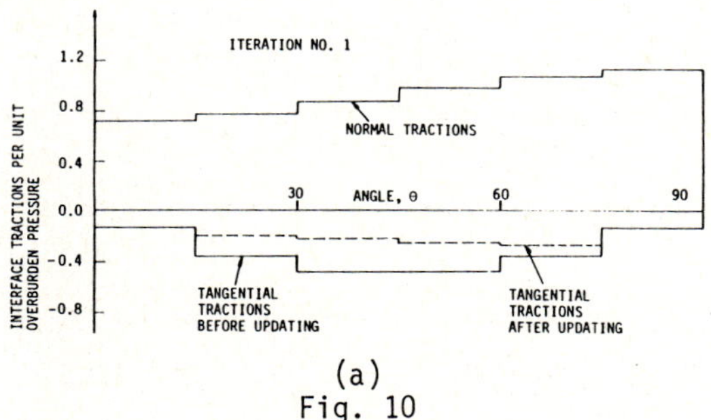

(a)
Fig. 10

                                        K.-J. Bathe & A. Chaudhary

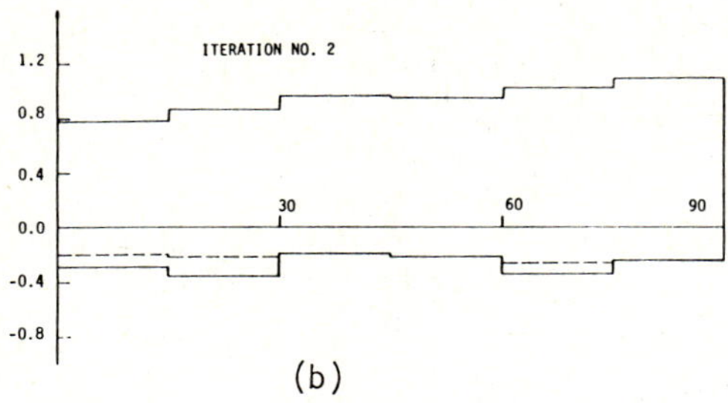

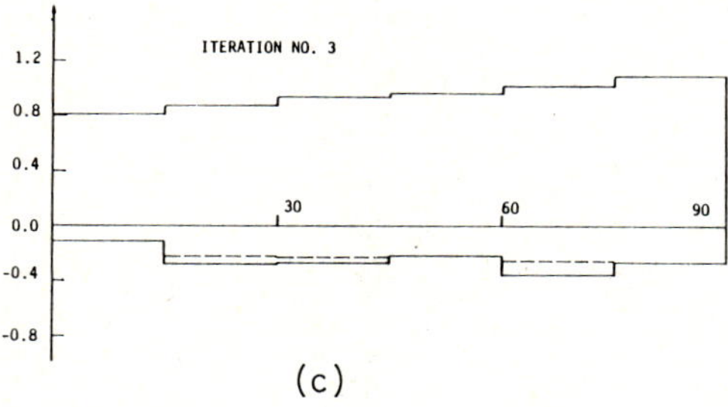

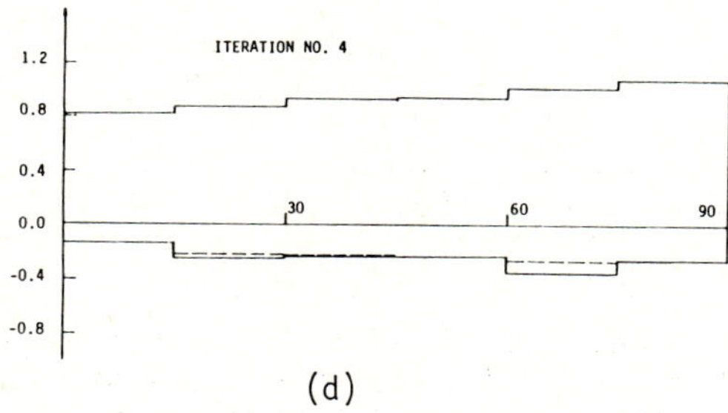

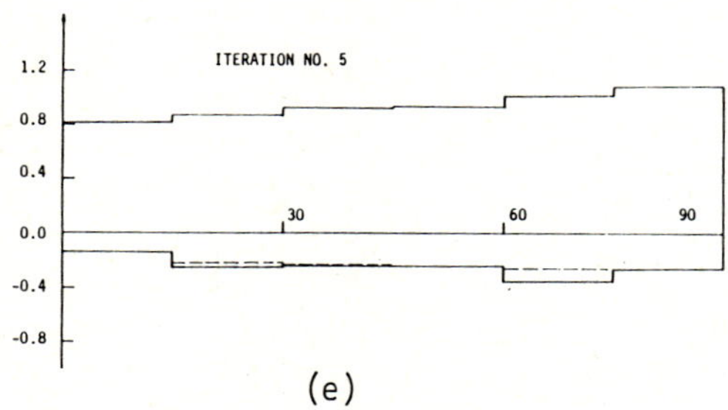

Figure 10   Mean tractions, T_t^j/A_j and T_n^j/A_j, for mesh B in the iterations. One step to total load and five iterations to convergence.

ITERATION NO. 1
1.2
0.8
0.4
0.0
-0.4
-0.8
INTERFACE TRACTIONS PER UNIT OVERBURDEN PRESSURE
NORMAL TRACTIONS
30
ANGLE, θ
60
90
TANGENTIAL TRACTIONS BEFORE UPDATING
TANGENTIAL TRACTIONS AFTER UPDATING

(a)

ITERATION NO. 2
1.2
0.8
0.4
0.0
-0.4
-0.8
30
60
90

(b)

ITERATION NO. 3
1.2
0.8
0.4
0.0
-0.4
-0.8
30
60
90

(c)

ITERATION NO. 4
1.2
0.8
0.4
0.0
-0.4
-0.8
30
60
90

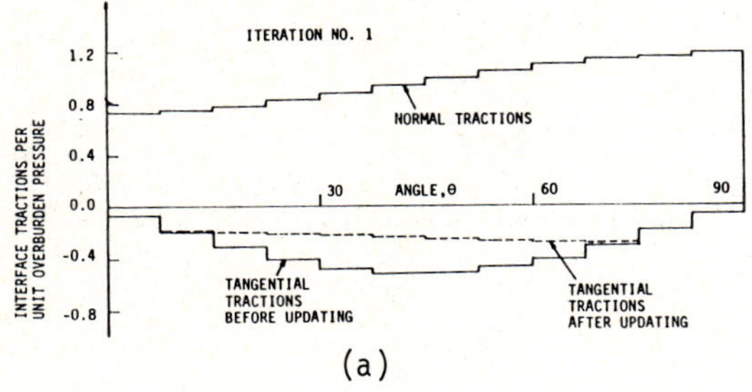

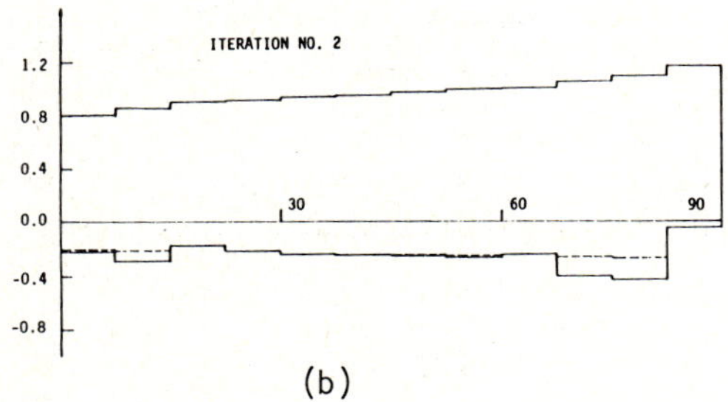

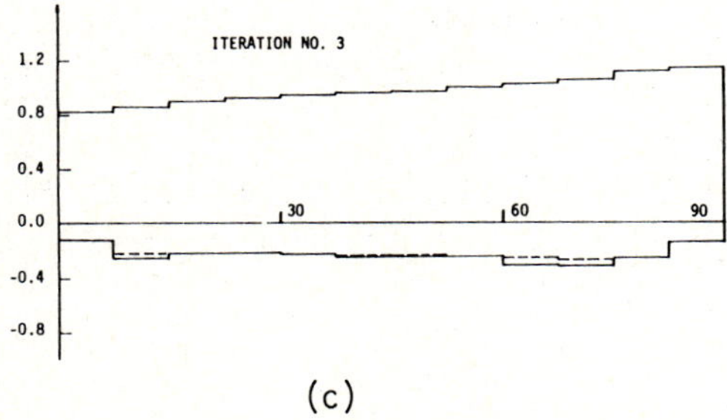

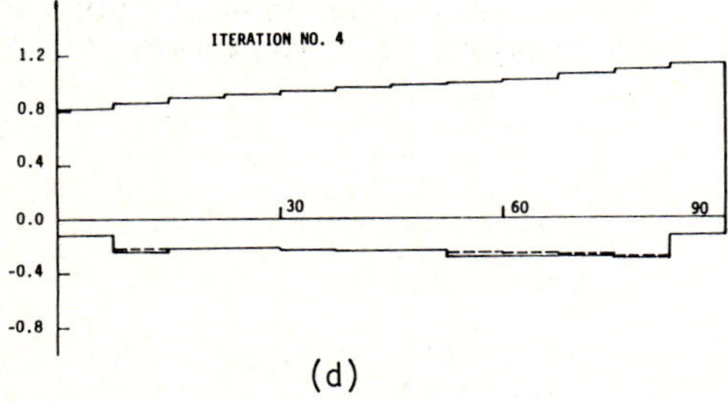

(d)

Fig. 11

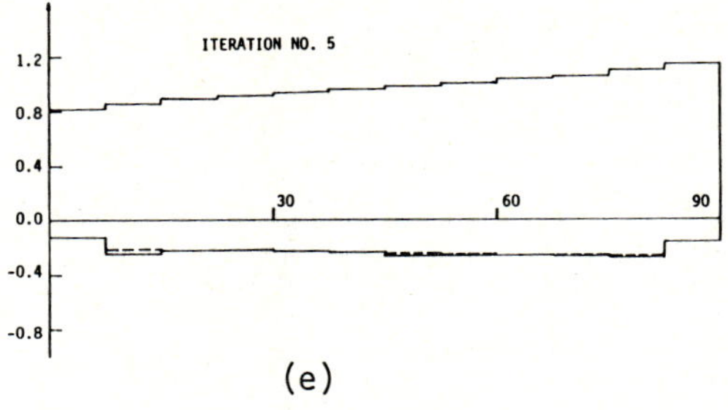

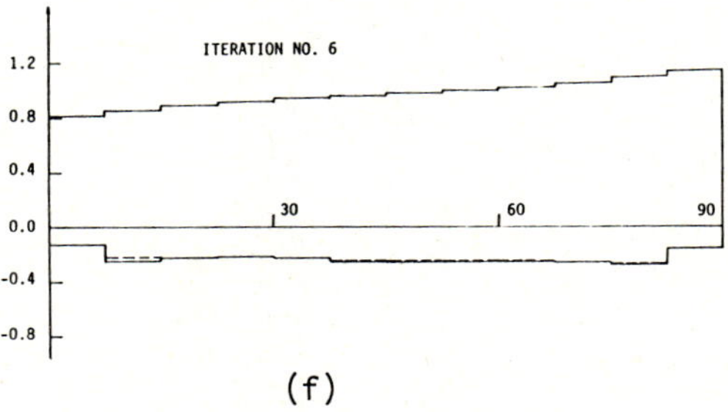

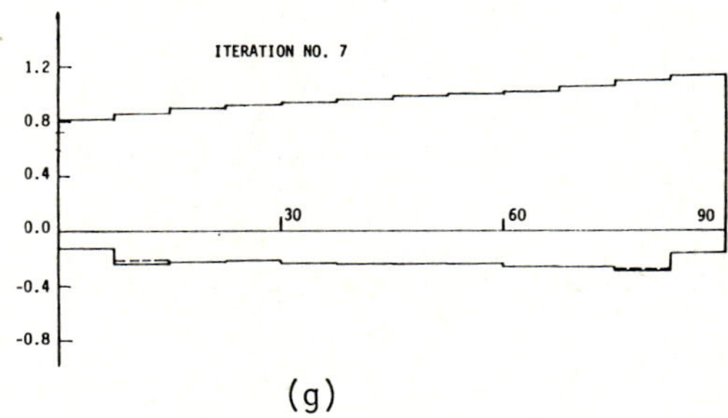

Figure 11   Mean tractions, T_t^j/A_j and T_n^j/A_j, for mesh C in the iterations. One step to total load and seven iterations to convergence.

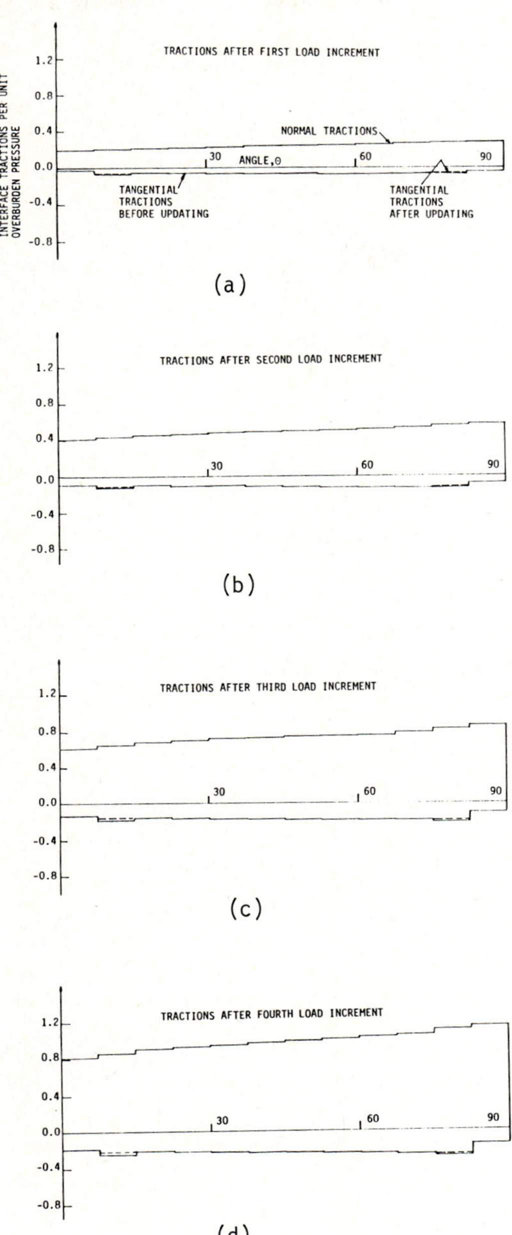

Figure 12 Mean tractions, T_t^j/A_j and T_n^j/A_j, for mesh C at convergence for each load step. Total load applied in four equal size load increments.

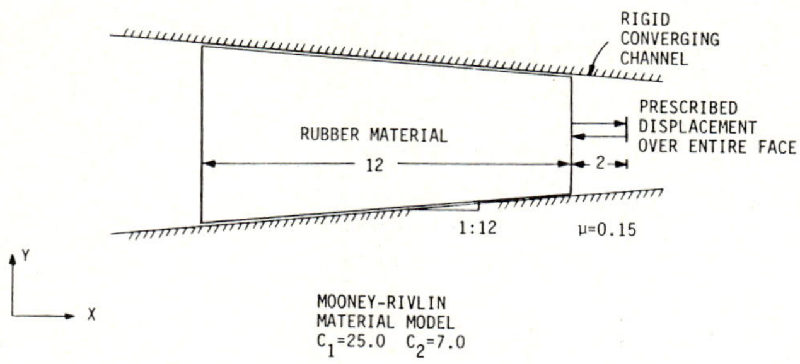

(a) Problem considered

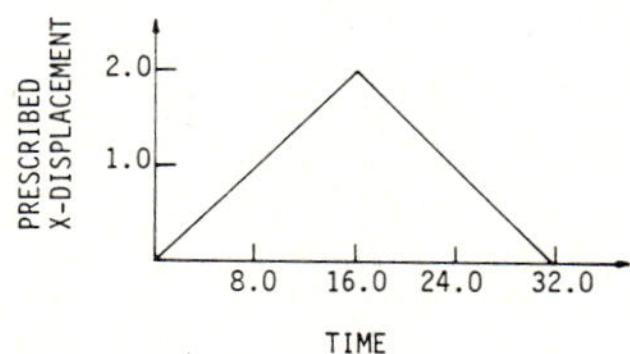

(b) Displacement history imposed on right face of sheet, $\Delta t = 0.5$

Figure 13 Rubber sheet analyzed

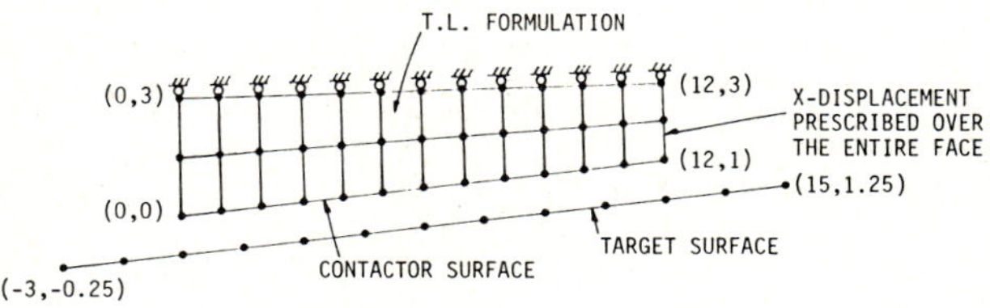

Figure 14 Finite element mesh used in analysis of rubber sheet; mesh B

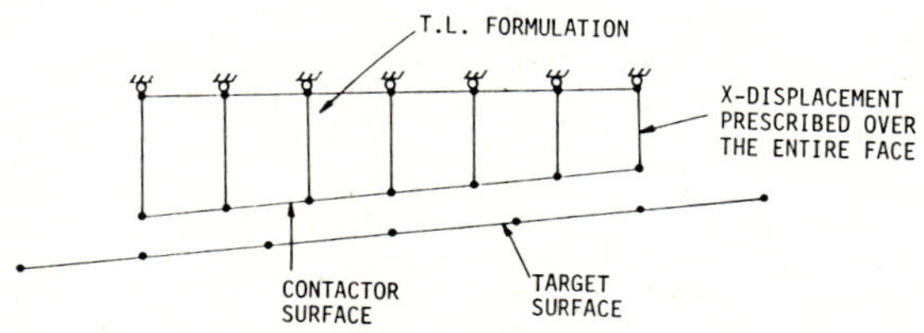

Figure 15 Finite element mesh used in analysis of rubber sheet; mesh A

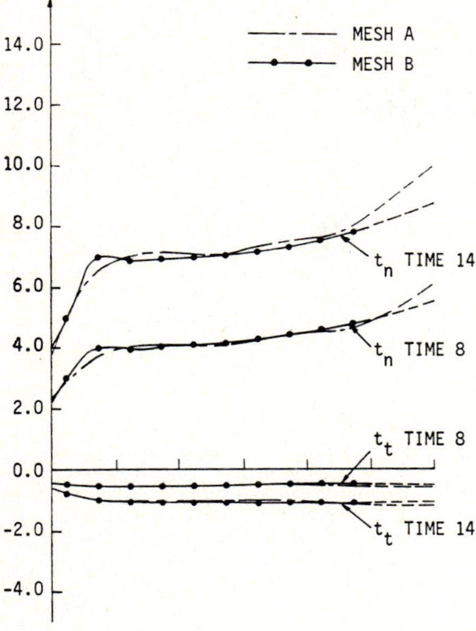

(a)   At times 8 and 14

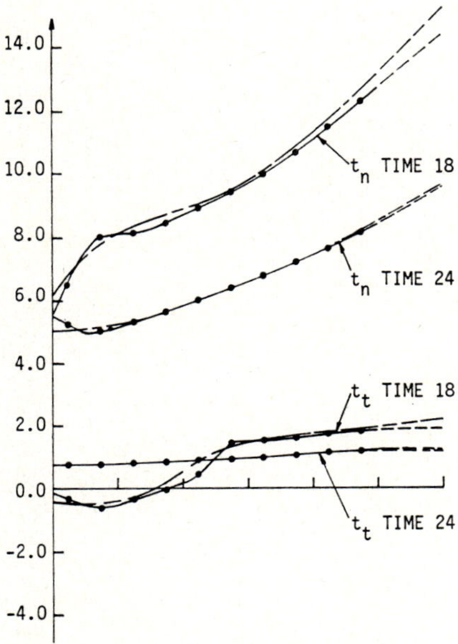

(b)   At times 18 and 24

Figure 16   Predicted tractions in analysis of rubber sheet

Unification of Finite Element Methods
H. Kardestuncer (Editor)
© Elsevier Science Publishers B.V. (North-Holland), 1984

CHAPTER 6

MIXED VARIATIONAL FINITE ELEMENT METHODS FOR INTERFACE PROBLEMS

J. Bielak & R.C. MacCamy

This paper presents a procedure for solving
interface problems: that is, situations in
which different partial differential equations
are to be solved in adjacent regions. One of
the regions is infinite in extent with homogeneous
equations. The other is finite but the equations
can be inhomogeneous. The method combines
variational finite element methods inside with
integral equation methods outside. A prototype
situation, that of electromagnetic theory, is
discussed.

1. Introduction.

We present here variational formulations for a class of
interface problems. These problems have the following form.
Let Ω be a bounded region in the plane with boundary Γ
and exterior Ω^+. Let p be a smooth, positive function in
$\overline{\Omega}$ and define the second order elliptic operator L by,

$$Lu = (pu_x)_x + (pu_y)_y. \qquad (1.1)$$

Let q and γ be functions defined on Ω and Γ respectively which are smooth but may be complex. Then given
(possibly complex) functions f and g on Γ and a nonnegative constant β_0 we seek u (possibly complex) such
that:

$$Lu + qu = 0 \quad \text{in} \quad \Omega; \quad \Delta u + \beta_0^2 u = 0 \quad \text{in} \quad \Omega^+$$

$$u^- = u^+ + f, \quad \gamma u_n^- = u_n^+ + g \quad \text{on} \quad \Gamma \qquad (1.2)$$

u satisfies a radiation condition in Ω^+ if $\beta_0 > 0$

u is bounded as $|x| \to \infty$ for $\beta_0 = 0$.*

Here the plus and minus denote limits from Ω^+ and Ω. We
call this problem (P_{β_0}).

(*) More generally one can allow $u \sim A \log |x|$ as $|x| \to \infty$,
A given.

150
J. Bielak & R.C. MacCamy

In the last section we indicate how such problems arise
in the study of two dimensional electromagnetic fields for
various choices of p, q, γ and β_0. In particular, we
indicate that the problem with $\beta_0 = 0$ is of considerable
importance. The problem P_{β_0} also arises in the study of
two-dimensional elastodynamics as discussed in [1]. In all
these applications one has,

$$q = \beta^2, \quad \beta > 0 \quad \text{or} \quad q = i\alpha^2, \quad \alpha > 0. \tag{1.3}$$

Our object is to give a variational formulation which
satisfies two conditions:
(1) One has to work only over Ω and Γ.
(2) All boundary conditions are natural.
The main problem is, of course, to account for the exterior
region. We do this by exploiting the fact that the exterior
equation has constant coefficients. Thus we can invoke the
ideas of boundary integral equations for exterior problems.
The necessary results are collected in section two.

In section three we show how to use the results from the
exterior problem to find problems which are equivalent to (P_{β_0})
in which one satisfies (1.2), and has non-local boundary
conditions connecting u and its normal derivative on Γ.
These transformed problems are then given variational formulations. We obtain a whole family $(VP)^\delta$ of these variational problems depending on a parameter δ, $0 \leq \delta \leq 1$.
In sections four and five we discuss the numerical implementation of our variational problems with finite elements.

Our methods are not completely new. The ideas bear
some resemblance to the use of hybrid methods for interior
Dirichlet problems [2]. They have some elements in common
with [3] and [4] and the work in [5] is, in effect, the
special case $(VP)^1$ for a simpler problem. Complete proofs
of the results quoted here can be found in [1].

2. <u>The exterior problem</u>.

We consider the equation,

$$\Delta u + \beta_0^2 u = 0, \qquad \beta_0 \geq 0. \tag{2.1}$$

We put,

$$g_{\beta_0}(x,y) = \frac{i}{4} H_0^{(2)}(\beta_0|x-y|) \text{ for } \beta_0 > 0; \quad g_0(x,y) = \frac{1}{2\pi} \log|x-y|, \tag{2.2}$$

where $H_0^{(2)}$ is the Hankel function of second kind and order
zero. Thus g_{β_0} is a fundamental solution for (2.1) and g_{β_0}
satisfies a radiation condition for $\beta_0 > 0$. Note, however,
that $g_0(x,y)$ becomes logarithmically infinite as $|x| \to \infty$.
We use g_{β_0} to define simple and double layers, $S[\varphi]$ and $D[\varphi]$
with density φ:

$$S[\varphi](x) = \int_\Gamma \varphi(y) g_{\beta_0}(x,y) \, dS_y; \quad D[\varphi](x) = \int_\Gamma \varphi(y) \frac{\partial}{\partial n_y} g_{\beta_0}(x,y) \, dS_y.$$
$$\tag{2.3}$$

For smooth curves Γ and functions φ the properties of $\mathcal{S}$ and $\mathcal{D}$ are well known. They satisfy (2.1) in Ω and in Ω^+ and the radiation conditions in Ω^+. We define integral operators S, N and D on Γ by,

$$S[\varphi](x) = \int_\Gamma \varphi(y) S_{\beta_0}(x,y) \, dS_y, \qquad S_{\beta_0}(x,y) = g_{\beta_0}(x,y) \Big|_{x \in \Gamma}$$

$$N[\varphi](x) = \int_\Gamma \varphi(y) N_{\beta_0}(x,y) \, dS_y, \qquad N_{\beta_0}(x,y) = \frac{\partial}{\partial n_x} g_{\beta_0}(x,y) \Big|_{x \in \Gamma}$$

$$D[\varphi](x) = \int_\Gamma \varphi(y) D_{\beta_0}(x,y) \, dS_y, \qquad D_{\beta_0}(x,y) = \frac{\partial}{\partial n_y} g_{\beta_0}(x,y) \Big|_{x \in \Gamma}.$$

$$(2.4)$$

The kernels N_{β_0} and D_{β_0} are continuous while S_{β_0} has a logarithmic singularity when $x = y$. Moreover, one has the symmetry properties,

$$S_{\beta_0}(x,y) = S_{\beta_0}(y,x); \quad N_{\beta_0}(x,y) = D_{\beta_0}(y,x). \tag{2.5}$$

One has, then, the well known limit relations:

$$\mathcal{S}[\varphi]^\pm = S[\varphi], \quad \left(\frac{\partial \mathcal{S}[\varphi]}{\partial n}\right)^\pm = \pm \frac{1}{2}\varphi + N[\varphi]$$

$$\mathcal{D}[\varphi]^\pm = \mp \frac{1}{2}\varphi + D[\varphi]. \tag{2.6}$$

The layers can be used to obtain representations for solutions of (2.1). We have, for any solution of (2.1) in Ω,

$$v = \mathcal{D}[v^-] - \mathcal{S}[v_n^-] \quad \text{in} \quad \Omega. \tag{2.7}$$

If v satisfies a radiation condition in Ω^+ we have,

$$v = \mathcal{S}[v_n^+] - \mathcal{D}[v_n^+] \quad \text{in} \quad \Omega^+ \quad \text{if} \quad \beta_0 > 0. \tag{2.8}$$

The representations in Ω^+ for $\beta_0 = 0$ is a little more complicated. The result is that if v is a solution of (2.1) which is bounded at infinity then,

$$v = \mathcal{S}[v_n^+] - \mathcal{D}[v^+] + C[v], \quad \int_\Gamma v_n^+ \, ds = 0, \tag{2.9}$$

where $C[v]$ is a constant.

From (2.7) and $(2.6)_3$ we obtain, for a solution of (2.1) in Ω,

$$\frac{1}{2} u^- = D[u^-] - S[u_n^-]. \tag{2.10}$$

Similarly, for $\beta_0 > 0$, and a solution of (2.1) in Ω^+, satisfying the radiation condition,

$$\frac{1}{2} u^+ = S[u_n^+] - D[u^+]. \tag{2.11}$$

Finally, for $\beta_0 = 0$, and a solution of (2.1) in Ω^+, bounded at infinity,

$$\frac{1}{2} u^+ = S[u_n^+] - D[u^+] + C[u], \qquad \int_\Gamma u_n^+ \, ds = 0. \tag{2.12}$$

Equations (2.10) and (2.11) can be used to establish existence theorems for the solution of Neumann problems. If $u_n^-(u_n^+)$ is specified on Γ then (2.10) ((2.11)) becomes an integral equation which can be solved for $u^-(u^+)$. Then (2.7) ((2.8)) yields a solution of (2.1) in Ω (Ω^+) with the specified normal derivative.

There is also a procedure to solve the Dirichlet problem which we will need. This procedure, developed in [6], is as follows. To find a solution of (2.1) in Ω (Ω^+) with $u^-(u^+)$ specified take,

$$u = S[\chi], \tag{2.13}$$

with χ satisfying,

$$S[\chi] = u^-(u^+). \tag{2.14}$$

Then, by $(2.6)_2$,

$$u_n^{\overline{+}} = \mp \frac{1}{2} \chi + N[\chi]. \tag{2.15}$$

The method just given has to be modified for the exterior problem with $\beta_0 = 0$. The appropriate procedure is,

$$u = S[\chi] + C; \qquad S[\chi] + C = u^+, \qquad \int_\Gamma \chi \, ds = 0.^* \tag{2.16}$$

<u>Remark</u>: There exists a countable infinity of values of β_0, β_0^n, for which the above integral equation methods will fail. We will assume that β_0 is not one of those values.

3. <u>The variational problems.</u>

As a first step in obtaining our variational procedures for the interface problems let us rephrase that problem using the results of section two. Suppose u is a solution of P_{β_0}, $\beta_0 > 0$ and put $\varphi = \gamma u_n^-$. Then we will have $u_n^+ = \varphi - g$ by (1.2). We substitute this into (2.11) to obtain.

$$\frac{1}{2} u^+ + D[u^-] - S[\varphi] = -S[g]. \tag{3.1}$$

(*) If u is allowed to have the behavior $u \sim A \log|\underset{\sim}{x}|$ in (P_0) then one replaces 0 here by A.

But now we can use the transition conditions again to rewrite (3.1) as,

$$\frac{1}{2} u^- + D[u^-] - S[\varphi] = \frac{1}{2} f - S[g] + D[f].\qquad(3.2)$$

Thus P_{β_0} is equivalent to the following problem which we denote by P^1. Find u and φ such that,

$$Lu + qu = 0 \quad \text{in} \quad \Omega, \qquad \gamma u_n^- = \varphi \text{ on } \Gamma \qquad(3.3)$$
$$\frac{1}{2} u^- + D[u^-] - S[\varphi] = \frac{1}{2} f - S[g] + D[f] \equiv F \quad \text{on} \quad \Gamma.$$

We give a variational formulation of P^1. Multiply $(3.3)_1$ by $\bar{v}$, integrate by parts and use $(3.3)_2$ to obtain,

$$-\int_\Omega (p\nabla u \cdot \nabla \bar{v} - qu\bar{v})\,dx + \int_\Gamma \frac{p}{\gamma} \varphi \bar{v}\,ds = 0.\qquad(3.4)$$

Then multiply $(3.3)_3$ by $\bar{\psi}$ and integrate over Γ:

$$\int_\Gamma (\frac{1}{2} u^- + D[u^-] - S[\varphi])\bar{\psi}\,ds = \int_\Gamma F\bar{\psi}\,ds.\qquad(3.5)$$

Our variational problem is then to find u and φ such that (3.4) and (3.5) hold for any (v, ψ).

For $\beta_0 = 0$ the method has to be modified. Instead of (3.5) we obtain,

$$\int_\Gamma (\frac{1}{2} u^- + D[u^-] - S[\varphi] + C)\bar{\psi}\,ds = \int_\Gamma F\bar{\psi}\,ds\qquad(3.5')$$

and we have to add the condition,

$$\int_\Gamma \varphi\,ds = \int_\Gamma g\,ds.\qquad(3.6)$$

Here we have to find u, φ and C so that (3.4), (3.5') and (3.6) hold for any (v, ψ).

Let us give a notation for the problem (3.4), (3.5) $(\beta_0 \neq 0)$. Let $U = (u, \varphi)$, $V = (v, \psi)$ and put

$$a^1(U,V) = -\int_\Omega (p\nabla u \cdot \nabla \bar{v} - qu\bar{v})\,dx + \int_\Gamma \frac{p}{\gamma} \varphi \bar{v}\,ds$$

$$+ \int_\Gamma (\frac{1}{2} u^- + D[u^-] - S[\varphi])\bar{\psi}\,ds\qquad(3.7)$$

$$\mathcal{F}^1(V) = \int_\Gamma F\bar{\psi}\,ds.$$

Then our variational problem is: find U such that for all V,

$$a^1(U,V) = \mathcal{F}^1(V)\qquad\qquad(VP)^1.$$

Remark: In the applications (see the last section) the functions f and g are usually u_0^+, $u_{0,n}^+$ for some function u_0 satisfying $\Delta u + \beta_0^2 u = 0$ in all space. For such a function we have by (2.10), $\frac{1}{2} f - D[f] + S[g] = 0$. It follows from $(3.3)_3$ that $F = f = u_0^+$.

We obtain a second variational problem by using $(2.13)-$ (2.15). We use (2.13) in Ω^+ with χ to be determined. Then from (2.14), (2.15) and the transition conditions to obtain,

$$u_n^- = \tfrac{1}{2}\chi + N[\chi] + g, \qquad u^- = S[\chi] + f \quad \text{on} \ \Gamma. \tag{3.8}$$

Then (P_{β_0}) is equivalent to the problem, denoted by $\wp^0$, find u and χ such that,

$$Lu + qu = 0 \quad \text{in} \ \Omega, \qquad \gamma u_n^- = \tfrac{1}{2}\chi + N[\chi] + g \quad \text{on} \ \Gamma$$

$$u^- = S[\chi] + f \quad \text{on} \ \Gamma. \tag{3.9}$$

We again give a variational formulation obtaining,

$$-\int_\Omega (p\nabla u \cdot \overline{\nabla v} - qu\overline{v})\,dx + \int_\Gamma \frac{p}{\gamma}(\tfrac{1}{2}\chi + N[\chi])\overline{v}\,ds = -\int_\Gamma \frac{p}{\gamma} g\,\overline{v}\,ds \tag{3.10}$$

$$\tfrac{1}{2}\int_\Gamma (u^- - S[\chi])\,\overline{\zeta}\,ds = \tfrac{1}{2}\int_\Gamma f\,\overline{\zeta}\,ds. \tag{3.11}$$

For $\beta_0 = 0$ we have to replace (3.11) by,

$$\tfrac{1}{2}\int_\Gamma (u^- - S[\chi] - C)\,\overline{\zeta}\,ds = \tfrac{1}{2}\int_\Gamma f\,\overline{\zeta}\,ds, \qquad \int_\Gamma \chi\,ds = 0. \tag{3.11'}$$

We introduce a notation analogous to (3.7). We put, for $\beta_0 > 0$, $U = (u,\chi)$, $V = (v,\zeta)$

$$G^0(U,V) = -\int_\Omega (p\nabla u \cdot \overline{\nabla v} - qu\overline{v})\,dx + \int_\Gamma \frac{p}{\gamma}(\tfrac{1}{2}\chi + N[\chi])\overline{v}\,ds$$

$$+ \tfrac{1}{2}\int_\Gamma (u^- - S[\chi])\,\overline{\zeta}\,ds, \tag{3.12}$$

$$\mathcal{F}^0(V) = -\int_\Gamma g\,\overline{v}\,ds + \tfrac{1}{2}\int_\Gamma f\,\overline{\zeta}\,ds.$$

Then (3.10) and (3.11) are,

$$G^0(U,V) = \mathcal{F}^0(V) \tag{VP^0}.$$

We want to study the form of the variational problems a little more closely. For simplicity let us assume that $\gamma = p$. In the applications this is usually true or else it can be achieved by changing variables. Let us write (3.7) and (3.8) in the obvious notations,

$$G^1(U,V) = G^1((u,\varphi),(v,\psi)) = A_{11}(u,v) + A_{12}^1(\varphi,v) + A_{21}^1(u,\psi) + A_{22}^1(\varphi,\psi)$$

$$G^0(U,V) = G^0((u,\chi),(v,\psi)) = A_{11}(u,v) + A_{12}^0(\chi,v) + A_{21}^0(u,\psi) + A_{22}^0(\chi,\psi) \tag{3.13}$$

We want to demonstrate the symmetries here, using (2.5). First we have, by $(2.5)_1$,

$$A_{22}^0(\chi,\psi) = -\int_\Gamma S[\chi]\,\overline{\psi}\,ds = -\int_\Gamma S(\overline{\psi})\,\chi\,ds = A_{22}^1(\overline{\psi},\overline{\chi}). \qquad (3.14)$$

Next, $(2.5)_2$ yields,

$$A_{12}^0(\chi,v) = \int_\Gamma (\chi + N[\chi])\,\overline{v}\,ds = \int_\Gamma (\overline{v} + D[\overline{v}])\,\chi\,ds = A_{21}^1(\overline{v},\overline{\chi}). \quad (3.15)$$

In the same way,

$$A_{21}^0(u,\psi) = A_{12}^1(\overline{\psi},\overline{u}). \qquad (3.16)$$

We can combine $(3.13)-(3.16)$ in the formula,

$$G^0(U,V) = G^0((u,\varphi),(v,\psi)) = G^1((\overline{v},\overline{\psi}),(\overline{u},\overline{\varphi})) = G^1(\overline{V},\overline{U}). \,(3.17)$$

Clearly one has the choice of using $(VP)^1$ or $(VP)^0$. The advantage of $(VP)^1$ is that it yields u_n^- directly as part of the solution. Its disadvantage is that it makes the computation of the external field a little complicated. One must determine u^- from the interior solution, compute u^+ and u_n^- from the transition conditions and then do the two integrations in (2.8). $(VP)^0$ yields the external field more readily with the single integration (2.12) but requires another integration, (2.15), to obtain u_n^-.

We observe that there is really a whole family of variational problems $(VP)^\delta$, $0 < \delta < 1$. We simply multiply $(VP)^1$ by δ and $(VP)^0$ by $(1-\delta)$ and add. Then if we put $U = (u,\varphi,\chi)$, $V = (v,\psi,\zeta)$ and

$$G^\delta(U,V) = \delta G^1((u,\varphi),(v,\psi)) + (1-\delta)G^0((u,\chi),(v,\zeta))$$

$$F^\delta(V) = \delta F^1((v,\psi)) + (1-\delta)F^0((v,\zeta))$$

$$(3.18)$$

we have the variational problems, find U such that for any V,

$$G^\delta(U,V) = F^\delta(V) \qquad\qquad (VP)^\alpha.$$

One can check that for $\delta = 1/2$ we have the symmetry relation,

$$G^{1/2}(\overline{V},\overline{U}) = G^{1/2}(U,V). \qquad (3.19)$$

We return to this relation in the next sections.

An analysis of the problems P^1 and P^0, as well as the variational problems $(VP)^1$ and $(VP)^0$, is presented in [1], for the case $\beta_0 > 0$. (The case $\beta_0 = 0$ can be treated similarly.) We review the results briefly. There are some technical conditions. We have indicated in section two that a countable infinity of β_0's must be avoided. Further, if q in $(1.2)_1$ is real and positive in Ω then it could happen that the problem $Lu + qu = 0$ in Ω, $pu_n = 0$ on Γ could have non-zero solutions. We assume that Ω is such that this cannot happen. Then the following facts have been established.

1. Suppose $f \in H_r(\Gamma)$ and $g \in H_{r-1}(\Gamma)$ for some $r \geq 1/2$.
Then (P^1) has a unique (generalized) solution (u,φ) with
$u \in H_{r+1/2}(\Omega)$, $\varphi \in H_{r-1/2}(\Gamma)$, $u^- \in H_r(\Gamma)$, $u_n^- \in H_{r-1}(\Gamma)$. If
one computes u^+ and u_n^+ from the transition conditions
then (2.8) yields a (classical) solution of $(1.2)_2$ in Ω^+,
satisfying the radiation condition and with $u \in H$, $\mathrm{loc}(\Omega^+)$.
The combined function is a (generalized) solution of (P_{β_0}).

2. Under the same conditions (P^0) has a unique generalized
solution (u,χ), with same regularity; (2.13) yields a solution
of $(1.2)_2$ with the same regularity and the combined function
yields a (generalized) solution of (P_{β_0}).

3. For $f \in H_{1/2}(\Gamma)$ and $g \in H_{-1/2}(\Gamma)$ $(VP)^1$ $((VP)^0)$ have
unique solutions (u,φ) $((u,\chi))$ with $u \in H_1(\Omega)$ and
$\varphi(\chi) \in H_{-1/2}(\Gamma)$.

   Results 1 and 2 are established by using known facts
about boundary value problems in Ω to reduce P^1 (P^0) to an
equation of Riesz-Schauder type for $\varphi(\chi)$ on the space
$H_{-1/2}(\Gamma)$. Then one can use the uniqueness of solutions of
(P_{β_0}) to show the homogeneous Riesz-Schauder equations have
only the trivial solution.

   In order to prove result 3 one has to establish coercivity
results of the form

$$\sup_{V \neq 0} \mathrm{Re}\, \frac{G^1(U,V)}{\|V\|} \geq k\|U\|, \qquad \sup_{V \neq 0} \mathrm{Re}\, \frac{G^0(U,V)}{\|V\|} \geq k\|V\|, \qquad (3.20)$$

where $\|U\|^2 = \|u\|^2_{H_1(\Omega)} + \|\varphi\|^2_{H_{-1/2}(\Gamma)}$. The estimates (3.20) can
be established by considering the adjoint variational problems
for $(VP)^1$ and $(VP)^0$, respectively. It turns out that because
of (3.17) the adjoint of $(VP)^1$ $((VP)^0)$ is essentially $(VP)^0$
$((VP)^1)$; hence one has a symmetric argument.

4. <u>Approximate variational problems.</u>

   In order to implement the variational problems numerically one introduces finite dimensional approximate spaces. We
illustrate with $(VP)^1$; the others are analogous. According
to result 3 in section 3, $(VP)^1$ has a solution
$(u,\varphi) \in H_1(\Omega) \times H_{-1/2}(\Gamma)$. We introduce families of subspaces,

$$S^{h\Omega} \subset H_1(\Omega), \qquad S^{h\Gamma} \subset H_{-1/2}(\Gamma). \qquad (4.1)$$

These are to be finite dimensional and to depend on parameters
h_Ω and h_Γ. We put $S^h = S^{h\Omega} \times S^{h\Gamma}$. Then our approximate
variational problem is:
Find $U^h = (u^h,\varphi^h) \in S^h$ such that for any $V^h = (v^h,\psi^h) \in S^h$,

$$G(U^h,V^h) = \mathfrak{J}(V^h). \qquad (AVP)^1$$

$(AVP)^1$ is equivalent to sets of algebraic equations. Let

$(\omega_1^h,\ldots,\omega_{N_{h_\Omega}}^h)$, $(\sigma_1^h,\ldots,\sigma_{N_{h_\Gamma}}^h)$ be bases for S^{h_Ω} and S^{h_Γ}.

Then we have

$$u^h = \sum_{i=1}^{N_{h_\Omega}} u_i^h \omega_i^h, \quad \varphi^h = \sum_{i=1}^{N_{h_\Gamma}} \varphi_i^h \sigma_i^h \tag{4.2}$$

and $(AVP)^1$ is equivalent to the algebraic equations

$$\underset{\sim}{A}_{11}\underset{\sim}{u}^h + \underset{\sim}{A}_{12}\underset{\sim}{\varphi}^h = \underset{\sim}{0}; \quad \underset{\sim}{A}_{21}\underset{\sim}{u}^h + \underset{\sim}{A}_{22}\underset{\sim}{\varphi}^h = \underset{\sim}{q}^h \tag{4.3}$$

where,

$$\underset{\sim}{u}^h \in \mathbb{R}^{N_{h_\Omega}}; \quad (\underset{\sim}{u}^h)_i = u_i^h; \quad \underset{\sim}{\varphi}^h \in \mathbb{R}^{N_{h_\Gamma}}, \quad (\underset{\sim}{\varphi}^h)_i = \varphi_i^h$$

$$\underset{\sim}{q}^h \in \mathbb{R}^{N_{h_\Gamma}}, \quad (\underset{\sim}{q}^h)_i = \int_\Gamma F \sigma_i^h\, ds \tag{4.4}$$

and the matrices are determined by

$$(\underset{\sim}{A}_{11})_{ij} = A_{11}(\omega_i^h,\omega_j^h), \quad (\underset{\sim}{A}_{12})_{ij} = A_{12}(\sigma_i^h,\omega_j^h)$$

$$\tag{4.5}$$

$$(\underset{\sim}{A}_{21})_{ij} = A_{21}(\omega_i^h,\sigma_j^h), \quad (\underset{\sim}{A}_{22})_{ij} = A_{22}(\sigma_i^h,\sigma_j^h).$$

We shall say a little more about numerical implementations in the next section. Here we want to review some further theoretical results from [1]. The results require the following approximation properties of the spaces S^{h_Ω} and S^{h_Γ}:

(A.1) There exists a constant γ_1 and an integer $k > 1$ such that for any $w \in H_\ell(\Omega)$, $1 \le \ell \le k$ there is a $w^{h_\Omega} \in S^{h_\Omega}$ with

$$\|w - w^{h_\Omega}\|_{H_r(\Omega)} \le \gamma_1 (h_\Omega)^{\ell-r} \|w\|_{H_\ell(\Omega)}, \quad 0 \le r \le \ell.$$

(A.2) There exists a constant $\gamma_2 > 0$ and a $k^1 > 1/2$ such that for any $\phi \in H_{\ell'}(\Gamma)$, $-1/2 \le \ell' \le k^1$ there is a $\phi^{h_\Gamma} \in S^{h_\Gamma}$ with

$$\|\phi - \phi^{h_\Gamma}\|_{H_s(\Gamma)} \le \gamma_2 (h_\Gamma)^{\ell'-s} \|\phi\|_{H_s(\Gamma)} \quad -\frac{1}{2} \le s \le \ell'.$$

The following results are established in [1]. Put $h = h_\Omega + h_\Gamma$. Then if h is sufficiently small:

(1) Equations (4.5) have a unique solution.

(2) Suppose $U = \{u,\varphi\}$ is the solution of $(VP)^1$ and

$u \in H_{1+\epsilon}(\Omega)$, $\varphi \in H_{-1/2+\epsilon}(\Gamma)$ with $\epsilon < \min(k,k^1)$ in (A.1) and (A.2). Then there exists a constant c, independent of h such that,

$$\|u - u^h\|_{H_1(\Omega)} + \|\varphi - \varphi^h\|_{H_{-1/2}(\Gamma)} \leq ch^\epsilon. \qquad (4.6)$$

As an example of the meaning of the above result one can take $S^{h\Omega}$ to consist of piecewise linear functions in Ω ($k = 2$) and $S^{h\Gamma}$ to consist of piecewise constant functions on Γ ($k^1 = 1$). Suppose then that the solution of (VP)^1 has $u \in H_2(\Omega)$ and $\varphi \in H_{1/2}(\Gamma)$ ($\epsilon = 1$). Then take $\epsilon = 1$ in (4.6) and get

$$\|u - u^h\|_1(\Omega) + \|\varphi - \varphi^h\|_{-1/2}(\Gamma) \leq ch. \qquad (4.7)$$

Thus we obtain order h convergence in the natural norm for (VP)^1. One can also show that there is a constant c^1 independent of the choice of h such that

$$\|u - u^h\|_0(\Omega) \in c^1 h^2. \qquad (4.8)$$

The proofs of the above results proceed in several stages. One shows first that the coercivity results (3.20) hold for any $U^h \in S^h$ when V is restricted to V^h. This is done by first using (3.20) to get a $V \in S$ which makes the inequalities valid and then using regularity results to show that V can be approximated with a $V^h \in S^h$. These coercivity results on S^h enable one to establish optimality; that is, to show that U is approximated by U^h, in the natural norm, as well as it is possible to approximate U, in that norm, by elements of S^h. Then one invokes (A.1) and (A.2). The L_2 estimate (4.8) is obtained via the Aubin-Nitsche trick.

5.   <u>Implementation of the numerical procedure.</u>

For purposes of illustration we now discuss the actual implementation of the finite element method described in the preceding section for the case in which Ω is the unit circle and f and g are given by u_0^+ and $u_{0,n}^+$, respectively, where u_0 represents a harmonic incident plane wave field. The material in Ω is homogeneous and u_0 is symmetric with respect to a diameter. The problem has been solved exactly in [7]. We divide the region Ω into circular sectors, with wedges around the origin, and consider a piecewise linear approximation for u in the polar coordinates r and θ; φ is taken to be a piecewise constant function on Γ. With this approximation the elements of the matrices $\underset{\sim}{A}_{11}$ and $\underset{\sim}{A}_{12}$ and the load vector g^h can be evaluated explicitly by direct integration. For $\underset{\sim}{A}_{21}$ we integrate numerically with standard Gauss-Legendre formulas since the kernel D_{β_0} in (2.4) that enters into the bilinear form A_{21} in (4.5) is continuous. Due to the logarithmic singularity of g_{β_0} which appears in the bilinear form A_{22} we use a modified Gauss-Legendre formula [8] that accommodates this singularity explicitly in evaluating the elements of $\underset{\sim}{A}_{22}$. $\underset{\sim}{A}_{11}$ is an $N_{h_\Omega} \times N_{h_\Omega}$ matrix with the elements indicated in Ω. $\underset{\sim}{A}_{12}$ and $\underset{\sim}{A}_{21}$ are $N_{h_\Omega} \times N_{h_\Gamma}$ and $N_{h_\Gamma} \times N_{h_\Omega}$, respectively, and $\underset{\sim}{A}_{22}$ will be $N_{h_\Gamma} \times N_{h_\Gamma}$. The last three matrices will be full. However, since the diameters of the elements inside Ω and the lengths of the intervals on Γ are about the same and of size h, then we see that $N_{h_\Omega} = N_{h_\Gamma}^2$. Thus, although the matrices $\underset{\sim}{A}_{12}$, $\underset{\sim}{A}_{21}$ and $\underset{\sim}{A}_{22}$ are full their size is much smaller than that of $\underset{\sim}{A}_{11}$ and an effective numerical procedure is still possible.

<u>Remarks</u>: (i) Note that while $\underset{\sim}{A}_{11}$ and $\underset{\sim}{A}_{22}$ are symmetric matrices, $\underset{\sim}{A}_{12}$ and $\underset{\sim}{A}_{21}$ are not generally the transpose of each other. Therefore, the system (4.3) is, in general, asymmetric. (ii) The form of equations (4.3) permits condensation. Suppose we are primarily concerned with the interior region. Then we may eliminate $\underset{\sim}{\varphi}^h$ and consider the system,

$$(\underset{\sim}{A}_{11} + \underset{\sim}{B}_{11})\underset{\sim}{u}^h = \underset{\sim}{r}^h \tag{5.1}$$

where $\underset{\sim}{B}_{11} = -1/2 A_{12}A_{22}A_{21}$ and $\underset{\sim}{r}^h = -A_{12}A_{22}^{-1}\underset{\sim}{q}^h$. The matrix $\underset{\sim}{B}_{11}$ has nonzero elements only for nodes on the boundary Γ. Thus it represents the impedance of the exterior region Ω^+, and constitutes, in effect, a discretized nonlocal absorbing boundary. $\underset{\sim}{r}^h$ represents the corresponding effective forcing function. Although $\underset{\sim}{B}_{11}$ is in general an asymmetric matrix it turned out to be symmetric for the present problem. (iii) The condensation procedure requires that the matrix $\underset{\sim}{A}_{22}$ be inverted; thus, it is not valid for values of β_0 for which the operator $S[\varphi]$ in (2.4) cannot be inverted. Direct solution of the complete system $(4.3)_1$, however, was possible for values of β_0 approaching these critical values. (iv) Other condensation schemes are clearly possible; see [1] for a scheme that is applicable if one is mainly concerned with the exterior region. (v) A symmetric discretized formulation of the general problem P is always possible by using the variational formulation $VP^{1/2}$ in view of the symmetry relationship (3.19) provided one chooses real basis functions. The price we pay for this symmetry is that the system (4.3) is replaced by a set of similar structure of $N_{h_\Omega} + 2N_{h_\Gamma}$ equations instead of the $N_{h_\Omega} + N_{h_\Gamma}$ in (4.3). See [1] for details. After condensation, however, the corresponding system leads to equations of the form,

$$(\underset{\sim}{A}_{11} + \frac{1}{2}\underset{\sim}{B}_{11} + \frac{1}{2}\underset{\sim}{B}_{11}^T)\underset{\sim}{u}^h = \underset{\sim}{t}^h$$

wnich is similar to (5.1) and is clearly symmetric. $\underset{\sim}{t}^h$ is the corresponding effective forcing function.

A comparison between the exact and approximate values of
u at the center of the circle is shown in Table 1 for dif-
ferent values of Nh_Γ ($Nh_\Omega = N^2_{h_\Gamma}$) for several combinations
of the system parameters. The results tend to confirm our
theoretical estimates that the convergence is of order h^2
for the elements used.

6. <u>Two dimensional electromagnetic problems.</u>

Many of the problems of electromagnetic problems can be
idealized in the following way. One has a field everywhere
in space, which we think of as filled with air. One introduces dielectric or metallic obstacles and seeks to determine
both the fields induced in the obstacles and the distortion
of the original field outside. This is an <u>interface</u> problem:
one has different sets of Maxwell's equations in air and in
the obstacles and transition conditions across the boundaries.

The above problems are usually considered for the case
of time-periodic fields of a single frequency and this is
the case we consider. (By taking inverse Fourier transforms
one can, in principle, solve time dependent problems from
the periodic case.)

The variables to be determined are the electric and
magnetic fields $\mathcal{E}$ and $\mathcal{H}$. These satisfy Maxwell's equations.
We write down these equations when the material in question
is either a dielectric or non-ferromagnetic metal in the time
periodic case with frequency ω. Further, we render the
position variables x non-dimensional by dividing by a
representative length a. The equations are:

$$\text{curl } \mathcal{E} = i\omega\mu a\mathcal{H}, \qquad \text{curl } \mathcal{H} = K\mathcal{E}, \tag{6.1}$$

where

$$K = -i\omega\epsilon a \quad \text{for dielectrics}, K = \sigma a \quad \text{for metal.} \tag{6.2}$$

Let us first do some scaling in the problem. Air is a
dielectric with permeabilities μ_0, ϵ_0. We introduce dimensionless fields H and E by writing,

$$\mathcal{H} = h_0 H, \qquad \mathcal{E} = i\omega\mu_0 a h_0 E. \tag{6.3}$$

Then (6.1) becomes,

$$\text{curl } E = \frac{\mu}{\mu_0} H; \qquad \text{curl } H = k E, \tag{6.4}$$

where

$$k = \omega^2 \mu_0 \epsilon a^2 = \beta^2 \quad \text{for dielectrics,}$$

$$k = i\omega\mu_0 \sigma a^2 = i\alpha^2 \quad \text{for metal.} \tag{6.5}$$

The parameters k, α and β are dimensionless. We allow for the possibility that the obstacles are inhomogeneous so that k, α and β can depend on position.

We can now describe the interface problem. Let Ω denote the obstacle region and Ω^+ its exterior. Then in Ω^+ we have $\mu = \mu_0$, $k = \beta_0^2 = \omega^2 a^2 \mu_0 \epsilon_0$. Thus we have,

$$\text{curl } \underset{\sim}{E} = \underset{\sim}{H}, \qquad \text{curl } \underset{\sim}{H} = \beta_0^2 \underset{\sim}{E} \text{ in } \Omega^+$$

$$\text{curl } \underset{\sim}{E} = (\tfrac{\mu}{\mu_0}) \underset{\sim}{H}, \quad \text{curl } \underset{\sim}{H} = k\underset{\sim}{E} \text{ in } \Omega . \tag{6.6}$$

The transition conditions across $\Gamma = \partial\Omega$ are that the tangential components of $\underset{\sim}{E}$ and $\underset{\sim}{H}$ are continuous, that is, if $\underset{\sim}{n}$ is the normal to Γ:

$$\underset{\sim}{n} \times \underset{\sim}{E}^+ = \underset{\sim}{n} \times \underset{\sim}{E}^-; \quad n \times \underset{\sim}{H}^+ = n \times \underset{\sim}{H}^- \text{ on } \Gamma, \tag{6.7}$$

where the plus and minus denote limits from Ω^+ and Ω.

The problem is to be driven by an incident field $\underset{\sim}{E}^0, \underset{\sim}{H}^0$ satisfying $(6.6)_1$ in all space.* The differences $\underset{\sim}{E} - \underset{\sim}{E}^0$ and $\underset{\sim}{H} - \underset{\sim}{H}^0$ are to satisfy a radiation condition. If we let $\underset{\sim}{E}$ and $\underset{\sim}{H}$ represent the scattered fields in Ω^+ then we still have equations (6.6) but (6.7) is replaced by,

$$\underset{\sim}{n} \times \underset{\sim}{E}^- = \underset{\sim}{n} \times \underset{\sim}{E}^+ + \underset{\sim}{n} \times \underset{\sim}{E}^0; \quad n \times \underset{\sim}{H}^+ = n \times \underset{\sim}{H}^- + n \times \underset{\sim}{H}^0 \text{ on } \Gamma, \tag{6.8}$$

with $\underset{\sim}{E}$ and $\underset{\sim}{H}$ satisfying a radiation condition.

We now specialize the geometry. We suppose that the obstacles consist of cylinders of uniform cross section Ω parallel to the z-axis. Then we limit ourselves to fields $\underset{\sim}{E}$, $\underset{\sim}{H}$ (and $\underset{\sim}{E}^0, \underset{\sim}{H}^0$) which depend only on x and y, not z. It can be shown that all such fields are combinations of fields of the following type:

<u>Transverse magnetic</u> (TM): $\underset{\sim}{E} = E(x,y) \hat{k}; \underset{\sim}{H} = H^1(x,y) \hat{\imath} + H^2(x,y) \hat{\jmath}$

<u>Transverse electric</u> (TE): $\underset{\sim}{H} = H(x,j) \hat{k}; \underset{\sim}{E} = E^1(x,j) \hat{\imath} + E^2(x,j) \hat{\jmath}.$

$$\tag{6.9}$$

Let us determine the structure of such fields.

(TM) We have,

$$E_y \hat{\imath} - E_x \hat{\jmath} = \frac{\mu}{\mu_0}(H^1 \hat{\imath} + H^2 \hat{\jmath}), \qquad H^2_x - H^1_y = kE. \tag{6.10}$$

(*) One can allow $\underset{\sim}{E}^0$ and $\underset{\sim}{H}^0$ to have singularities in Ω^+. In fact, when $\beta_0 = 0$ one must have singularities in order to obtain a non-trivial problem.

We introduce a function u by the formulas,

$$u_y = \frac{\mu}{\mu_0} H^1, \qquad u_x = - \frac{\mu}{\mu_0} H^2. \tag{6.11}$$

Then $(6.10)_1$ is satisfied if $E = u$ and $(6.10)_2$ is satisfied
if

$$\left(\frac{\mu_0}{\mu} u_x\right)_x + \left(\frac{\mu_0}{\mu} u_y\right)_y = - ku. \tag{6.12}$$

Conversely, if u satisfies (6.12) and we put $E = u$ and
define H^1, H^2 by (6.11) we have a solution of (6.10).
Observe that on the surface Γ of our cylinder we have, for TM
fields,

$$\underset{\sim}{n} \times \underset{\sim}{E} = \underset{\sim}{\tau} u, \qquad \underset{\sim}{n} \times \underset{\sim}{H} = \frac{\mu_0}{\mu} u_\nu \hat{k} \tag{6.13}$$

where $\underset{\sim}{\tau}$ and $\underset{\sim}{\nu}$ are the unit tangent and normal to the
boundary of Ω in the x-y plane.

(TE) We have,

$$E_x^2 - E_y^1 = \frac{\mu}{\mu_0} h, \qquad H_y \hat{i} - H_x \hat{j} = k(E^1 \hat{i} + E^2 \hat{j}). \tag{6.14}$$

This time we define u by,

$$u_y = kE^1, \qquad u_x = - kE^2, \qquad H = u, \tag{6.15}$$

so that

$$\left(\frac{u_x}{k}\right)_x + \left(\frac{u_y}{k}\right)_y = - \frac{\mu}{\mu_0} u. \tag{6.16}$$

Instead of (6.13) we have,

$$n \times \underset{\sim}{E} = \frac{1}{k} u_\nu \hat{k}, \qquad n \times \underset{\sim}{H} = \underset{\sim}{\tau} u. \tag{6.17}$$

We can now obtain four different problems, all fitting
the framework discussed in the preceding sections. We again
let Ω^+ denote the exterior of Ω in the x-y plane. The
exterior region is air.

(I)   <u>Dielectric cylinder - TM fields</u>.

$$u_{xx} + u_{yy} = -\beta_0^2 u \quad \text{in} \quad \Omega^+, \qquad \left(\frac{\mu_0}{\mu} u_x\right) + \left(\frac{\mu_0}{\mu} u_y\right)_y = -\beta^2 u \quad \text{in} \quad \Omega$$

$$u^- = u^+ + u_0^+; \qquad \frac{\mu_0}{\mu} u_\nu^- = u_\nu^+ + u_{0,\nu} \quad \text{on} \quad \Gamma.$$

(II)   <u>Dielectric cylinder - TE fields</u>.

$$u_{xx} + u_{yy} = -\beta_0^2 u \quad \text{in} \quad \Omega^+; \qquad \left(\frac{u_x}{\beta^2}\right)_x + \left(\frac{u_y}{\beta^2}\right)_y = - \frac{\mu_0}{\mu} u \quad \text{in} \quad \Omega$$

$$u^- = u^+ + u_0^+, \quad \frac{\beta_0^2}{\beta^2}\, u_\nu^- = u_\nu^+ + u_{0,\nu} \quad \text{on} \quad \Gamma.$$

(III) <u>Metallic cylinder</u> - TM <u>fields</u>.

$$u_{xx} + u_{yy} = -\beta_0^2 \quad \text{in} \quad \Omega^+; \quad \left(\frac{\mu_0}{\mu}\, u_x\right)_x + \left(\frac{\mu_0}{\mu}\, u_y\right)_y = -i\alpha^2 u \quad \text{in} \quad \Omega$$

$$u^- = u^+ + u_0^+; \quad \frac{\mu_0}{\mu}\, u_\nu^- = u_\nu^+ + u_{0,\nu} \quad \text{on} \quad \Gamma.$$

(IV) <u>Metallic cylinder</u> - TE <u>fields</u>.

$$u_{xx} + u_{yy} = -\beta_0^2 u \quad \text{in} \quad \Omega^+; \quad \left(\frac{u_x}{\alpha^2}\right)_x + \left(\frac{u_y}{\alpha^2}\right)_y = -i\,\frac{\mu_0}{\mu} \quad \text{in} \quad \Omega$$

$$u^- = u^+ + u_0^+, \quad -\frac{i\beta_0^2}{\alpha^2}\, u_\nu^- = u_\nu^+ + u_{0,\nu} \quad \text{on} \quad \Gamma.$$

<u>Remarks</u>: 1. For any materials except ferromagnetic ones there is only a small variation in μ. Hence it is not a bad approximation to assume $\mu/\mu_0 = 1$.

2. Although the theory can be carried through for any choice of the parameters there are really only two important cases. At low (say, 60 cycle) frequencies the parameter β for a dielectric is very small while the parameter α for metals is $0(1)$. At higher frequencies, say $\omega = 0(10^{10})$ the parameter β is $0(1)$ but the parameter α is very large. Thus the dielectric problems are both meaningful at higher frequencies. For the metallic cylinder problem at low frequencies one can, with small error, put $\beta_0 = 0$. This is what is usually done with a statement that one "neglects displacement current in air". This is the origin of our problem (P_0). At higher frequencies the usual approximation is that the metal has "infinite conductivity" in which case one simply solves an exterior boundary value problem with $\mathcal{E}_{\text{tang}}$ equal to zero on the obstacle. Thus problems III and IV are really meaningful only if $\beta_0 = 0$.

<u>Acknowledgement</u>. This work was supported by the National Science Foundation under Grants CEE-8210859 (J.B.) and MCS-8219675 (R.C. MacC.).

Table 1. Relative Displacement at Origin

γ	$\dfrac{q}{p\beta_0^2}$	β_0	N_{h_Γ}						Exact	
			5		10		20			
			Re	Im	Re	Im	Re	Im	Re	Im
6	1/2	$.25\pi$	0.8807	0.1421	0.8802	0.1401	0.8800	0.1398	0.8800	0.1398
		$.50\pi$	0.5327	0.0249	0.5335	0.0225	0.5334	0.0222	0.5333	0.0222
		π	0.1343	-0.2314	0.1367	-0.2301	0.1369	-0.2295	0.1368	-0.2293
1/6	2	$.25\pi$	1.8209	-0.0394	1.8051	-0.0419	1.8003	-0.0421	1.7983	-0.0420
		$.50\pi$	-0.5755	1.7429	-0.5650	1.7012	-0.5619	1.6856	-0.5606	1.6785
		π	-2.3700	-0.5224	-1.8809	-0.6694	-1.7531	-0.6850	-1.7050	-0.6859

References

[1]   Bielak, J. and MacCamy, R.C., An exterior interface problem in two-dimensional elastodynamics, Quart. of Appl. Math. 41 (1983) 143-160.

[2]   Fix, G.J., Hybrid finite element methods, in: Noye, John (ed.), Numerical Simulation of Fluid Motion (North-Holland, Amsterdam, 1978).

[3]   MacCamy, R.C. and Marin, S.P., A finite element method for exterior interface problems, Int. Jrnl. Math. and Math. Anal. 3 (1980) 311-350.

[4]   Aziz, A.K. and Kellogg, R.B., Finite element analysis of a scattering problem, Math. of Comp. 37 (1981) 261-272.

[5]   Johnson, C. and Nedelec, J.C., On the coupling of boundary integral and finite element methods, Math. of Comp. 35 (1980) 1063-1079.

[6]   Hsiao, G. and MacCamy, R.C., Solutions of boundary value problems by integral equations of the first kind, SIAM Review 15 (1973) 687-705.

[7]   Trifunac, M.D., Surface motion of a semi-cylindrical alluvial valley for incident plane SH waves, Bull. Seism. Soc. Am 61 (1971) 1755-1770.

[8]   Harris, C.G. and Evans, W.A.B., Extension of numerical quadrature formulae to cater for end point singular behavior over finite intervals, Int. J. Comp. Maths. 6B (1977) 219-227.

Unification of Finite Element Methods
H. Kardestuncer (Editor)
© Elsevier Science Publishers B.V. (North-Holland), 1984

CHAPTER 7

PRECONDITIONED ITERATIVE METHODS FOR NONSELFADJOINT OR INDEFINITE ELLIPTIC BOUNDARY VALUE PROBLEMS

J.H. Bramble & J.E. Pasciak

We consider a Galerkin-Finite Element approximation
to a general linear elliptic boundary value problem
which may be nonselfadjoint or indefinite. We
show how to precondition the equations so that the
resulting systems of linear algebraic equations
lead to iteration procedures whose iterative
convergence rates are independent of the number of
unknowns in the solution.

1.   <u>INTRODUCTION</u>.

In recent years, the application of iterative methods to
preconditioned linear systems has been extremely successful in a variety
of complex physical applications [3,16]. Many articles are available
in the literature which report on the favorable performance of such
methods [3,6,10,12].

The two aspects of a resulting algorithm consist of the
preconditioner and the underlying iterative method [1,8,12]. Various
iterative methods, the most popular being the conjugate gradient (CG) and
certain normal forms of the CG method, have been considered extensively
both from a theoretical and an experimental viewpoint (see [10] and the
references therein). It has been demonstrated that, in general,
iterative algorithms with the same theoretical convergence rates

converge, in practice, at about the same rate[1]. The question of choosing
an appropriate preconditioner is much more difficult. The
preconditioner must in some way be similar to the inverse of the system
which is being solved. Consequently, the evaluation of the
preconditioner usually requires the solution of a system of equations and
so if the method is to result in an improvement of computational
efficiency, the preconditioner must have some property which makes it
easier to solve than the original system. The iterative convergence
rate of the algorithm is extremely sensitive to the choice of
preconditioner. Indeed, the choice of a more appropriate
preconditioner may reduce the number of iterations by an order of
magnitude or more in a given problem.

In this paper we illustrate some techniques for analysing
preconditioned iterative methods for nonsymmetric problems. We will
discuss the problem of choosing an appropriate preconditioner and study
two different iterative algorithms. Typical finite element
discretization of an elliptic boundary value problem leads to a matrix
problem

$$(1.1) \qquad\qquad Mc = d.$$

where M is the "stiffness" matrix associated with the discretization
and is nonsingular and c and d are vectors. We seek a
preconditioner M_1^{-1} such that M_1 is symmetric positive definite,
$(M_1)^{-1}$ is easier to compute than $(M)^{-1}$, and $(M_1)^{-1}$ "approximates
in some sense" $(M)^{-1}$. System (1.1) can of course be replaced by the
equivalent system

$$(1.2) \qquad M^t M_1^{-1} M_1^{-1} Mc = M^t M_1^{-1} M_1^{-1} d .$$

The matrix $M' \equiv M^t M_1^{-1} M_1^{-1} M$ is symmetric positive definite and the
first algorithm is defined by applying the conjugate gradient method
to (1.2). Alternatively, (1.1) is equivalent to the problem

$$(1.3) \qquad M_1^{-1} M^t M_1^{-1} Mc = M_1^{-1} M^t M_1^{-1} d.$$

---

[1] The number of iterations to reach a desired accuracy may vary by at most
a factor of five [6,10].

The matrix $M'' \equiv M_1^{-1} M^t M_1^{-1} M$ although not usually symmetric, is a symmetric operator with respect to the inner product defined by

$$<<w,v>> = (M_1 w) \cdot v .$$

The CG method can be applied to (1.3) in the $<<\cdot,\cdot>>$ inner product and leads to Algorithm II of Section 2. Our analysis suggests that the preconditioned iterative method based on (1.3) is more robust than that based on (1.2) since results for (1.2) require additional hypotheses. In fact, we have not been able to obtain results for the scheme based on (1.2) unless the elements used in the methods are of "quasi-uniform" size.

We shall present two general theorems which can be used to derive certain discrete stability estimates. Such estimates lead to bounds on the iterative convergence rates of algorithms for finding the solution of matrix equations resulting from the finite element discretization of elliptic boundary value problems which may be nonsymmetric and/or indefinite. We show how these general results can be applied in a finite element approximation to the Poincaré problem. Both strategies depend upon a priori stability estimates for the continuous problem and use the approximation properties of the discretization to derive the stability estimate for the matrix problems.

The first theorem leads to a strategy which uses a positive definite symmetric problem as a preconditioner for a more complicated nonsymmetric and/or indefinite problem. The problem of the efficient solution of positive definite problems, although not completely solved, has been extensively researched. For example, matrices corresponding to positive definite symmetric problems often have certain diagonal dominance properties which imply that various sparse matrix packages [9,11] can be used for their solution. Also, there are "fast solver" algorithms available for certain elliptic problems on a variety of domains [5,14,15]. Our analytical results guarantee that the iterative convergence rate for our algorithms is independent of the number of unknowns in the system. Thus the cost of convergence to a given accuracy grows linearly with the size of the problem.

The first strategy is applicable to, for example, problems where the differential operator A can be decomposed into a symmetric positive definite operator L and a compact (but not small) perturbation B. The

operators A, L, and B are approximated by discrete operators A_h,
L_h, and B_h derived by finite elements. The discrete approximation
to the solution u of the original problem is defined as the solution
of

$$(1.4) \qquad (L_h + B_h)U = F.$$

Problem (1.4) can be replaced by the equivalent problem

$$(1.5) \qquad L_h^{-1}(L_h + B_h)U = L_h^{-1} F .$$

We derive the appropriate stability estimates for (1.5) which guarantee
that the CG method applied, with respect to $<\!\!<\cdot,\cdot>\!\!>$, to (1.3) converges
at a rate independent of the number of unknowns in the discretization. In
addition, the stability results yield immediately estimates for the
discretization error $u-U$.

   We give a second theorem which, under additional hypotheses,
provides another stability estimate. This estimate, under a further
restriction, can be used to show that the CG method applied to (1.2)
converges to the solution of (1.2) at a rate which is independent of the
number of unknowns in the discretization.

   An outline of the remainder of the paper is as follows. In Section 4
we describe two conjugate gradient algorithms for matrix problems.
Section 3 gives some preliminaries and notation to be used in the paper.
In Section 4 we state the type of estimates needed to guarantee rapid
convergence for some iterative methods for solving nonsymmetric and/or
indefinite problems. Two theorems used to derive the stability estimates
are given in Section 5. In Section 6 we apply the theorems to a finite
element approximation of a general elliptic boundary value problem.
Finally in Section 7 we apply a stability estimate to bound the
discretization error.

2.   <u>CONJUGATE GRADIENT ALGORITHMS</u>.

   We describe the algorithms which result from applying the conjugate
gradient method to the preconditioned systems (1.2) and (1.3). In either
case we assume that we are given an initial approximation c_0 to the
solution c of (1.1) and the iterative algorithm produces a sequence of

iterates c_i for $i > 0$. We stop the iterative procedure when the residual error $d-Mc$ becomes sufficiently small. We note that applying the conjugate gradient method to preconditioned systems as illustrated in the following algorithms is not novel however we include the details for completeness.

Applying the conjugate gradient method to (1.2) gives the following algorithm:

ALGORITHM I. $\quad M' = M^t M_1^{-1} M_1^{-1} M$

(1) Define $\quad r_0 = p_0 = M^t M_1^{-1} M_1^{-1} (d-Mc_0)$.

(2) For $i \geq 0$ define

$$\alpha_i = \frac{r_i \circ p_i}{(M' p_i) \circ p_i}$$

$$c_{i+1} = c_i + \alpha_i p_i$$

$$r_{i+1} = r_i - \alpha_i M' p_i$$

$$\beta_i = \frac{(M' r_{i+1}) \circ p_i}{(M' p_i) \circ p_i}$$

$$p_{i+1} = r_{i+1} - \beta_i p_i \quad .$$

Applying the conjugate gradient method in the $<<\cdot,\cdot>>$ inner product to (1.3) gives the following algorithm:

ALGORITHM II. $\quad M'' = M_1^{-1} M^t M_1^{-1} M$.

(1) Define $\quad r_0 = p_0 = M_1^{-1} M^t M_1^{-1}(d-Mc_0)$.

(2) For $i \geq 0$ define

$$\alpha_i = \frac{(M_1 r_i) \circ p_i}{(M_1 p_i) \circ (M'' p_i)}$$

$$c_{i+1} = c_i + \alpha_i p_i$$

$$r_{i+1} = r_i - \alpha_i M'' p_i$$

$$\beta_i = \frac{(M_1 r_{i+1}) \circ (M'' p_i)}{(M_1 p_i) \circ (M'' p_i)}$$

$$p_{i+1} = r_{i+1} - \beta_i p_i \; .$$

3.   PRELIMINARIES AND NOTATION.

Throughout this paper we shall be concerned with solving boundary value problems on a bounded domain Ω contained in R^2 with boundary Γ . To state our stability estimates, we shall make use of various spaces of functions defined on Ω . The space $L^2(\Omega)$ is the collection of square integrable functions on Ω ; that is, a function $f(x)$ defined for (x,y) in Ω is in $L^2(\Omega)$ if

$$\int_\Omega f(x,y)^2 \; dxdy < \infty \quad .$$

The $L^2(\Omega)$ inner product is defined by

$$(f,g) \equiv \int_\Omega f(x,y) \; g(x,y)dxdy \quad \text{for} \quad f, \; g \in L^2(\Omega).$$

We shall also use the Sobolev space $H^1(\Omega)$. Loosely, a function f is in $H^1(\Omega)$ if $f, \frac{\partial f}{\partial x}$ and $\frac{\partial f}{\partial y}$ are all in $L^2(\Omega)$. Thus for functions in $H^1(\Omega)$, we can define the Dirichlet form by

$$D(f,g) \equiv \int_\Omega \left(\frac{\partial f}{\partial x} \; \frac{\partial g}{\partial x} + \frac{\partial f}{\partial y} \; \frac{\partial g}{\partial y}\right) dxdy \quad .$$

We shall also denote the $L^2(\Gamma)$ inner product by

$$<f,g> \equiv \int_\Gamma fg \; ds \; .$$

For any positive integer r, the Sobolev space of $L^2(\Omega)$-functions whose r^{th} order partial derivatives belong to $L^2(\Omega)$ will be denoted by $H^r(\Omega)$.

We also let C and C_i for $i \geq 0$ denote positive constants. The values of C and C_i may be different in different places however C and C_i shall always be independent of the mesh parameter h defining the approximation method. Thus C and C_i will always be independent of the number of unknowns in the discretization.

To define the approximation of later sections we shall need a collection of finite element approximation subspaces $\{S_h\}$, $0 < h \leq 1$, contained in $H^1(\Omega)$. Typically, finite element approximation subspaces are defined by partitioning the domain Ω into subregions of size h and defining S_h to be the set of functions which are continuous on Ω and piecewise polynomial when restricted to the subregions (see [4,7,17] for details). For example, one could partition Ω into triangles of size h and define S_h to be the functions which are continuous on Ω and linear on each of the triangles. Alternatively, Ω could be partitioned into rectangles and S_h could be defined to be the functions which are continuous on Ω and bilinear on each of the rectangles.

4. <u>ESTIMATES FOR THE CONJUGATE GRADIENT METHOD</u>.

Our analysis of iterative algorithms for preconditioned systems is based on stability estimates for the continuous or nondiscrete problem and the error estimates between the continuous solutions and their discrete approximations. To study the properties of the solutions of boundary value problems in partial differential equations, it is natural to consider operators in their basis free representations since complete sets of basis functions are usually too complex to be of much practical value. Consequently, it is natural to think of the process of solving for the discrete solution of the finite element equations as a basis free operator on the finite element subspace S_h of $H'(\Omega)$. We represent differential and solution operators by the notation A, B, L, or T whereas their discrete counterparts shall be respectively denoted A_h, B_h, L_h and T_h.

The CG method can be applied to find the solution X of the problem

$$(4.1) \qquad\qquad L_h X = Y$$

where L_h is a symmetric positive definite operator with respect to some inner product (cf. [13]). The CG algorithm requires an initial guess X_0 and produces an approximation X_n to X after n iterative steps. It is well known that

$$(4.2) \qquad \|X - X_n\|_H \leq 2 \left(\frac{\sqrt{\gamma}-1}{\sqrt{\gamma}+1} \right)^n \|X - X_n\|_H$$

where γ is the condition number for L_h and is defined to be the
ratio of the largest eigenvalue of L_h to the smallest. We note that
if L_h satisfies the inequality

$$(4.3) \qquad C_0 \|W\|_H^2 \leq (L_n W, W)_H \leq C_1 \|W\|_H^2 \qquad \text{for all} \quad W \in S_n,$$

where $(\cdot, \cdot)_H$ denotes the H-inner product, then the condition
number γ is bounded by C_1/C_0. Thus estimates of the type (4.3)
in conjunction with (4.2) lead to convergence estimates for the CG
method applied to (4.1).

The problem of finding the finite element solution in the examples
of later sections can be reduced to solving for the solution X of a
nonsingular operator equation

$$(4.4) \qquad A_h X = Y$$

where A_h is a nonsymmetric and/or nonpositive operator on S_h. We
shall first precondition the system, multiply by the adjoint and
then apply the CG method in the appropriate inner product.

We assume that we have a symmetric positive definite operator
T_h defined on S_h for a preconditioner. The types of preconditioners
for which we can get analytic results will be described in later sections.

We note that problem (4.4) can be replaced by the problem of
finding X in S satisfying

$$(4.5) \qquad A_h^* T_h T_h A_h X = A_h^* T_h T_h Y$$

where A_h^* is the $L^2(\Omega)$ - adjoint of A_h. The CG method with respect
to the $L^2(\Omega)$ inner product can be used to solve (4.5). The
convergence rate of the resulting algorithm is bounded by (4.2) in
the $L^2(\Omega)$ norm where γ is bounded by C_1/C_0 for any C_0 and
C_1 satisfying

$$(4.6) \qquad C_0 \|W\|^2_{L^2(\Omega)} \;\underline{\le}\; \|T_h A_h\, W\|^2_{L^2(\Omega)} \;\underline{\le}\; C_1 \|W\|^2_{L^2(\Omega)} \qquad \text{for all}\quad W \in S_h \;.$$

In certain applications, estimate (4.6) can be used to derive bounds on the iterative convergence rate of Algorithm I.

Alternatively, problem (4.4) is also equivalent to the problem of finding X in S_h satisfying

$$(4.7) \qquad\qquad T_h\, A_h^* \, T_h\, A_h\, X = T_h\, A_h^*\, T_h\, Y \;.$$

The operator $B \equiv T_h\, A_h^*\, T_h\, A_h$ is symmetric positive definite in the inner product $(T_h^{-1} W, V)$. Applying the CG method to the solution of (4.7) in this inner product gives an algorithm which converges at a rate described by (4.2) where $\gamma \underline{\le} C_1/C_0$ for any C_0 and C_1 satisfying

$$(4.8) \quad C_0(T_h^{-1} W, W) \underline{\le} (T_h\, A_h W,\, A_h W) \underline{\le} C_1(T_h^{-1} W, W) \qquad \text{for all}\quad W \in S_h \;.$$

In applications, estimate (4.8) is used to derive iterative convergence rates for Algorithm II.

5. STABILITY THEOREM.

In this section we give general results which can be used to derive estimates of the form (4.6) and (4.8).

<u>Theorem 1</u>.   Let R be a continuous operator and R_h be its discrete approximation. Assume that the following stability and error estimates hold:

$$(5.1) \qquad \|\theta\|_{H^1(\Omega)} \underline{\le} C\{\|(I+R_h)\,\theta\|_{H^1(\Omega)} + \|\theta\|_{L^2(\Omega)}\} \qquad \text{for all}\quad \theta \in S_h.$$

For any $\varepsilon > 0$ there exists C_ε such that

$$(5.2) \qquad \|\phi\|_{L^2(\Omega)} \underline{\le} C_\varepsilon \|(I+R)\phi\|_{L^2(\Omega)} + \varepsilon\|\phi\|_{H^1(\Omega)} \qquad \text{for all}\quad \phi \in H^1(\Omega).$$

$$(5.3) \qquad \|(R-R_h)\phi\|_{L^2(\Omega)} \underline{\le} Ch\, \|\phi\|_{H^1(\Omega)} \qquad \text{for all}\quad \phi \in H^1(\Omega).$$

Then there exists $h_0 > 0$ such that for $h < h_0$

$$(5.4) \qquad \|\theta\|_{H^1(\Omega)} \leq C\|(I+R_h)\theta\|_{H^1(\Omega)} \qquad \text{for all} \quad \theta \, \varepsilon \, S_h \ .$$

<u>Remark 1</u>. Estimate (5.4) combined with

$$(5.5) \qquad \|(I+R_h)\theta\|_{H^1(\Omega)} \leq C\|\theta\|_{H^1(\Omega)} \qquad \text{for all} \quad \theta \, \varepsilon \, S_h$$

guarantees a uniform (independent of h) iterative convergence rate for the CG iteration for the solution of

$$(I+R_h)^* (I+R_n)U = F$$

where $*$ denotes the adjoint with respect to the $H^1(\Omega)$ inner product. In our finite element applications, $I+R_h = T_h A_h$ and

$$C_0\|\theta\|^2_{H^1(\Omega)} \leq (T_h^{-1}\,\theta,\theta) \leq C_1\|\theta\|^2_{H^1(\Omega)} \qquad \text{for all} \quad \theta \, \varepsilon \, S_h \ .$$

Thus (5.4) and (5.5) will imply (4.8) for the particular examples of the next section.

<u>Theorem 2</u>.   Let T^1 and T^2 be continuous operators and T_h^1 and T_h^2 be their corresponding discrete approximations. Assume that the following three estimates hold:

$$(5.6) \qquad C_0\|T^1 u\|_{L^2(\Omega)} \leq \|T^2 u\|_{L^2(\Omega)} \leq C_1\|T^1 u\|_{L^2(\Omega)} \qquad \text{for all} \quad u \, \varepsilon \, L^2(\Omega).$$

$$(5.7) \qquad \|(T^i-T_h^i) u\|_{L^2(\Omega)} \leq Ch^2 \|u\|_{L^2(\Omega)} \qquad \text{for all} \quad u \, \varepsilon \, L^2(\Omega) \ .$$

$$(5.8) \qquad \|(T_h^i)^{-1} U\|_{L^2(\Omega)} \leq Ch^2\|U\|_{L^2(\Omega)} \qquad \text{for all} \quad U \, \varepsilon \, S_h \ .$$

for $i = 1,2$. Then

$$(5.9) \qquad C_0\|U\|^2_{L^2(\Omega)} \leq \|T_h^2(T_h^1)^{-1} U\|^2_{L^2(\Omega)} \leq C_1\|U\|^2_{L^2(\Omega)} \qquad \text{for all} \quad U \, \varepsilon \, S_h \ .$$

<u>Remark 2</u>. Estimate (5.8) is an inverse property for the operator T_h^1 and in applications is derived from the hypothesis that the mesh elements are of "quasi uniform" size. Estimate (5.9) coincides with (4.6) when $A_h = \left(T_h^1\right)^{-1}$.

<u>Remark 3</u>. The proofs of the above two theorems are simple and consequently will not be included.

6.   THE POINCARÉ PROBLEM.

To illustrate our approach we consider a finite element approximation of the Poincaré problem in this section. We consider the following model problem:

$$-\Delta u + \frac{\partial u}{\partial x} + Ku = f \quad \text{in} \quad \Omega$$

(6.1)

$$\frac{\partial u}{\partial \eta} + \beta \frac{\partial u}{\partial \tau} + \gamma u = 0 \quad \text{on} \quad \Gamma$$

where $\Delta = \dfrac{\partial^2}{\partial x^2} + \dfrac{\partial^2}{\partial y^2}$, η and τ are respectively the normal and

tangential directions along Γ. For simplicity we have considered constant coefficients in defining the differential equation as well as the boundary condition. Our results and iterative algorithms extend to variable coefficient problems without any complications. We also assume that the solution of (6.1) exists and is unique.

The finite element approximation to (6.1) can then be defined by the Galerkin technique. Multiplying (6.1) by an arbitrary function ϕ and integrating by parts shows that the solution u satisfies

$$(6.2) \qquad D(u,\phi) + \left(\frac{\partial u}{\partial x},\phi\right) + K(u,\phi) + \langle \beta \frac{\partial u}{\partial \tau} + \gamma u,\phi \rangle = (f,\phi) \ .$$

The finite element approximation U to u is then defined to be the function U in S_h which satisfies

$$(6.3) \qquad D(U,\theta) + \left(\frac{\partial U}{\partial x},\theta\right) + K(U,\theta) + \langle \beta \frac{\partial U}{\partial \tau} + \gamma U,\theta \rangle = (f,\theta) \qquad \text{for all} \ \theta \in S_h .$$

178   *J.H. Bramble & J.E. Pasciak*

Equation (6.3) can be used to derive a system of equations of the form
(1.1) defining the discrete solution U, i.e., using a basis for S_h,
(6.3) gives N equations for the N unknowns defining U in that
basis.

   To describe iterative methods for the solution of (6.3) and/or
the corresponding matrix system, we shall need to use some operator
notation. First, we consider the Neumann problem

$$w - \Delta w = f \quad \text{in} \quad \Omega$$

(6.4)

$$\frac{\partial w}{\partial \eta} = 0 \quad \text{on} \quad \Gamma$$

Given a function f in $L^2(\Omega)$, the solution w of (6.4) is in $H^2(\Omega)$
if as we shall assume, Γ is sufficiently smooth. We denote the
solution operator T as the map which takes f to $Tf \equiv w$. T is a
bounded map of $L^2(\Omega)$ into $H^2(\Omega)$. The finite element approximation to
(6.4) is the function W in S_h satisfying

(6.5) $D(W,\theta) + (W,\theta) = (f,\theta)$ for all $\theta \in S_h$.

The discrete solution operator T_h can then be defined as the map which
takes f to $T_h f \equiv W$. T_h is a map from $L^2(\Omega)$ onto S_h and the
following convergence estimate is well known (cf. [2]):

(6.6) $\|(T_h - T)f\|_{H^1(\Omega)} \leq Ch\|f\|_{L^2(\Omega)}$.

In a similar manner, we can define solution operators for the following
variational problems:

$$D(\chi,\phi) + (\chi,\phi) = (\frac{\partial z}{\partial x},\phi) + (k-1)(z,\phi)$$

and

$$D(\psi,\phi) + (\psi,\phi) = \langle \beta \frac{\partial \omega}{\partial \tau} + \gamma\omega,\phi \rangle$$

We define the solution operators $R^1 z \equiv \chi$ and $R^2 \omega \equiv \psi$. The corresponding finite element approximations are given by the solutions X and Y in S_h satisfying

$$D(X,\theta) + (X,\theta) = (\tfrac{\partial z}{\partial x},\theta) + (K-1)(z,\theta) \quad \text{for all} \quad \theta \in S_h$$

and

$$D(Y,\theta) + (Y,\theta) = \langle \beta \tfrac{\partial \omega}{\partial \tau} + \gamma\omega,\theta \rangle \quad \text{for all} \quad \theta \in S_h \, ,$$

respectively. The discrete solution operators are then defined by $R_h^1 z \equiv X$ and $R_h^2 \omega \equiv Y$ and the following convergence estimates hold:

$$(6.7) \qquad \|(R_h^1 - R^1)z\|_{L^2(\Omega)} \leq Ch\|z\|_{H^1(\Omega)} \quad .$$

and

$$(6.8) \qquad \|(R_h^2 - R^2)\omega\|_{L^2(\Omega)} \leq Ch\|\omega\|_{H^1(\Omega)} \quad .$$

In terms of operators, problem (6.1) is equivalent to

$$(I + R^1 + R^2)u \equiv TA\,u = Tf \ .$$

The existence and uniqueness properties of solutions of (6.1) can be used to show that for any $\varepsilon > 0$ there is a constant C_ε such that

$$(6.9) \qquad \|\phi\|_{L^2(\Omega)} \leq C_\varepsilon \|(I+R^1+R^2)\phi\|_{L^2(\phi)} + \varepsilon\|\phi\|_{L^2(\Omega)} \quad .$$

The discrete estimate

$$(6.10) \qquad \|\theta\|_{H^1(\Omega)} \leq C\{\|(I+R_h^1 + R_h^2)\theta\|_{H^1(\Omega)} + \|\theta\|_{L^2(\Omega)}\} \quad \text{for all} \quad \theta \in S_h$$

is immediate from the definition of R_h^i. Problem (6.3) can be stated in terms of operators as

$$(I + R_h^1 + R_h^2)U \equiv T_h A_h\,U = T_h\,f \ .$$

Applying Theorem 1 we get the following stability estimate:

$$(6.11) \qquad C_0 \|W\|^2_{H^1(\Omega)} \leq \|T_h A_h W\|^2_{H^1(\Omega)} \leq C_1 \|W\|^2_{H^1(\Omega)} \qquad \text{for all } W \in S_h.$$

The second inequality in (6.11) can be easily derived from the definitions. The constants C_0 and C_1 in (6.10) are independent of the mesh size h. Now it is easy to check that

$$(6.12) \quad (T_h^{-1} W, V) = D(W,V) + (W,V) \quad \text{for all } W, V \in S_h.$$

Comparing (6.12), (6.11), (4.7) and (4.8) implies that the CG method applied to

$$(6.13) \qquad T_h A_h^* T_h A_h U = T_h A_h^* T_h f$$

converges with a reduction per iteration which can be bounded independently of the number of unknowns.

Let M and M_1 respectively denote the "stiffness" matrices corresponding to (6.3) and (6.5) in a given basis $\mathbb{B} = \{\mathbb{B}_i\}_{i=1}^N$ for S_h. If the coefficients of a function W in S_h in terms of the basis $\mathbb{B}$ are represented by the vector c then

$$d = M_1^{-1} M^t M_1^{-1} M c$$

gives the coefficients of $T_h A_h^* T_h A_h W$ in terms of $\mathbb{B}$. Consequently, the sequence of vectors c_i generated by Algorithm II gives the coefficients of the sequence of functions generated by the CG method applied to (6.13). Thus the iterative convergence estimates for the CG method applied to (6.13) imply iterative convergence rates for Algorithm II.

The above procedure is an example of an iterative convergence analysis in $H^1(\Omega)$. We also note that if T_h^1 is another discrete operator on S_h which is spectrally equivalent to T_h in the sense that

$$(6.14) \qquad C_0(T_h W,W) \leq (T_h^1 W,W) \leq C_1(T_h W,W) \quad \text{for all } W \in S_h$$

then T_h can be replaced by T_h^1 in (6.11).

We next consider an iterative analysis in $L^2(\Omega)$ based on Theorem 2. Let $T^1: L^2(\Omega) \to H^2(\Omega)$ denote the solution operator for problem (6.1) with $\beta = 0$, i.e., $T^1 f \equiv u$. The solution operator T^1 satisfies an estimate of the form

$$(6.15) \qquad C_0 \|T^1 f\|_{L^2(\Omega)} \leq \|Tf\|_{L^2(\Omega)} \leq C_1 \|T^1 f\|_{L^2(\Omega)} \, .$$

We have restricted to the case of $\beta = 0$ since (6.15) is well known in that case. Assume that both T^1 and T can be approximated in the same finite element subspaces and let T_h^1 and T_h denote the corresponding discrete solution operators. The following convergence estimates are well known for a wide class of finite element applications [27]:

$$(6.16) \qquad \|(T_h^i - T^i)f\|_{L^2(\Omega)} \leq Ch^2 \|f\|_{L^2(\Omega)} \, .$$

We finally assume that the inverse properties

$$(6.17) \qquad \|(T_h^i)^{-1}\theta\|_{L^2(\Omega)} \leq Ch^{-2} \|\theta\|_{L^2(\Omega)}, \quad \theta \in S_h,$$

are also satisfied. Estimates of the type (6.17) can usually be derived from inverse assumptions for the subspaces. Applying Theorem 2 gives that

$$(6.18) \qquad C_0\|W\|_{L^2(\Omega)} \leq \|T_h(T_h^1)^{-1} W\|_{L^2(\Omega)} \leq C_1\|W\|_{L^2(\Omega)} \qquad \text{for all } W \in S_h \, .$$

Estimate (6.18) guarantees that the CG method applied in $L^2(\Omega)$ for the solution of

$$(6.19) \qquad A_h^* T_h T_h A_h X = A_h^* T_h T_h f$$

where $A_n = (T_n^1)^{-1}$, will converge to the solution X at a rate which is independent of the number of unknowns in S_h. The resulting algorithm does not however correspond to Algorithm I. To guarantee rapid iterative

convergence rates for Algorithm I we must make additional assumptions.
Again we use the basis $\mathcal{B}$ for S_h . If $W \varepsilon S_h$ we denote by C_W the
coefficients of W in the basis $\mathcal{B}$. We require that

$$(6.20) \qquad C_0(C_W \cdot C_W) \leq (W,W)_{L^2(\Omega)} \leq C_1(C_W \cdot C_W) \quad \text{for all} \quad W \varepsilon S_h .$$

Estimate (6.20) states that the Gram or mass matrix is "equivalent"
to the coordinate inner product. Combining (6.19) and (6.20) implies

$$(6.21) \qquad C_0 \, |c|^2 \leq |M_1^{-1} M \, c|^2 \leq C_1 |c|^z$$

for all N dimensional vectors c. Estimate (6.21) is finally an
estimate which can be applied to guarantee uniform iterative convergence
rates for Algorithm I.

7.   AN ESTIMATE FOR THE DISCRETIZATION ERROR.

In order to estimate the discretization error u-U with u and U
defined by (6.2) and (6.3) respectively, we introduce the $H^1(\Omega)$-
projection P_h onto S_h . It is defined for $v \varepsilon H^1(\Omega)$ by

$$(7.1) \qquad D(P_h v,\theta) + (P_h v,\theta) = D(v,\theta) + (v,\theta), \quad \text{for all} \quad \theta \varepsilon S_h .$$

It is well known that P_h satisfies

$$(7.2) \qquad \|(I-P_h)v\|_{H^1(\Omega)} \leq Ch^{r-1}\|v\|_{H^r(\Omega)}$$

for $v \varepsilon H^r(\Omega)$ and some r > 1 which depends on the choice of
S_h(cf. [2,7]). In view of (7.2), to estimate u-U we need only consider
P_h u-U . Hence we apply (5.4) to obtain

$$\|P_h u-U\|_{H^1(\Omega)} \leq C\|(I+R_h)(P_h u-U)\|_{H^1(\Omega)} ,$$

with $R_h = R_h^1 + R_h^2$. From the definitions of R^1, R_h^1, R^2 and R_h^2 we
see that

$$(I+R_h)(P_h u-U) = P_h(R^1+R^2)(P_h-I)u .$$

Hence

$$\|P_h u - U\|_{H^1(\Omega)} \leq C\|P_h(R^1+R^2)(P_h-I)u\|_{H^1(\Omega)}$$

from which it follows immediately that

$$(7.3) \qquad \|P_h u - U\|_{H^1(\Omega)} \leq C\|(I-P_h)u\|_{H^1(\Omega)} \quad .$$

Thus using (7.2) we obtain the estimate for the discretization error,

$$\|u-U\|_{H^1(\Omega)} \leq Ch^{r-1}\|u\|_{H^r(\Omega)} \quad .$$

REFERENCES.

[1] O. Axelsson; A class of iterative methods for finite element equations , Comp. Methods Appl. Mech. Engng., V. 9, pp. 123-137.

[2] I. Babuška and A.K. Aziz; Part I. Survey lectures on the mathematical foundations of the finite element method , The Mathematical Foundations of the Finite Element Method with Applications to Partial Differential Equations, A.K. Aziz, ed. Academic Press, New York, 1972.

[3] J.H. Bramble and J.E. Pasciak; An efficient numerical procedure for the computation of steady state harmonic currents in flat plates , COMPUMAG conf., Genoa, 1983.

[4] J.H. Bramble, J.E. Pasciak, and A.H. Schatz; Preconditioners for interface problems on mesh domains, preprint.

[5] B.L. Buzbee, F.W. Dorr, J.A. George, and G.H. Golub; The direct solution of the discrete Poisson equation on irregular regions , SIAM J. Numer. Anal., V. 8, 1971, pp. 722-736.

[6] R. Chandra; Conjugate gradient methods for partial differential equations, Yale University, Dept. of Comp. Sci. Report No. 129, 1978.

[7] P.G. Ciarlet; The finite element method for elliptic problems, North-Holland, Amsterdam, 1978.

[8] P. Concus, G. Golub, and D. O'Leary , A generalized conjugate gradient method for the numerical solution of elliptic partial differential equations , in Sparse Matrix Computation, J. Bunch and D. Rose, eds., Academic Press, New York, 1976, pp. 309-322.

[9]   S.C. Eisenstat, M.C. Gursky, M.H. Schultz, A.H. Sherman; Yale
      sparse matrix package, I. the symmetric codes, Yale Univ. Dept.
      of Comp. Sci. Report No. 112.

[10] H. Elman; Iterative methods for large, sparse, nonsymmetric
      systems of linear equations, Yale Univ. Dept. of Comp. Sci.
      Report No. 229, 1978.

[11] A. George and J.W.H. Liu; User Guide for SPARSPAK, Waterloo
      Dept. of Comp. Sci. Report No. CS-78-30.

[12] J.A. Meijerink and H.A. Van der Vorst; An iterative solution method
      for linear systems of which the coefficient matrix is a symmetric
      M-matrix , Math. Comp. 1973, V. 31, pp. 148-162.

[13] W.M. Patterson; Iterative methods for the solution of a linear
      operator equation in Hilbert space - A survey, lecture notes in
      mathematics, Springer-Verlag, No. 394, 1974.

[14] W. Proskurowski and O. Widlund; On the numerical solution of
      Helmholtz's equation by the capacitance matrix method , Math. Comp.,
      V. 20, 1976, pp. 433-468.

[15] A.H. Schatz; Efficient finite element methods for the solution of
      second order elliptic boundary value problems on piecewise smooth
      domains, Proceedings of the conference Constructive methods for
      singular problems , November 1983, Oberwolfach, West Germany,
      P. Grisvard, W. Wendland and J. Whiteman, editors, Springer-Verlag
      lecture notes in mathematics, to appear.

[16] J. Simkin and C.W. Trowbridge; On the use of the total scalar
      potential in the numerical solution of field problems in
      electromagnetics , Inter. J. Numer. Math. Eng., 1979, V. 14,
      pp. 423-440.

[17] O.C. Zienkiewicz; The finite element method , 3rd edition,
      McGraw-Hill, 1977.

CHAPTER 8

ON THE UNIFICATION OF FINITE ELEMENTS & BOUNDARY ELEMENTS

C.A. Brebbia

This paper reviews some of the applications of boundary element
methods for the solution of engineering problems. The paper
considers how the new technique relates to classical finite elements,
by reviewing the fundamentals of mechanics, in particular virtual
work and associated principles. This approach gives a common basis
for all approximate techniques and helps to understand the relationship between finite and boundary element method. The paper stresses
the range of applications for which the boundary element method can
give accurate results and be computationally efficient.

1. INTRODUCTION

In the last few years the applications of boundary integral equations in
engineering have undergone important changes. The brave attempts during
the sixties and early seventies pioneers such as Jawson [1], Symm [2],
Massonet [3], Hess [4], Cruse [5] and few others, have now borne fruit
in the newly developed boundary element method. In this way boundary
integral equations have become an engineering tool rather than a mathematical method with important but rather restrictive applications.

Since the early 1960's a small group at Southampton University in England
started working on the applications of integral equations to solve stress
analysis problems. Some of this work has been reported at the first
international Conference on Variational Methods in Engineering, held there
in 1972 [6]. More is expected to be presented during the second Conference
reconvened for 1985. These Conferences are held to discuss the different
techniques of engineering analysis and how they are interrelated. The
importance of the BIE presentations during the 1st Conference is that this
was the first time that boundary integral equations were interpreted as a
variational technique. The work at Southampton was continued throughout the
seventies through a series of theses mainly concerned with boundary integral
solutions of elastostatic problems. At the same time new developments
in finite elements started to find their way into boundary integral equations and the problem of how to relate the technique to other approximate
solutions was solved using weighted residuals [7]. This work at Southampton
University culminated around 1978 when the first book was published with
the title "Boundary Elements" [8]. The work was expanded to encompass time
dependent and non-linear problems in two subsequent books [9],[10], one of
them very recently published [10]. The importance of this work is that it
stresses the common principles and fundamentals relationships governing

the different techniques, rather than trying to set the boundary element
method as a completely separate computational technique.

Five important international conferences have already been held on the
topic of boundary elements in 1978 (Southampton) [11], 1980 (Southampton)
[12], 1981 (California) [13], 1982 (Southampton) [14], 1983 (Hiroshima)
[15] and the next one is to be held in July 1984 on board the Queen
Elizabeth II cruiser. The frequency of the meetings and the increasing
number of papers presented at each of them is evidence of the healthy
growth of the new method. In addition, a series of state of the art books
are regularly published to highlight the main developments of the technique
[16][17][18].

The success and rapid acceptance of the new technique is due to some
important advantages over classical finite elements, which are better
understood by reviewing the main characteristics of the method. The
boundary element method as understood nowadays is a reduction technique
based on boundary integral equation formulations and interpolation function
of the type used in finite elements. The main characteristic of the
method is that it reduces the dimensionality of the problem by one and
hence produces a much smaller system of equations and more important for
the practicing engineer, considerable reductions in the data required to
run a problem. The latter advantage is making boundary elements a
favourite for many mechanical engineering problems when the numerical
model has to be interfaced with mesh generators and other CAD facilities.

In addition the numerical accuracy of the method is generally greater than
that of finite elements, which have led many engineers to use BEM for
problems such as fracture mechanics and others where stress concentration
can occur. This accuracy is due to using a mixed formulation type of
approach for which all boundary values are obtained with similar degree of
accuracy. In this respect BEM is closely related to the mixed formulations
pioneered by Reissner [19] and excellently explained and generalized by
Washizu [20] and Pian and Tong [21]. The method is also well suited to
problem solving with infinite domains such as those frequently occurring in
soil mechanics and hydrodynamics, and for which the classical domain
methods are unsuitable. A boundary solution is formulated in terms of
influence functions obtained by applying a fundamental solution. If the
solution is suitable for an infinite domain no outer boundaries need to
be defined.

It is now generally accepted that the best way of formulating boundary
elements for general engineering problems is by using weighted residual
techniques, as shown in references [7],[8] and [10]. This formulation
closely relates the BEM to the variational methods and to the original
interpretation of virtual work proposed by Bernoulli.   It also allows
for complicated non-linear and time dependent problems to be properly
formulated, without need to find an integral expansion beforehand.

The term boundary element now also implies that the surface of the domain
is divided into a series of elements over which the functions under
consideration vary in accordance with some interpolation functions, in
much the same way as in finite elements. By contrast with past integral
equations formulations – which were restricted to concentrated sources –
these variations permit the proper description of curved surfaces in
addition to working with more accurate higher order interpolation
functions.

Summarizing, after years of research and development the boundary element method has emerged as a powerful mathematical tool for the solution of a large variety of engineering problems. The acceptance of the technique amongst practicing engineers is mainly due to the following advantages:

i)   Simple data preparation, which considerably reduces the amount of manpower required to run a problem

ii) More accurate results, which makes the technique especially attractive for stress concentration problems, fracture mechanics application and others. This increased accuracy also allows the designer to work with coarser meshes than in finite elements with further reduction in manpower.

iii) Definition of system and interpretation of results become easier which permits a better interfacing to surface modelling and other CAD systems.

iv) Problems with infinite domains can be solved accurately, which makes the method well suited for applications such as soil mechanics and hydrodynamics.

2. FUNDAMENTAL PRINCIPLES

In what follows we will consider problems in linear elasticity for which the problem can be expressed in function of a set of equilibrium equations and another set of compatibility relations, related together by constitutive laws. These equations will be written using the indicial notation. Dynamic loading will not be considered explicitly but it can be easily included using D'Alembert's hypothesis, i.e. by considering that at a given time the dynamic and static forces are in equilibrium. This simple but brilliant idea facilitates the dynamic analysis.

The approximate methods of solution used in engineering analysis have all a common basis not only given by the fundamental equations of physics but also by the fact that the actual approximations can be interpreted using the principle of virtual work. The application of this principle in different ways gives rise to the diverse techniques of engineering analysis. It is important to point out that the principle itself is a fundamental idea based on philosophical and physical intuition rather than higher mathematics. In this respect it is interesting to remark that the principle has been discussed since the beginning of western civilization and is related to the 'potentialities' of physical systems as discussed by Aristotle [22].

From classical antiquity onward the principle has been frequently applied and several well known fields of mathematics related to it, such as the Calculus of Variations, Functional Analysis, Distribution Theory, etc. These mathematics, although impressive, should not distract us from the elegance, simplicity and generality of the original virtual work statement.

In this section we will try to point out how the principle of virtual work can be used to generate models in solid mechanics. This is first done by assuming that the same physical equations apply to two different states, one is the 'actual' and the other is the 'virtual' state. The actual state is usually defined in terms of an approximation in the practice. The products of these two states give rise to virtual work statements. This section will attempt to classify these statements depending on which type of relationships are identically satisfied and which are to be imposed on the approximate functions. These formulations are as well known as virtual displacements and virtual forces, but can also be some type of mixed or hybrid approach. We will particularly consider the possibility of producing generalized formulations and taking them to the boundary, as it is due in boundary elements.

The simplicity of the virtual work approach allows for the formulation of
very general approximate models, valid even for non linear and time dependent
problems. The formulation of different techniques - including boundary
elements - becomes then independent of the existence or otherwise of a
functional or integral statement. These formulations will not be discussed
here, but the interested reader is referred to [23].

VIRTUAL WORK

The Virtual Work principle can be interpreted as the work done by one
state ('actual') over another ('virtual'). This work can be expressed in
different ways, depending on the variables under consideration. For
instance if one is dealing with displacements and body and traction forces
one can write the following virtual work statement

$$\int b_k \, u_k^* \, d\Omega + \int_\Gamma t_k \, u_k^* \, d\Gamma = \int b_k^* \, u_k \, d\Omega + \int_\Gamma t_k^* \, u_k \, d\Gamma \qquad (1)$$

Notice that the work has been defined in terms of the usual inner product,
i.e. the multiplication of the variables integrated over the domain and
external surface. b_k are the body forces, t_k the surface tractions
and u_k the displacement components. The virtual field is indicated by an
asterisk.

The same principle can also be expressed in terms of the internal work,
which gives,

$$\int \sigma_{jk} \, \varepsilon_{jk}^* \, d\Omega = \int \sigma_{jk}^* \, \varepsilon_{jk} \, d\Omega \qquad (2)$$

σ_{jk} and ε_{jk} are the stress and strain components respectively.

Still more interestingly, virtual work could be given as a relationship
between compatibility equations and stress functions. If the compatibility
relationships are expressed by the R_k components of a compatibility vector
and the associated stress function χ_k one can write,

$$\int R_k \, \chi_k^* \, d\Omega = \int R_k^* \, \chi_k \, d\Omega \qquad (3)$$

These three statements are equally valid and they can even be added to find
an extended version of virtual work as we will see soon.

This presentation of virtual work has some advantages over the more
classical variational type of approach as we will see shortly. The classical approach originated with Bernoulli constraint equations, usually
requires the definition of some Lagrangian multipliers to generalize the
principles. Our approach instead is much simpler.

VIRTUAL DISPLACEMENTS

It is now easy to deduce different versions of the virtual work principle
by applying the above equations. Let us start with identity (1) integrating by parts the surface integral on the right hand side. In order to do

this we can use the well known Gauss theorem and for linear strain-displacement relations — which we accept are identically satisfied — obtain,

$$\int (\sigma^*_{jk,j} + b^*_k)\, u_k\, d\Omega + \int \sigma^*_{jk}\, \varepsilon_{jk}\, d\Omega - \int b_k\, u^*_k\, d\Omega - \int_\Gamma t_k\, u^*_k\, d\Gamma = 0 \qquad (4)$$

where $t_k = n_j\, \sigma_{jk}$; n_j are the direction cosines of the normal with respect to x_j axis. If furthermore we accept i) reciprocity as given by equation (2), ii) that the virtual displacements u^*_k are such that the displacements boundary condition on Γ_1 ($u_k = \bar{u}_k$ on Γ_1 where $\bar{u}_k$ are known values) are identically satisfied, i.e. $u^*_k \equiv 0$ on Γ_1 and iii) that the virtual field satisfies equilibrium, one finds,

$$\int \sigma_{jk}\, \varepsilon^*_{jk}\, d\Omega = \int b_k\, u^*_k\, d\Omega + \int_{\Gamma_2} \bar{t}_k\, u^*_k\, d\Gamma \qquad (5)$$

which is the usual expression for virtual work. Notice that the other part of the boundary Γ_2 is that on which the traction boundary conditions are prescribed, i.e. $t_k = \bar{t}_k$ on Γ_2.

Another form of virtual displacements can be obtained by integrating by parts the left hand side integral in (5). This gives

$$\int (\sigma_{jk,j} + b_k)u^*_k\, d\Omega = \int_{\Gamma_2} (t_k - \bar{t}_k)u^*_k\, d\Gamma \qquad (6)$$

The above statement is equivalent to (5) provided that we accept that the strain-displacement equations and constitutive relationships are identically satisfied.

The above restrictions to virtual work give rise to the possibility of defining a functional called total potential energy, composed of two parts, i.e. the internal strain energy function,

$$U(\varepsilon) = \frac{1}{2} \int \sigma_{jk}(\varepsilon)\, \varepsilon_{jk}\, d\Omega = \frac{1}{2} \int \varepsilon_{\ell m}\, d_{\ell mjk}\, \varepsilon_{jk}\, d\Omega \qquad (7)$$

and the potential of the loads (assuming they are conservative

$$\Omega = - \int_{\Gamma_2} \bar{t}_k\, u_k\, d\Gamma - \int b_k\, u_k\, d\Omega \qquad (8)$$

The total potential energy is then

$$\Pi(u) = U + \Omega \qquad (9)$$

Equilibrium statements (5) or (6) for instance are now defined by the 'variation' of Π, i.e.

$$\Pi^* = U^* + \Omega^* = 0 \qquad (10)$$

190   *C.A. Brebbia*

Notice that Potential energy is function of the displacements and strains.

As it is well known this principle is the basis of the stiffness finite element formulations.

Principle of Virtual Forces

The converse of the Principle of Virtual Displacements is the Principle of Virtual Forces which can be described in several different ways. In this paper we will start by using the virtual work relationship (equation (2)),

$$\int \sigma_{ij}^{*} \, \varepsilon_{ij} \, d\Omega = \int \sigma_{ij} \, \varepsilon_{ij}^{*} \, d\Omega \tag{11}$$

Accepting that $\varepsilon_{ij}^{*} = \tfrac{1}{2}(u_{i,j}^{*} + u_{j,i}^{*})$ we can transform the right hand side term of (11) into,

$$\int \sigma_{ij} \, \varepsilon_{ij}^{*} \, d\Omega = -\int (\sigma_{jk,j})u_{k}^{*} \, d\Omega + \int_{\Gamma} t_{k} \, u_{k}^{*} \, d\Gamma \tag{12}$$

Furthermore accepting that the σ state satisfies the equilibrium equations

$$\sigma_{jk,j} + b_{k} = 0$$
$$t_{k} = n_{j} \, \sigma_{jk} \tag{13}$$

one can write (12) as,

$$\int \sigma_{ij} \, \varepsilon_{ij}^{*} \, d\Omega = \int b_{k} \, u_{k}^{*} \, d\Omega + \int t_{k} \, u_{k}^{*} \, d\Gamma \tag{14}$$

In order to eliminate u_{k}^{*} from (11) we can use another reciprocity relationship – equation (1) – this gives,

$$\int \sigma_{ij} \, \varepsilon_{ij}^{*} \, d\Omega = \int u_{k} \, b_{k}^{*} \, d\Omega + \int u_{k} \, t_{k}^{*} \, d\Gamma \tag{15}$$

Hence equation (11) can be written as,

$$\int \sigma_{ij}^{*} \, \varepsilon_{ij} \, d\Omega = \int b_{k}^{*} \, u_{k} \, d\Omega + \int t_{k}^{*} \, u_{k} \, d\Gamma \tag{16}$$

The unknown boundary displacements on Γ_{2} can be eliminated by stipulating that the t_{k}^{*} components vanish there and (16) becomes

$$\int \sigma_{ij}^{*} \, \varepsilon_{ij} \, d\Omega = \int b_{k}^{*} \, u_{k} \, d\Omega + \int_{\Gamma_{1}} t_{k}^{*} \, \bar{u}_{k} \, d\Gamma \tag{17}$$

where the bar on u_k components indicates that these values are known.

We now want to demonstrate that equation (17) will produce as stationarity requirement, the compatibility equations. In order to do so consider the b_k^* body forces which, if the virtual stresses satisfy equilibrium, (i.e. $\sigma_{jk,j}^* + b_k^* = 0$) give rise to,

$$\int b_k^* u_k \, d\Omega = -\int (\sigma_{jk,j}^*) u_k \, d\Omega =$$

$$= \int_\Omega \sigma_{jk}^* \{\tfrac{1}{2}(u_{j,k} + u_{k,j})\} \, d\Omega - \int_\Gamma t_k^* u_k \, d\Gamma$$

Noticing that $t_k^* \neq 0$ only on Γ_1 one can substitute (18) into equation (17) and obtain,

$$\int \sigma_{ij}^* \{\varepsilon_{ij} - \tfrac{1}{2}(u_{i,j} + u_{j,i})\} d\Omega + \int_{\Gamma_1} (u_k - \bar{u}_k) t_k^* \, d\Gamma = 0 \tag{19}$$

The stationarity requirements are compatibility, i.e.

$$\varepsilon_{ij} = \tfrac{1}{2}(u_{i,j} + u_{j,i}) \tag{20}$$

plus the associated displacement boundary conditions

$$u_k = \bar{u}_k \qquad \text{on } \Gamma_1 \tag{21}$$

For the principle of virtual displacements instead the requirements were equilibrium (equation (6)). This means that if one has the equilibrium or the strain displacements set of equations one can derive the other set from one of the two principles (i.e. virtual displacements or forces). This allows us to produce a consistent set of equations which in some cases may be difficult to find otherwise.

Some researchers have applied these ideas to deduce a consistent set of equations for shell theory for instance.

One can· now define a functional called complementary energy, composed of two parts, i.e. the internal complementary energy, given by

$$W(\sigma) = \tfrac{1}{2} \int \sigma_{jk} \, \varepsilon_{ij}(\sigma) \, d\Omega \tag{22}$$

$$= \tfrac{1}{2} \int \sigma_{jk} \, c_{ij\ell m} \, \sigma_m \, d\Omega$$

plus the potential of the surface forces assuming that the $\bar{u}_k$ displacements are not functions of t_k, i.e.

$$\Omega_c = - \int_{\Gamma_1} t_k \, \bar{u}_k \, d\Gamma - \int_{\Omega} b_k \, u_k \, d\Omega \tag{23}$$

The total complementary energy is then

$$\Pi_c = W + \Omega_c \tag{24}$$

The condition for stationarity is defined by

$$\Pi_c^* = W^* + \Omega_c^* = 0 \tag{25}$$

where the * stands as always for variation. This principle is the basis of the flexibility formulations. Notice that

$$W^*(\sigma) = \int \sigma_{ij}^* \, c_{jk\ell m} \, \sigma_{\ell m} \, d\Omega \tag{26}$$

$$= \int \sigma_{ij}^* \, \varepsilon_{ij} \, d\Omega$$

GENERALIZED PRINCIPLE

A great deal of confusion still exists regarding the so called generalized principles, which are based on a generalization of the previous two cases. Frequently researchers deduced them from an extension of the Principle of Minimum Potential Energy. (see Washizu [20] and Reissner [19]). We will now see that this is not really necessary and they can easily be obtained from considerations of virtual work.

Let us first write the principle of virtual work for the case of virtual displacements but without the restrictions that they have to be identically zero on the Γ_1 part of the boundary. In this case we have

$$\int \sigma_{jk}(\varepsilon) \, \varepsilon_{jk}^* \, d\Omega = \int b_k \, u_k^* \, d\Omega + \int_{\Gamma_2} \bar{t}_k \, u_k^* \, d\Gamma + \int_{\Gamma_1} t_k \, u_k^* \, d\Gamma \tag{27}$$

Then we write the expression for virtual forces also without the restriction that $t_k^* \equiv 0$ on Γ_2 and assuming that the strains are functions of stress, i.e. $\varepsilon_{jk} = c_{jk\ell m} \, \sigma_{\ell m}$. This gives

$$\int \sigma_{jk}^* \, c_{jk\ell m} \, \sigma_{\ell m} \, d\Omega = \int b_k^* \, u_k \, d\Omega + \int_{\Gamma_2} t_k^* \, u_k \, d\Gamma + \int_{\Gamma_1} t_k^* \, \bar{u}_k \, d\Gamma \tag{28}$$

One can now replace the first integral on the right hand side by

$$\int b_k^* u_k \, d\Omega = - \int (\sigma_{jk,j}^*) u_k \, d\Omega$$

$$= \int \sigma_{jk,j}^* (\tfrac{1}{2}(u_{i,j} + u_{j,i})) d\Omega - \int_\Gamma t_k^* u_k \, d\Gamma \tag{29}$$

In what follows we will accept that $\varepsilon_{ij} = \tfrac{1}{2}(u_{i,j} + u_{j,i})$. Hence we can subtract (28) from (27) taking into consideration (29) and obtain

$$\int \{\sigma_{jk} \varepsilon_{jk}^* + \sigma_{jk}^* \varepsilon_{jk} - \sigma_{jk}^* c_{jk\ell m} \sigma_{\ell m}\} \, d\Omega =$$

$$= \int b_k u_k^* \, d\Omega + \int_{\Gamma_2} \bar{t}_k u_k^* \, d\Gamma + \int_{\Gamma_1} t_k u_k^* \, d\Gamma$$

$$+ \int_{\Gamma_1} (u_k - \bar{u}_k) t_k^* \, d\Gamma$$

This is a well known 'generalized' expression. In order to investigate the stationarity conditions associated with it we can carry out an integration by parts and obtain the following equation,

$$\int \{\{\sigma_{jk,j} + b_k\} u_k^* + (c_{jk\ell m} \sigma_{\ell m} - \varepsilon_{jk}) \sigma_{jk}^*\} \, d\Omega \tag{31}$$

$$- \int_{\Gamma_2} (t_k - \bar{t}_k) u_k^* \, d\Gamma - \int_{\Gamma_1} (\bar{u}_k - u_k) t_k^* \, d\Gamma = 0$$

The stationary conditions are   i) the equilibrium equations in Ω; ii) the stress boundary conditions on Γ_2 ; iii) the stress-strain relationships and iv) the displacement boundary conditions on Γ_1.

Remembering the definition of W (equation (22)) the second integral in (31) can be written as,

$$\int \left\{ \frac{\partial W}{\partial \sigma_{jk}} - \varepsilon_{jk} \right\} \sigma_{jk}^* \, d\Omega \tag{32}$$

(where W is the complementary strain energy density). Notice that u_k and t_k components in (31) are <u>independent.</u>

With these concepts in mind we can now propose an energy functional for the generalized principle, i.e.

$$\Pi_G(u,\sigma) = \int \{\sigma_{jk}\varepsilon_{jk} - W(\sigma)\}d\Omega - \int b_k u_k \; d\Omega$$

$$- \int_{\Gamma_2} t_k u_k \; d\Gamma - \int_{\Gamma_1} t_k(u_k - \bar{u}_k)d\Gamma \tag{33}$$

The condition of stationarity which produces expression (31) is

$$\Pi_G^* = 0 \tag{34}$$

This model requires expansions for both stresses and displacements and gives rise to the mixed formulation of boundary elements.

HYBRID MODEL

Another interesting development in recent years has been the study of the so-called "hybrid" models. In this case we can start with expression (31) but select stress functions σ which identically satisfy the equilibrium equation, i.e.

$$\sigma_{jk,j} + b_k = 0 \tag{35}$$

Hence equation (31) reduces to

$$\int \left\{\frac{\partial W}{\partial \sigma_{jk}} - \varepsilon_{jk}\right\}\sigma_{jk}^* \; d\Omega = \int_{\Gamma_2} u_k^*(t_k - \bar{t}_k)d\Gamma + \int_{\Gamma_1} t_k^*(\bar{u}_k - u_k)d\Gamma \tag{36}$$

Integrating by parts the term in ε_{jk} and assuming that $\sigma_{jk,k}^* \equiv 0$ one finds,

$$\int W^* \; d\Omega = \int_{\Gamma_2} (t_k - \bar{t}_k)u_k^* \; d\Gamma + \int_{\Gamma_2} t_k^* u_k \; d\Gamma + \int_{\Gamma_1} \bar{u}_k t_k^* \; d\Gamma \tag{37}$$

Note that expressions for the u_k displacements are required only on the boundaries. This means that the only expressions needed on the volume are the stress σ_{jk} which has to satisfy the equilibrium equations.

If the displacement boundary conditions are made to satisfy the boundary conditions on Γ_1, equation (37) can be written

$$\int W^* \; d\Omega = - \int_{\Gamma_2} \bar{t}_k u_k^* \; d\Gamma + \int_{\Gamma} (t_k u_k^* + u_k t_k^*)d\Gamma \tag{38}$$

which is the form usually presented in the literature [20]. The integral

on the left hand side of this equation produces a flexibility matrix.
After certain manipulations this can be transformed into a stiffness matrix
and used in the same manner as the matrices deduced using the principle of
virtual displacements.

BOUNDARY SOLUTIONS

To obtain the type of boundary solutions used in boundary elements one can
also start with equation (31) but this time satisfying identically the
stress-strain relations. Here we have,

$$\int \{\sigma_{jk,j} + b_k\} u_k^* \, d\Gamma = \int_{\Gamma_2} (t_k - \bar{t}_k) u_k^* \, d\Gamma + \int_{\Gamma_1} (\bar{u}_k - u_k) t_k^* \, d\Gamma \tag{39}$$

The aim is now to try to reduce the solution to the boundary. One possibility is to propose σ_{jk} functions which satisfy the equilibrium equations
as we have done in the case of "hybrid" formulations. Another is to find
virtual work field u_k^*, σ_k^* which are in equilibrium. To this end we can
integrate twice by parts the first integral in (39) which gives

$$\int (\sigma_{jk,j}^*) u_k \, d\Omega + \int b_k \, u_k^* \, d\Omega =$$

$$= \int_{\Gamma_1} t_k \, u_k^* \, d\Gamma + \int_{\Gamma_2} \bar{t}_k \, u_k^* \, d\Gamma + \int_{\Gamma_1} \bar{u}_k \, t_k^* \, d\Gamma + \int_{\Gamma_2} u_k \, t_k^* \, d\Gamma \tag{40}$$

One can now look for functions such as the fundamental solutions which
satisfy the equilibrium equation, such that,

$$\sigma_{jk,j}^* + \Delta_\ell^i = 0 \tag{41}$$

where Δ_ℓ^i is the Dirac delta function and represents a unit load at the
point 'i' acting in the ℓ direction. This solution is sometimes called
Kelvin's solution and will produce for each direction 'ℓ' the following
equation

$$u_\ell^i + \int_\Gamma u_k \, t_k^* \, d\Gamma = \int b_k \, u_k^* \, d\Omega + \int_\Gamma t_k \, u_k^* \, d\Gamma \tag{42}$$

u_ℓ^i represents the displacement at 'i' in the 'ℓ' direction. Notice that
for simplicity we have added together the two types of boundary, $\Gamma = \Gamma_1 + \Gamma_2$
u_k^* and t_k^* are components of the fundamental solution, i.e. displacement and
tractions due to a unit concentrated load at the point 'i' acting in the
'ℓ' direction. If we consider unit forces acting in the three directions,
equation (42) can be written as,

$$u_\ell^i + \int_\Gamma u_k \, t_{\ell k}^* \, d\Gamma = \int_\Gamma t_k \, u_{\ell k}^* \, d\Gamma + \int b_k \, u_{\ell k}^* \, d\Omega \qquad (43)$$

where $t_{\ell k}^*$ and $u_{\ell k}^*$ represent the tractions and displacements in the k direction due to unit forces acting in the ℓ direction. Equation (43) is valid for the particular point 'i' where those forces are applied.

Expression (43) gives rise to the so called direct boundary element method which is described in detail in reference [10]. For the purposes of this paper it is important to point out that the u_k and t_k unknowns are all defined on the boundary. Furthermore if the domain term in body forces - which does not contain any unknown - is taken to the boundary, one only needs to compute boundary integrals which effectively reduces the dimensionality of the problem by one. Several ways in which the body force term can be taken to the boundary are described in reference [10] and [24].

The boundary element as described above is basically a point collocation technique as the fundamental solutions are applied at different points 'i' on the boundary. It is also possible to distribute these fundamental solutions over portions of the boundary or elements but in this case a double integration will be required which complicates the solution and reduces the efficiency of the new technique. This technique is described by Wendland in reference [25].

SUMMARY

Table I summarizes the main characteristics of each of the five engineering analysis methods discussed earlier, together with the most usual type of statement which gives origin to the techniques to be described in detail in part 3 of this paper.

The generalized functional Π_G can be considered as the starting point for all the formulations. Notice that there are many other stationary conditions that we could include in Π_G but they have not yet produced practical methods of engineering analysis. By contrast the five formulations shown in the table are well known in engineering, although most of the engineering codes are based on the displacement formulation. More recently a substantial number of codes have started to appear based on boundary methods using the fundamental solution due to Kelvin or similar.

Although Table I helps to understand the common basis of these methods of engineering analysis it is necessary, if one wishes to combine them, to see the form that the element matrices take for each of the formulations. This analysis is carried out in the next section.

3. THE DISCRETE ELEMENT METHODS

In this section we will try to deduce the matrices corresponding to the different methods seen in Section 2, starting with the simpler - the finite element displacement method - for completeness.

i) Displacement Model

In this case one starts with the following expression,

TABLE I
GENERALIZED FUNCTIONAL

$$\Pi_G = \int \{\sigma_{jk,j} + b_k\}u_k^* \, d\Omega + \int \left\{\frac{\partial W}{\partial \sigma_{jk}} - \varepsilon_{jk}^*\right\} d\Omega - \int_{\Gamma_2} (t_k - \bar{t}_k)u_k^* \, d\Gamma - \int_{\Gamma_1} (\bar{u}_k - u_k)t_k^* \, d\Gamma$$

METHOD	STATIONARY CONDITIONS	INDEPENDENT VARIABLES	IDENTICALLY SATISFIES	USUAL STATEMENT
DISPLACEMENT	$\sigma_{jk,j} + b = 0$ $t_k = \bar{t}_k$ on Γ_2	u	$\dfrac{\partial W}{\partial \sigma_{jk}} = \varepsilon_{jk}$ $u_k = \bar{u}_k$ on Γ_1	$\int \sigma_{jk}\varepsilon_{jk}^* d\Omega = \int_{\Gamma_2} \bar{t}_k u_k^* d\Gamma + \int b_k u_k^* d\Omega$
BOUNDARY (with fundamental solution)	$\sigma_{jk,j} + b_k = 0$ $t_k = \bar{t}_k$ on Γ_2 $u_k = \bar{u}_k$ on Γ_1	u , t (boundary only)	$\dfrac{\partial W}{\partial \sigma_{jk}} = \varepsilon_{jk}$	$u_\ell^i + \int_\Gamma u_k t_{\ell k}^* d\Gamma = \int_\Gamma t_k u_{\ell k}^* d\Gamma + \int b_k u_{\ell k}^* d\Omega$
MIXED	$\sigma_{jk,j} + b_k = 0$ $\dfrac{\partial W}{\partial \sigma_{jk}} = \varepsilon_{jk}$ $t_k = \bar{t}_k$ on Γ_2; $u_k = \bar{u}_k$ on Γ_1	u , σ		$\int\{\sigma_{jk}\varepsilon_{jk}^* + \sigma_{jk}^*\varepsilon_{jk} - W^*\}d\Omega =$ $= \int_{\Gamma_2} \bar{t}_k u_k^* d\Gamma + \int_{\Gamma_1} t_k u_k^* d\Gamma + \int_{\Gamma_1} (u_k - \bar{u}_k)t_k^* d\Gamma + \int b_k u_k^* d\Omega$
HYBRID	$\dfrac{\partial W}{\partial \sigma_{jk}} = \varepsilon_{jk} \quad t_k = \bar{t}_k$ on Γ_2; $u_k = \bar{u}_k$ on Γ_1 (or part)	u(Boundary only σ	$\sigma_{jk,j} + b_k = 0$	$\int W^* d\Omega = -\int_{\Gamma_2} \bar{t}_k u_k^* d\Gamma + \int_\Gamma (t_k u_k^* + u_k t_k^*)d\Gamma$
FLEXIBILITY	$\dfrac{\partial W}{\partial \sigma_{jk}} = \varepsilon_{jk}$ $u_k = \bar{u}_k$ on Γ_1	σ	$\sigma_{jk,j} + b_k = 0$ $t_k = \bar{t}_k$ on Γ_2	$\int \sigma_{jk}^*\varepsilon_{ij}\, d\Omega = \int_{\Gamma_1} t_k^* \bar{u}_k \, d\Gamma + \int b_k^* u_k d\Omega$

$$\int \sigma_{jk}\, \varepsilon^*_{jk}\, d\Omega = \int_{\Gamma_2} \bar{t}_k\, u^*_k\, d\Gamma + \int b_k\, u^*_k\, d\Omega \tag{44}$$

which can be rendered in matrix form as follows,

$$\int \underset{\sim}{\sigma}^T\, \underset{\sim}{\varepsilon}^*\, d\Omega = \int \underset{\sim}{\bar{t}}^T\, \underset{\sim}{u}^*\, d\Gamma + \int \underset{\sim}{b}^T\, \underset{\sim}{u}^*\, d\Omega \tag{45}$$

We can now propose to use displacement functions such that

$$\underset{\sim}{u} = \underset{\sim}{\phi}^T\, \underset{\sim}{u}_e \qquad ; \qquad \underset{\sim}{u}^* = \underset{\sim}{\phi}^T\, \underset{\sim}{u}^*_e \tag{46}$$

where the ϕ are the interpolation functions for the displacements over one element and u_e the nodal unknowns. Differentiating we obtain the strains, i.e.

$$\underset{\sim}{\varepsilon} = \underset{\sim}{B}\, \underset{\sim}{u}_e \tag{47}$$

and accepting the strain-stress relationships we can write,

$$\underset{\sim}{\sigma} = \underset{\sim}{D}\, \underset{\sim}{\varepsilon} \tag{48}$$

$\underset{\sim}{D}$ is the matrix of elastic constant.

Under these conditions equation (45) becomes,

$$\underset{\sim}{u}_e^{*,T}\, \underset{\sim}{K}_e\, \underset{\sim}{u}_e = \underset{\sim}{u}_e^{*,T}\, \underset{\sim}{P}_e \tag{49}$$

where,

$$\underset{\sim}{K}_e = \int \underset{\sim}{B}^T\, \underset{\sim}{D}\, \underset{\sim}{B}\, d\Omega$$
$$\underset{\sim}{P}_e = \int \underset{\sim}{\phi}\, \underset{\sim}{\bar{t}}\, d\Gamma + \int \underset{\sim}{\phi}\, \underset{\sim}{b}\, d\Omega \tag{50}$$

As the virtual displacements are arbitrary, equation (49) can be written simply as

$$\underset{\sim}{K}_e\, \underset{\sim}{u}_e = \underset{\sim}{P}_e \tag{51}$$

The functions $\underset{\sim}{u}$ in (46) are assumed to be admissible and ϕ part of a complete set of functions. Notice that they have to satisfy identically the boundary conditions $u_k = \bar{u}_k$ on Γ_1 including the interelement surfaces.

ii)   <u>Mixed Models</u>

Here we start with the complete Π^*_G expression, i.e.

$$\int \{\sigma_{jk}\,\varepsilon^*_{jk} + \sigma^*_{jk}\,\varepsilon_{jk} - W^*\}d\Omega =$$

$$= \int_{\Gamma_2} \bar{t}_k\,u_k\,d\Gamma + \int_{\Gamma_1} t_k\,u^*_k\,d\Gamma + \int_{\Gamma_1} (u_k-\bar{u}_k)t^*_k\,d\Gamma + \int b_k\,u^*_k\,d\Omega \tag{52}$$

which can be written in matrix form as,

$$\int \{\underset{\sim}{\sigma}^T\,\underset{\sim}{\varepsilon}^* + \underset{\sim}{\varepsilon}^T\,\underset{\sim}{\sigma}^* - W^*\}d\Omega = \int_{\Gamma_2} \bar{\underset{\sim}{t}}^T\,\underset{\sim}{u}^*\,d\Gamma + \int_{\Gamma_1} \underset{\sim}{t}^T\,\underset{\sim}{u}^*d\Gamma \tag{53}$$

$$+ \int_{\Gamma_1} (\underset{\sim}{u}-\bar{\underset{\sim}{u}})^T\,\underset{\sim}{t}^*\,d\Gamma + \int \underset{\sim}{b}^T\,\underset{\sim}{u}^*\,d\Omega$$

We now adopt expressions for both displacements and stress over an element, i.e.

$$\underset{\sim}{u} = \underset{\sim}{\phi}^T\,\underset{\sim}{u}_e \quad ; \quad \underset{\sim}{\sigma} = \underset{\sim}{\psi}^T\,\underset{\sim}{\sigma}_e \tag{54}$$

(Notice that we will assume displacement and stress continuity here for simplicity. Otherwise extra "jump" terms should be included in Π^*_G).

The expressions for $\underset{\sim}{\varepsilon}$, $\underset{\sim}{t}$ and $\underset{\sim}{W}$ become

$$\underset{\sim}{\varepsilon} = \underset{\sim}{B}\,\underset{\sim}{u}_e \quad ; \quad \underset{\sim}{t} = \underset{\sim}{N}\,\underset{\sim}{\sigma} = \underset{\sim}{N}\,\underset{\sim}{\psi}^T\,\underset{\sim}{\sigma}_e \tag{55}$$

$$W^* = \underset{\sim}{\sigma}^{*,T}_e (\underset{\sim}{\psi}\,\underset{\sim}{C}\,\underset{\sim}{\psi}^T)\underset{\sim}{\sigma}_e$$

where C is the elastic compliance matrix ($\underset{\sim}{C} = \underset{\sim}{D}^{-1}$). N is a matrix of direction cosines on the boundary. Substituting (55) into (53) we obtain,

$$\underset{\sim}{u}^{*,T}_e\,\underset{\sim}{A}^T_e\,\underset{\sim}{\sigma}_e + \underset{\sim}{\sigma}^{*,T}_e\,\underset{\sim}{A}\,\underset{\sim}{u}_e - \underset{\sim}{\sigma}^{*,T}_e\,\underset{\sim}{F}\,\underset{\sim}{\sigma}_e = \underset{\sim}{u}^{*,T}_e\,\underset{\sim}{P}_e + \underset{\sim}{\sigma}^{*,T}\,\underset{\sim}{Q}_e \tag{56}$$

where,

$$\underset{\sim}{A}^T_e = \int \underset{\sim}{B}^T\,\underset{\sim}{\psi}\,d\Omega - \int_{\Gamma_1} \underset{\sim}{\phi}^T\,\underset{\sim}{N}\,\underset{\sim}{\psi}^T\,d\Gamma$$

$$\underset{\sim}{P}_e = \int \underset{\sim}{\phi}\,\underset{\sim}{b}\,d\Omega + \int_{\Gamma_2} \underset{\sim}{\phi}\,\bar{\underset{\sim}{t}}\,d\Gamma$$

$$\underset{\sim}{F}_e = \int \underset{\sim}{\psi}\,\underset{\sim}{C}\,\underset{\sim}{\psi}^T\,d\Omega$$

$$\underset{\sim}{Q}_e = - \int_{\Gamma_1} \underset{\sim}{\psi} \; \underset{\sim}{N}^T \; \underset{\sim}{\bar{u}} \; d\Gamma$$

One can then assemble all the elements together and obtain the following matrices for the whole structure,

$$\underset{\sim}{U}^{*,T}(\underset{\sim}{A}^T \underset{\sim}{\sigma} - \underset{\sim}{P}) + \underset{\sim}{\sigma}^{*,T}(\underset{\sim}{A} \; \underset{\sim}{U} - \underset{\sim}{F} \; \underset{\sim}{\sigma} - \underset{\sim}{Q}) = \underset{\sim}{0} \qquad (58)$$

where $\underset{\sim}{U}$ and $\underset{\sim}{\sigma}$ are the nodal stresses and displacements for the whole structure. The final system of equations can be written as,

$$\begin{bmatrix} -\underset{\sim}{F} & \underset{\sim}{A} \\[2ex] \underset{\sim}{A}^T & \underset{\sim}{0} \end{bmatrix} \begin{Bmatrix} \underset{\sim}{\sigma} \\[2ex] \underset{\sim}{U} \end{Bmatrix} = \begin{Bmatrix} \underset{\sim}{Q} \\[2ex] \underset{\sim}{P} \end{Bmatrix} \qquad (59)$$

Notice that the system is symmetric but not positive definite.

iii) <u>Hybrid Models</u>

In the case of hybrid models we can start with the following statement,

$$\int W^* \, d\Omega = - \int_{\Gamma_2} \bar{t}_k \, u_k^* \, d\Gamma + \int_\Gamma (t_k \, u_k^* + u_k \, t_k^*) d\Gamma \qquad (60)$$

which in matrix form can be written as

$$\int \underset{\sim}{\sigma}^{*,T} \underset{\sim}{C} \; \underset{\sim}{\sigma} \, d\Omega = - \int_{\Gamma_2} \underset{\sim}{t}^T \, \underset{\sim}{u}^* \, d\Gamma + \int_\Gamma (\underset{\sim}{t}^T \, \underset{\sim}{u}^* + \underset{\sim}{t}^{*,T} \, \underset{\sim}{u}) d\Gamma \qquad (61)$$

The $\underset{\sim}{u}$ vector refers to the boundary displacements only and the $\underset{\sim}{\sigma}$ needs to satisfy the equilibrium equations. Furthermore on the <u>external</u> Γ_1 boundaries the displacements usually identically satisfy the $u_k = \bar{u}_k$ boundary conditions. With these conditions in mind we can define,

$$\underset{\sim}{\sigma} = \underset{\sim}{\psi}^T \underset{\sim}{\sigma}_e$$

$$\underset{\sim}{u} = \underset{\sim}{\phi}^T \underset{\sim}{u}_e \qquad (62)$$

The boundary tractions are written in function of the nodal stresses

$$\underset{\sim}{t} = \underset{\sim}{N} \; \underset{\sim}{\psi}^T \underset{\sim}{\sigma}_e \qquad (63)$$

have many practical advantages as the hybrid finite elements are in many cases more accurate than the displacement models.

In spite of the many possibilities offered by the combination of different methods, few researchers use more than one technique at a time. Sometimes however, engineers are forced to look into some model combination because of the special characteristics of the problem. Cases such as offshore structures, buildings on soil foundations, etc. may require a finite element analysis coupled with a special analysis for the water or soil. Many of these analyses are nowadays carried out by coupling finite and boundary element solutions. These combinations and the use of approximate solutions will be discussed in this section.

In order to effect the combination one should first notice that the values of T in equation (72) are the actual surface tractions at the nodes. In finite elements instead these values are weighted as shown in the right hand side of equation (51) and the "integrated tractions" are concentrated at the nodes. These values are repeated by the vector P. It is now possible to relate the values of P and T through a distribution matrix M whose terms represent the weighting of the boundary values of the tractions by the interpolation functions, i.e.

$$P = M \, T \tag{73}$$

In order to combine the boundary element region with the finite element part, one can deduce a matrix which can be easily implemented in finite element codes. We start by transforming equation (72) and inverting G, i.e.

$$G^{-1}(H \, U \, - \, B) = T \tag{74}$$

Next one multiplies both sides by the distribution matrix M to obtain the weighted traction vectors, P of finite elements, as follows,

$$(M \, G^{-1} \, H)U = (M \, G^{-1} \, B) + M \, T \tag{75}$$

These terms can be redefined using

$$K' = M \, G^{-1} \, H$$
$$\tag{76}$$
$$P' = M \, T + M \, G^{-1} \, B$$

Hence equation (75) presents now the same form as finite elements, i.e.

$$K' \, U = P' \tag{77}$$

The main discrepancy that arises now between this formulation and finite element displacement models is that the matrix K' is generally asymmetric. The asymmetry is due to the approximation involved in the discretization process and the choice of the assumed solution. The matrix can be made symmetric by minimizing the square of the errors in the non-symmetric off-diagonal terms as the asymmetry is small in most practical applications.

This gives a new matrix whose coefficients are defined by

$$\underset{\sim}{K} = \tfrac{1}{2}(\underset{\sim}{K'} + \underset{\sim}{K'}'^{\,T})$$

(78)

This matrix can now be assembled with the finite element displacement
matrices as usual. The disadvantage of this technique is that the
equations within the boundary element region are all coupled, which gives
a full system of equations. Because of this it is sometimes preferable
to use approximate boundary elements. Typical cases where the researchers
may prefer the approximate ones are when the infinite medium is described
using boundary elements.

Approximate boundary elements are based on the assumption that sufficiently
far from the region under consideration – which can be assumed to be discretized using finite elements – the behaviour of the fundamental solution
can be approximated. This approximation results in a simple expression for
the boundary solution at the interface without involving the neighbouring
points. The full explanation of the way these solutions can be deduced is
given in references [26][27]. The methodology to determine these conditions can be applied to a large number of problems and it is interesting
to mention that when applied to certain classical problems, it produces
some well known equations, such as the Somerfeld radiation conditions

5. CONCLUSIONS

The purpose of this chapter has been to present the common basis of the
more important methods of solutions used in engineering analysis. These
methods which can be called DISCRETE ELEMENT TECHNIQUES encompass the
classical stiffness finite element formulation, mixed, hybrid and
flexibility techniques and the newly developed boundary element method.

In this chapter the common basis of all the techniques is stressed by
pointing out how the principle of virtual work can be used to generate
solid mechanics models. The simplicity of the virtual work approach allows
for the formulation of very general approximate methods, valid even for
nonlinear and time dependent problems. The formulation of the different
techniques – including boundary elements – becomes then independent of the
existence or otherwise of a known integral or variational statement.

A generalized principle is presented from which all the different models are
deduced. The principle has been presented using considerations of virtual
work only, rather than by extending the Principle of Minimum Potential
Energy. This new deduction may have important applications in other cases
as virtual work is more general than the energy principles. By identically
satisfying some of the stationary conditions implied by the generalized
functional one can obtain the different discrete element techniques.

The chapter ends by pointing out a way of combining finite and boundary
elements stressing the possibility of using approximate boundary elements
as described in references [10][26].

REFERENCES

[1]   JASWON, M.A. "Integral Equation Methods in Potential Theory, I".
      Proc. R. Soc. A., 1963, 275.

[2]   SYMM, G.T. "Integral Equation Methods in Potential Theory, II"
      Proc. R. Soc., A, 1963, 275.

[3]   MASSONET, C.E. "Numerical Use of Integral Procedures" in Stress
      Analysis, Zienkiewicz, O.C. and Holister, G.S. (Eds) Wiley, 1966.

[4]   HESS, J.L. and SMITH, A.M.O. "Calculation of Potential Flow about
      Arbitrary Bodies" Progress in Aeronautical Science, 8, Kuchemann, D
      (Ed.) Pergamon Press, 1967.

[5]   CRUSE, T. "Application of the Boundary-Integral Equation Solution
      Method in Solid Mechanics" in "Variational Methods in Engineering,
      Vol. II" (C.A. Brebbia and H. Tottenham, Eds) Southampton University
      Press, 1973 and 1975.

[6]   BREBBIA, C.A. and TOTTENHAM, H. (Eds) "Variational Methods in
      Engineering" (2 volumes) Southampton University Press, England, 1973.
      Reprinted 1975.

[7]   BREBBIA, C.A. "Weighted Residual Classification of Approximate Methods"
      Applied Mathematical Modelling, 2, September 1978.

[8]   BREBBIA, C.A. "The Boundary Element Method for Engineers" Pentech
      Press, London, Halstead Press, NY, 1978. Reprinted 1980,1982.

[9]   BREBBIA, C.A. and WALKER, S. "Boundary Element Techniques in
      Engineering" Butterworths, London, 1979.

[10] BREBBIA, C.A., TELLES, J. and WROBEL, L. "The Boundary Element
      Technique – Theory and Applications in Engineering" Springer-Verlag,
      Berlin and NY, 1984.

[11] BREBBIA, C.A. (Ed.) "Recent Advances in Boundary Element Methods"
      Proc. 1st Int. Conf. on BEM, Southampton University, 1978. Pentech
      Press, London, 1978.

[12] BREBBIA, C.A. (Ed.) "New Developments in Boundary Element Methods"
      Proc. 2nd Int. Conf. on BEM, Southampton University, 1980.
      CML Publications, Southampton, 1980.

[13] BREBBIA, C.A. (Ed.) "Boundary Element Methods" Proc. 3rd Int.
      Conf. on BEM, Salifornia, 1981. Springer-Verlag, Berlin & NY 1981.

[14] BREBBIA, C.A. (Ed.) "Boundary Element Methods in Engineering"
      Proc. 4th Int. Conf. on BEM, Southampton, 1982. Springer-Verlag,
      Berlin & NY, 1982.

[15] BREBBIA, C.A., FUTAGAMI, T. AND TANAKA, M. (Eds) "Boundary Elements"
      Proc. of 5th Int. Conf. in BEM, Hiroshima, 1983. Springer-Verlag,
      Berlin-Ny, 1983.

[16] BREBBIA, C.A. (Ed.) "Progress in Boundary Element Methods, Vol.1"
     Pentech Press, London, Halstead Press, NY, 1982.

[17] BREBBIA, C.A. (Ed.) "Progress in Boundary Element Methods, Vol.2"
     Pentech Press, London, Springer-Verlag, NY, 1983.

[18] BREBBIA, C.A. (Ed.) "Progress in Boundary Element Methods, Vol.3"
     Springer-Verlag, Berlin and NY, 1984.

[19] REISSNER, E. "A Note on Variational Principles in Elasticity"
     Int. J. Solids and Structures, 1, 1965, pp.93-95 and 357.

[20] WASHIZU, K. "Variational Methods in Elasticity and Plasticity"
     2nd Edition. Pergamon Press, NY, 1975.

[21] PIAN, T.H.H. and TONG, P. "Basis of Finite Elements for Solid
     Continua" Int. J. Num. Method. Engg., 1, 1969, pp.3-28.

[22] ARISTOTLE "Physics" Oxford University Press, 1942.

[23] BREBBIA, C.A. "Basic Principles" Opening Address. 5th Int. Conf.
     on BEM, Hiroshima, Nov. 1983. Published in "Boundary Elements"
     (C.A. Brebbia, ed.) Springer Verlag, Berlin - NY, 1983.

[24] DANSON, D. "Linear Isotropic Elasticity with Body Forces" Chapter
     in "Progress in Boundary Elements" Vol. 2 (C.A. Brebbia, Ed.)
     Pentech Press, London and Springer-Verlag, NY, 1983.

[25] WENDLAND, W. "Asymptotic Accuracy and Convergence" Chapter in
     "Progress in Boundary Element Methods" Vol. 1 (C.A. Brebbia, ed.)
     Pentech Press, London, Halstead Press, NY, pp.289-313, 1981.

[26] BREBBIA, C.A. "Coupled Systems" Section in "Handbook of Finite
     Elements" (H. Kardestuncer and J. Norrie, Eds.) MacGraw Hill, NY,
     1984.

[27] BREBBIA, C.A. "New Developments" Invited Lecture published in
     Proceedings of the International Conference on Finite Element Methods,
     August 1982, Shanghai, China. Published by Gordon & Breach, Science
     Publishers, NY, 1982.

CHAPTER 9

UNIFICATION OF FEM WITH LASER EXPERIMENTATION

H. Kardestuncer & R.J. Pryputniewicz

Unification of finite element methods with laser experimentation is presented. It is pointed out that most engineering problems contain regions in which finite element modeling encounters difficulties due to nonlinearities, irregular boundaries, ambiguous energy functionals, etc. Measurements obtained by laser experimentation, particularly in these regions, can be digitized and automatically incorporated into the finite element modeling to improve results. Unification is possible in solid mechanics, fluid mechanics, gas dynamics, heat transfer, and in an everincreasing number of other fields.

INTRODUCTION

Solution methodologies for engineering problems can, in general, be categorized as experimental, analytical, and numerical. In the recent past, the emphasis appears to have shifted from the first to the last. Certainly, each methodology has considerable advantage over the others for a given class of problems and each makes use of the others for verification of the results. In many cases, even the data furnished by one methodology has been utilized by the others. In spite of recent advances in number crunching equipment which have drawn considerable attention to numerical methodologies, in particular to the finite element methods, the importance of experimental, analytical, or semi-analytical methods has not diminished.

Today's demands for optimum and reliable design are, to a great extent, satisfied by application of finite element methods. In these applications, the finite element methods are used to solve problems for which exact solutions are nonexisting, or are very difficult to obtain. Also, the finite element methods are the only way to analyze complex three-dimensional structures, response of which to applied load system cannot be predicted in any other way. However, results obtained by the finite element methods are subject to the boundary conditions used, rely greatly on the accurate knowledge of material properties, depend on accurate representation of the object's geometry, and are sensitive to the shape and size of elements employed in modeling. All of the information necessary to "run" the finite element models is obtained either from published data (for example, material properties), from design specifications (object's geometry), or from experimental studies (boundary conditions, shape and size of elements).

As is often the case with new and powerful methods, the finite element method has been over-used, perhaps even misused. Only recently have we begun to realize that virtually all versions of FEM contain some shortcomings. As a result, the need for unifying (merging, coupling) FEM in the physical and time domain with other methods has begun to manifest itself (see, for instance, Kardestuncer (1975, 1978, 1979, 1980, 1982), and Zienkiewicz et al. (1977, 1980). Here we are interested in exploring the unification of laser experimentation with FEM in space and time simultaneously.

Other experimental techniques which can be used in conjunction with the finite element methods employ strain gauges, photoelastic procedures, etc. These experimental procedures,

although conventionally used, do not provide all the
information necessary to reliably model an object's response
to the applied load system by the finite element methods. For
example, strain gauges give only pointwise information for the
surface of the object directly under them; to obtain a
complete strain mapping a large number of strain gauges must
be bonded to the surface of interest. The procedure,
moreover, is invasive and interferes with the object's
performance. In photoelastic modeling, on the other hand, an
oversimplification is made because the object is formed in
plastic which has properties totally different from those of
the actual material the object is made of. Although such a
model, when observed under polarized light, is very useful in
identification of stress fields, it does not represent the
true response of the object to the applied force system.

Ideally, what is needed is an experimental method that would
provide necessary displacements and/or deformations at any
point on the investigated object. Also, the results should be
provided in three dimensions with high accuracy and precision
in such a way as not to interfere with the object's
performance.

Recent advances in the field of optoelectronics have led to
development of methods satisfying the above requirements.
These methods utilize lasers as a source of light and as such
can be called the <u>laser methods</u>. Although there are several
laser methods available today, of particular interest to
finite element analysis are: (i) hologram interferometry, (ii)
heterodyne hologram interferometry, (iii) laser speckle
metrology, (iv) fiber optic metrology, and (v) directed light
beam metrology. Each of these methods has certain
characteristics which make it particularly useful in specific
experiments. In general, however, all of these methods allow
highly accurate, precise, noninvasive, rapid determination of
the object's response to the applied load system.

In this paper, laser methods are described, including their representative applications, with particular emphasis on their unification with the finite element methods to improve the results.

UNIFICATION IN ERROR ANALYSIS

The most important issue in any approximate procedure in engineering is the accuracy of the results. How good are they? What are the upper bounds of errors? Such questions have always been asked though answers have not always been found. Nevertheless, problems were solved and systems were put into service. The easiest response to these questions has always been the use of a factor of safety (FS) big enough to accommodate all uncertainties. How big should it be has, of course, been another question. If it was big enough, the engineer was successful; if not, he was doomed.

An alternative approach to these questions has been to experiment (full scale, half scale, whatever) before putting the system into service. Recognizing that things designed and built yesterday are not as complex as those designed and built today, experimentation and the choice of FS were relatively easier tasks than they are today. In recent years, however, the availability of numerical tools (both in respect to methodology and equipment) has enabled engineers to design very complicated systems by successfully solving very difficult problems. Nevertheless, one question raised above still remains: how good are the results?

When we examine the finite element methods, for instance, we find that error sources are quite numerous. Basically, these sources can be categorized as mathematical modeling, discretization, and manipulation (Melosh and Utku (in print)). In addition, each of these has many subdivisions of error sources. To address all and come up with a generally acceptable methodology for error bounds might very well be the most difficult task in numerical methods today.

Some of the error sources are rather general--tool-dependent (i. e., they include errors due to equipment, methodology, solution procedures, etc.); others are very specific--problem-dependent (i. e., they include physical and geometric characteristics of the domain). The latter are more difficult to deal with.

Many have addressed the question of error bounds for problems of the first kind in finite element methods; in particular, are works by Babuska and Rheinboldt (1977, 1978, 1980), Szabo and Mehta (1978), Peano et al. (1979), Utku and Melosh (1984), and a very fine work on a posteriori error analysis by Kelly et al. (1983). Error bounds and controls for problems of the second kind, in particular for those which are time and path dependent, have yet to be established.

When it comes to the finite element methods, certainly h and p (mesh size and order of polynomials, respectively) are the more important (or, the easier) parameters to play with in estimating or even minimizing error bounds which are due to discretization only. The work by Kelly et al. (1983), cited earlier, estimates and minimizes error bounds based on information obtained during the solution process itself. Using two independent error measures consisting of an error indication and an error estimation, they establish certain criteria for where to refine a given mesh and when to stop adaptive processes. The programs developed using either or both (the latter, often referred to as the pony express policy, is claimed to be the better) are called self-adaptive processes because they require no interaction on the part of the user. Supposedly, it is also more practical and less expensive than theoretical a priori error estimates and classical approaches requiring multiple analyses.
Self-adaptive processes, however, are tested for linear and

self-adjoint boundary value problems only. One of the main
features of <u>a posteriori</u> error analysis presented by Kelly et
al. (1983) is that it involves local rather than global
computations. It also necessitates establishing an energy
functional beforehand for a given problem.

In spite of many useful properties of energy functionals,
there are a good many problems for which one can not come up
with a functional which is valid for all stages of the
problem. Furthermore, in self-adaptive processes, the
coefficients characterizing the physical domain must be
well-defined and their variations in respect to time and path
must be sufficiently smooth. Otherwise, codes developed based
on h- and p-processes will be insufficient. Nevertheless,
they cover at least one aspect of error analysis and
minimization. The fact remains, however, that development of
fully automatic self-adaptive processes is one of the most
crucial needs of finite element computations today. To
achieve this, one must not develop algorithms based on the
computed information alone. Instead, information based on
actual measurements made during the processing must also be
incorporated into the algorithm. These measurements
(observations) should be employed not for veryfying the
computed values of the unknown function (as is often done) but
for estimating and even controlling errors.

When discretizing the domain, engineers generally pay
attention to certain regions of the domain which are critical
or very sensitive to changes in h- and p-refinement
parameters. Localized error norms in these regions may
fluctuate drastically or even diverge as in the case of
ill-conditioned systems of equations. If u_c and u_m represent
the computed and measured values of the unknown function,
respectively, then the error can be defined as either
$e_c = u - u_c$ or $e_m = u - u_m$, where u is the correct answer.
Moreover, if the measurents are of very high precision, then
we suggest that the latter be employed for error estimates and

for adaptive processes in the critical regions of the domain.

In structural mechanics problems, the energy of the error
corresponding to a particular solution is given as

$$\eta^2 = -\int_\Omega e\, r\, d\Omega\,,\tag{1}$$

where r represents the residual forces (Kelly et al. (1983)).
Following the same procedure as in Kelly et al. (1983) one can
obtain the possible refinement on u by using a hierarchic mode
N_{p+1} (the finite element basis function for the polynomial of
degree p+1). Since the energy absorbed by this additional
mode is assumed to be directly proportional to the
corresponding force and inversely proportional to the
stiffness, the above equation for the ith element becomes

$$\eta_i^2 = \frac{\left[\int N_{p+1}\, r_i\, d\Omega_i\right]^2}{K_{ii}^{p+1}}\,.\tag{2}$$

Hence, this procedure suggests that among all the available
N_{p+1} , the one that gives the greatest error decrease should be
chosen as the new refinement. One should, however, make sure
that N_{p+1} is not orthogonal to r_i otherwise $\eta_i = 0$ indicating
that the estimate may be deceptive.

UNIFICATION IN EVALUATING ELEMENT MATRICES

Experimental techniques can also be used for direct evaluation
of the element stiffness and/or flexibility coefficients. In
particular, when an element's shape is irregular (i. e.,
possesses curved lines and surfaces, which is often the case
at the free boundaries) or when the material is anisotropic,
composite, nonlinear, or stratified, the computed stiffness
matrix, even using higher order isoparametric elements
(implying p-refinements), may not yield the accuracy desired.

     H. Kardestuncer & R.J. Pryputniewicz

The h-refinements for those elements would, on the other hand, increase the number of equations to be solved, thereby decreasing the accuracy of the results. In the case of solid elements, for instance, (whether one-, two-, or three-dimensional), the stiffness matrices can be obtained experimentally. For this we shall refer to Castigliano's second theorem in tensor notation (Kardestuncer (1977)). Thus,

$$p^{iq} = \frac{\partial W}{\partial u_{iq}} \ , \tag{3}$$

where

$$W = \tfrac{1}{2} p^{iq} u_{iq} \ , \qquad i = 1, ..., n \ , \qquad q = 1,2,3 \tag{4}$$

represents strain energy stored in the element. Substituting Eq. 4 into Eq. 3 and keeping in mind that

$$\frac{\partial u_{jr}}{\partial u_{iq}} = 1 \quad \text{for } i,q = j,r, \text{ respectively;} \quad \text{zero otherwise,}$$

the result is

$$p^{iq} = \frac{\partial p^{iq}}{\partial u_{jr}} u_{jr} \ , \tag{5}$$

which can be written as

$$p^{iq} = K^{ijqr} u_{jr} \ . \tag{6}$$

Kardestuncer (1969) has investigated the tensorial properties of Eq. 6 and the similarities with the following well-known tensorial equation in solid mechanics

$$\sigma^{iq} = E^{ijqr} \epsilon_{jr} \ . \tag{7}$$

Note that Eq. 7 is a physical equation (Hooke's Law) without geometry (i. e., direction but no distance) whereas Eq. 6 contains <u>geometry</u> as well as <u>physics</u>.

The bivalent version of the quadrivalent tensor on the right
hand side of Eq. 6 is the stiffness matrix of the element.
This equation indicates that the stiffness (or flexibility)
matrix coefficients can be determined by observing (measuring)
the variation of p^{iq} in respect to u_{jr} or vice versa. Note
that in this equation i and j represent the nodes (the
integration points in the standard FEM) of the element and q
and r are the directions of coordinate axes (local and
global). Today, there are many high precision instruments
that can evaluate the stiffness (or flexibility) matrix
coefficients of an element of any shape and material. Here,
we emphasize the use of laser technology for such evaluation.
Since the measurements are continuous (independent of time and
path), the stiffness matrix coefficients for those elements
(highly nonlinear both in respect to time and path) can be
determined at any increment of time and/or load. These
coefficients can then be incorporated into the global **K** prior
to the solution procedure.

Let us assume that the overall final stiffness matrix is
partitioned as follows

$$
\begin{bmatrix} \mathbf{p}_I \\ \hline \mathbf{p}_{II} \\ \hline \mathbf{p}_{III} \end{bmatrix}
=
\left[
\begin{array}{c|c|c}
\left[\mathbf{K}_{I,I}\right]_{c,m} & \left[\mathbf{K}_{I,II}\right]_{c,m} & \left[\mathbf{K}_{I,III}\right]_{c} \\
\hline
 & \left[\mathbf{K}_{II,II}\right]_{c,m} & \left[\mathbf{K}_{II,III}\right]_{c} \\
\hline
\text{SYMMETRIC} & & \left[\mathbf{K}_{III,III}\right]_{c}
\end{array}
\right]
\begin{bmatrix} \mathbf{u}_{I,c} \\ \hline \mathbf{u}_{II,c} \\ \hline \mathbf{u}_{III,m} \end{bmatrix}
\; , \tag{8}
$$

where the subscripts c and m identify the computed and
measured entities, respectively. Note that in certain
portions of **K**, the measured and computed elements are coupled
and identified with subscripts c,m. Equation 8 can be further

reduced to

$$
\begin{bmatrix}
\mathbf{p}_\mathrm{I} - \left[\mathbf{K}_{\mathrm{I,III}}\right]_c \mathbf{u}_{\mathrm{III},m} \\[2ex]
\mathbf{p}_\mathrm{II} - \left[\mathbf{K}_{\mathrm{II,III}}\right]_c \mathbf{u}_{\mathrm{III},m}
\end{bmatrix}
=
\begin{bmatrix}
\left[\mathbf{K}_{\mathrm{I,I}}\right]_{c,m} & \left[\mathbf{K}_{\mathrm{I,II}}\right]_{c,m} \\[2ex]
\left[\mathbf{K}_{\mathrm{II,I}}\right]_{c,m} & \left[\mathbf{K}_{\mathrm{II,II}}\right]_{c,m}
\end{bmatrix}
\begin{bmatrix}
\mathbf{u}_{\mathrm{I},c} \\[2ex]
\mathbf{u}_{\mathrm{II},c}
\end{bmatrix}
. \qquad (9)
$$

Since the left hand side of this equation contains all the
known entities (whether given, computed, or measured), its
solution is possible and will yield the unknown values of the
function at the nodes where no measurements have been taken.

Once we have determined $\mathbf{u}_{\mathrm{I},c}$ and $\mathbf{u}_{\mathrm{II},c}$, we can compute the
residual force vector as

$$
\mathbf{r} = \mathbf{p}_\mathrm{III} - \left[\mathbf{K}_{\mathrm{III,I}}\right]_c \mathbf{u}_{\mathrm{I},c} - \left[\mathbf{K}_{\mathrm{III,II}}\right]_c \mathbf{u}_{\mathrm{II},c} - \left[\mathbf{K}_{\mathrm{III,III}}\right]_c \mathbf{u}_{\mathrm{III},m} \neq 0 \qquad . \qquad (10)
$$

The components of this vector corresponding to element i can
then be utilized in Eq. 2 to determine the next refinement for
the adaptive processes presented by Kelly et al. (1983). We
shall now present various laser methods to obtain the
measurements mentioned above.

HOLOGRAM INTERFEROMETRY

The most useful of all methods of hologram interferometry, for
finite element applications, is the two-beam, off-axis method
(Pryputniewicz (1979, 1982a)). In this method, the laser beam
is divided into two beams, as shown in Fig. 1. One of the
beams is made to interact with the object, or scene being
recorded (the so-called object beam) while the other beam does
not interact with the object at all. In fact, the second beam
is a reference beam against which the object beam is recorded.
A setup for recording holograms is made so that the object
beam and the reference beam overlap in a given region of

space.   As a result of this, the two beams interfere with each
other and the resulting interference pattern is recorded in a
suitable medium (Smith (1977)).   The exposed medium, upon
processing, becomes a hologram.   The hologram is reconstructed
with the same setup that was used in recording, except that
now it is illuminated with the reference beam alone.   Of the
images produced during reconstruction, the most applicable to
finite element analysis is the virtual image.   To observe the
virtual image, the reconstruction should be viewed through the
hologram as if it were a window.   The image is seen in the
space which was occupied by the object while the hologram was
recorded, even though the original object had since been
removed. The image observed has all the visual characteristics
of the original object.   In fact, there is no visual test that
can differentiate between the two.

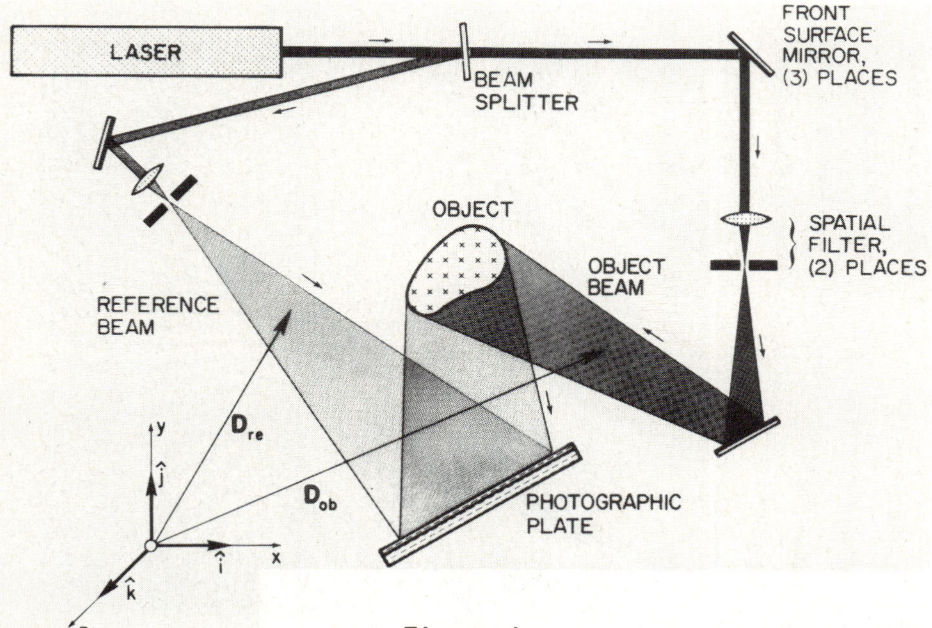

Figure 1.
Setup for recording and reconstruction of holograms.
Directions of propagation of the object beam and the
reference beam are defined by position vectors **D**
specified in respect to the x-y-z coordinate system.

There are three basic variations of hologram interferometry:
(i) real-time, (ii) time-average, and (iii) double-exposure.

<u>Real-time hologram interferometry</u> involves recording a single
exposure hologram as shown in Fig. 1, processing it, and
reconstructing it by illumination with the original reference
beam. The reconstructed image is superimposed onto the
original object which is also illuminated with the same beam
as used in recording the hologram. If the object is now even
slightly displaced and/or deformed, interferometric comparison
between the holographically reconstructed image and the new
state of the object occurs instantaneously (Fig. 2). The
particular advantage of the real-time method is that different
types of motion, dynamic as well as static, can be studied
with a single holographic exposure.

Figure 2.
Images obtained using real-time hologram interferometry:
studies of microcracks in porous, ceramic components.

In <u>time-average interferometry</u> a single holographic recording of an object undergoing a cyclic vibration is made. With the exposure time long compared to one period of the vibration cycle, the hologram effectively records an ensemble of images corresponding to the time-average of all positions of the object during its vibration. In the reconstruction of such a hologram, interference occurs between the entire ensemble of the recorded images, with the images recorded near zero velocity contributing most strongly. As such, images reconstructed from the time-average hologram have intensity distribution given by the zero-order Bessel function (Fig. 3). In the case of stroboscopic illumination of a vibrating object, however, cosinusoidal intensity distributions are obtained.

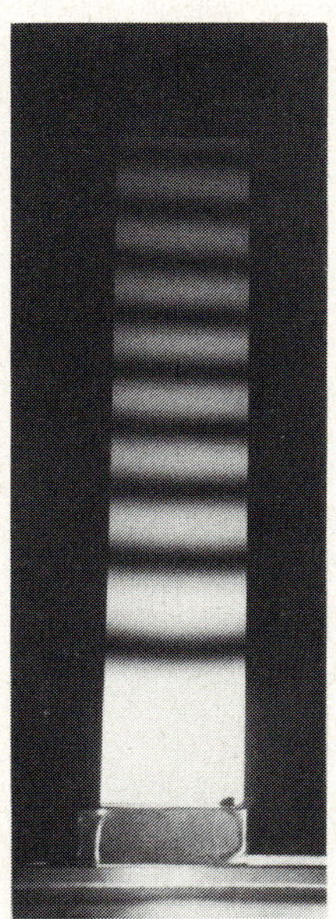

Figure 3.
Time-average hologram
of a vibrating
cantilever beam:
the first flexure
mode.

The <u>double-exposure hologram interferometry</u>, which can be considered to be a special case of the time-average method (where only two exposures of the object are made in the same medium), is the most widely used of all holographic methods. In this method, the object is displaced and/or deformed between the two exposures. Therefore, the object beam during the second exposure is different from that used in making the first exposure. During reconstruction of the double-exposure hologram, both object beams are faithfully reconstructed, forming images of the object's initial and final positions. Since these images are formed in coherent laser light, they interfere with each other forming a pattern of bright and dark

fringes resulting in cosinusoidal intensity variation of the
image (Fig. 4).   These fringes are a direct measure of changes
in the object's position and/or shape which occured between
the two exposures.

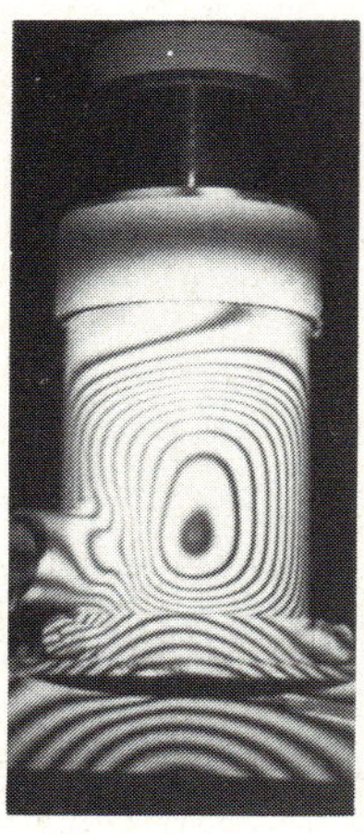

Figure 4.
Double-exposure hologram
of a hydraulic cylinder:
pressure between the
exposures was increased
from 5,100 psi to 5,800 psi.

QUANTITATIVE INTERPRETATION OF HOLOGRAMS

There are a number of methods dealing with interpretation of
the fringes observed within the holographically reconstructed
image (Stetson (1979), Schuman and Dubas (1979), Vest (1978),
Pryputniewicz (1980a), Pryputniewicz and Stetson (1976,
1980)).   The most general of these methods employs multiple
observations of the holographic image.   It results in
displacement vector **u** expressed as a product of the inverse of
the matrix formed by the sum of the projection matrices B with
the matrix representing the sum of the observed vectors $\mathbf{u}_{ob}$
(Stetson (1979), Pryputniewicz and Stetson (1980),
Pryputniewicz (1980a))

$$\mathbf{u} = \left[\sum_{i=1}^{n} B^i \right]^{-1} \left(\sum_{i=1}^{n} \mathbf{u}_{ob}^i \right) \quad . \tag{11}$$

In Eq. 11, i denotes the observation number with n being the total number of observations while $\mathbf{u}_{ob}$ is measured in the plane normal to the direction of observation and is defined by the corresponding B. The projection matrix B^i, for the ith

direction of observation, can be either one of the following two types. If the projection is made in a direction parallel to the direction of observation, then the projection is normal; if it is not, the projection is oblique. The normal projection is defined as a difference between the identity matrix and the matrix resulting from the dyadic product of the ith unit observation vector with itself. In the case of the oblique projection, the matrix is formed by the dyadic product of the object's surface unit normal vector with the unit vector defining the particular direction of observation.

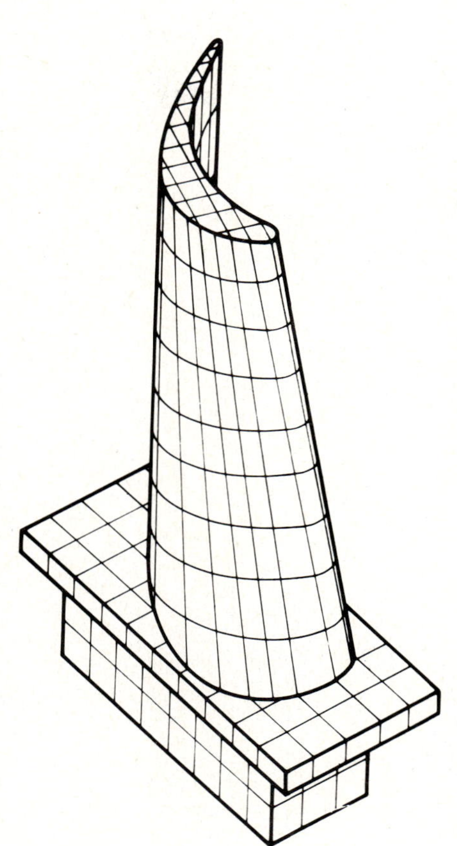

Figure 5.
Typical finite element
breakup of an airfoil.

Of particular interest in FEM modeling (Fig. 5) is the application of double-exposure hologram interferometry in determination of strains

and rotations (Pryputniewicz and Stetson (1976)). In this case, the strain-rotation matrix f is determined directly from the parameters S (defining illumination and observation

geometry) and S_f (defining shape and distribution of fringes
seen during reconstruction of a hologram)

$$f = \left[S^T S\right]^{-1}\left[S^T S_f\right] \quad . \tag{12}$$

Decomposition of the matrix f into the symmetric part e and
the antisymmetric part θ gives strains and rotations,
respectively.

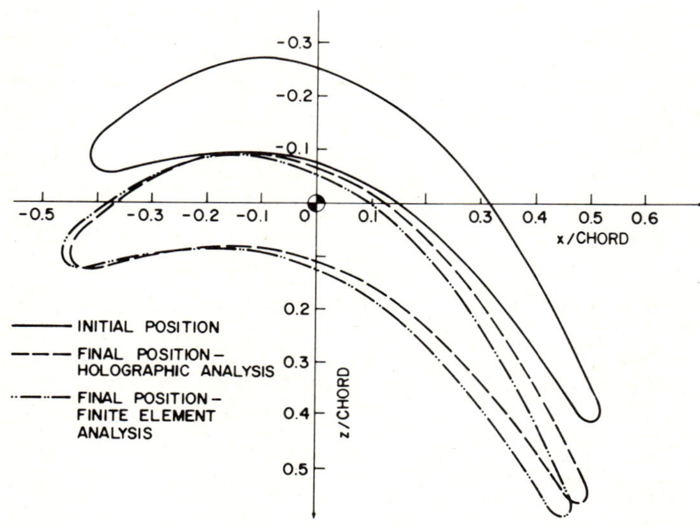

Figure 6.
Displacements of a radially loaded airfoil:
the finite element computations were subject to
the boundary conditions obtained from
the double-exposure holograms.

The matrix S_f appearing in Eq. 12 consists of fringe vectors,
one for each direction of observation, which can be computed
from the fringe patterns produced during reconstruction of the
holograms, that is,

$$\omega = \mathbf{S_f} \cdot \mathbf{D} \quad , \tag{13}$$

where ω is the fringe-locus function, constant values of which
define fringe loci on the object's surface, and **D** specifies
coordinates at which the specific values of ω were determined.
Knowledge of the fringe vector is essential in quantitative
interpretation of holograms (Fig. 6). The fringe vector, as
expressed in Eq. 13 and used in Eq. 12, is also helpful is
determining the system's optimum geometry for recording of
holograms.

It should be noted that analysis of holographically produced
fringes does not depend on material properties at all. In
fact, the holographic procedures are particularly suited for
quantitative determination of a material's constitutive
behavior.

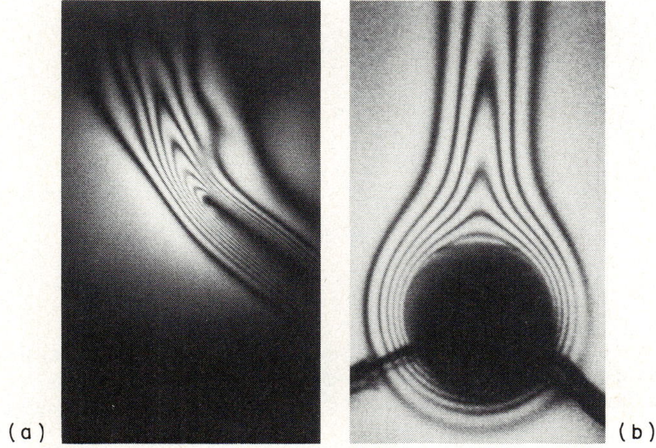

Figure 7.
Double-exposure holograms of: (a) heated inclined plate,
(b) heated horizontal rod.

Hologram interferometry is also very useful in heat transfer
studies. For example, Fig. 7 shows typical images recorded
during studies of heat transfer characteristics of flat and

curved surfaces. From reconstructed images, temperature
distributions can be determined to within a fraction of one
degree, anywhere within the image, without any interference
whatsoever with the studied "space".

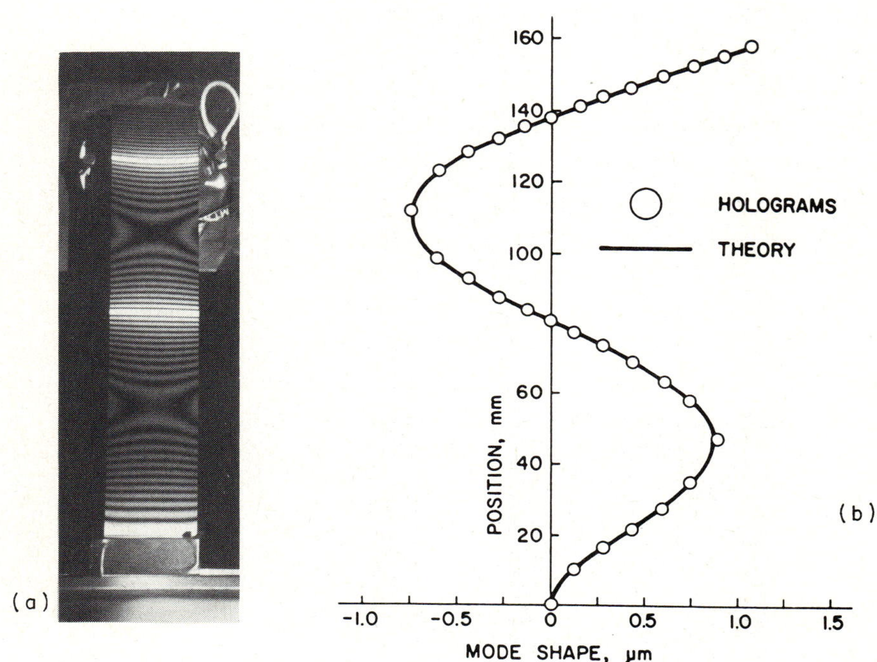

Figure 8.
Quantitative study of the vibrating beams: (a) the time
average hologram of the cantilever beam, (b)
displacements corresponding to the third flexure
mode shown in (a).

In the case of time-average hologram interferometry,
displacements are found from (Pryputniewicz (in print))

$$u_z = \frac{\lambda}{2\pi(\hat{S}_{2_z} - \hat{S}_{1_z})} |\omega_t| \quad , \tag{14}$$

where λ is the laser wavelength, $|\omega_t|$ is the argument of the

zero order Bessel function, while $\hat{S}_{1Z}$ and $\hat{S}_{2Z}$ are components
of the unit vectors defining directions of illumination and
observation, respectively. Typical results for a vibrating
cantilever beam are given in Fig. 8 showing good agreement
with the theoretical predictions. In the case shown, the beam
is vibrating in the third flexure mode as vividly depicted by
the hologram (Fig. 8-a) where nodes are demarcated by the
brightest fringes and antinodes by the darkest fringes; for
this mode, the theoretically predicted frequency was 1772 cps,
while that determined experimentally was 1733 cps.
Also, Fig. 8-b shows very good agreement between the mode
shape determined from the hologram and the mode shape
predicted by a theoretical model, which was developed to
simulate beam vibrations. However, it should be noted that
the theoretical computation was subject to the boundary
conditions which were provided from the results obtained
directly from the holograms. As shown in Fig. 8-b, the
maximum displacement of the beam, vibrating in the third
flexure mode, is 1.05 microns.

In the manner similar to that described above, mode shapes at
other frequencies can be studied using methods of hologram
interferometry.

HETERODYNE HOLOGRAM INTERFEROMETRY

Heterodyne hologram interferometry is similar to the
double-exposure hologram interferometry in that, there also,
two images of an object, at different states of stress, are
recorded in the same medium. However, each of these images is
recorded with a different reference beam, in such a way that
the reference beams can later be reconstructed independently
(Dandliker et al. (1976), Pryputniewicz (1982b)). This allows
introduction, during the reconstruction process, of a small
frequency shift between the two reconstructed and interfering
light fields, resulting in an intensity modulation at a beat

frequency of these light fields, for any point within the
interference pattern.

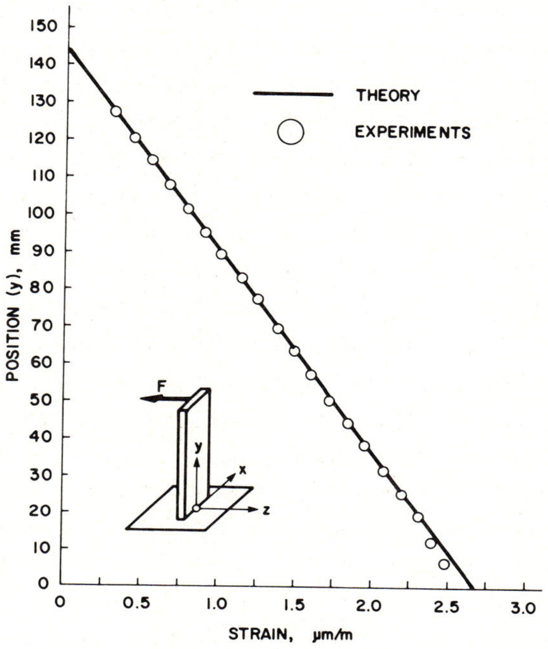

Figure 9.
Strains determined from a
heterodyne hologram of a loaded
cantilever beam.

The optical phase difference, corresponding to the
displacement and/or deformation recorded within the hologram
being reconstructed, is converted into the phase of the beat
frequency of the two interfering light fields. This phase, in
turn, is interpolated optoelectronically, resulting in
determination of fringe orders to within 1/1000 of one fringe.
This high accuracy in determining fringe orders leads to
determination of displacements to within 0.3 nm, and strains
to within 0.000,02 % (Pryputniewicz (1982a, 1982b)).

Representative results obtained using heterodyne hologram interferometry are shown in Fig. 9. In this case, a prismatic cantilever beam was loaded in the direction normal to its neutral plane, between the exposures of the heterodyne hologram. Resulting interferograms were scanned by placing a fiber-optic detector probe in the image plane formed by a lens placed between the hologram and the detector (Pryputniewicz (1982b). The resulting phase measurements were then processed using the equations relating them to parameters characterizing the system used to record, reconstruct, and scan the heterodyne hologram.

Figure 9 shows that the results obtained from the heterodyne holograms correlate very well with the theory. It should be noted that the measured strains ranged from 0.3 microns/m to 2.5 microns/m and were well below the values that can be reliably detected by conventionally used strain measuring devices. Also, the results presented in Fig. 9 were obtained without contacting the object at all, and without interfering with it in any other way. All measurements were made remotely by scanning the object's image, thus producing the results in a truly noninvasive manner.

SPECKLE METROLOGY

Any object illuminated with laser light will seem to have a granular appearance. That is, its surface will appear to be covered with fine randomly distributed light and dark irregular spots. If the observer moves, these spots appear to twinkle and move relative to the object. This phenomenon is caused by each point on the object scattering some light toward the observer. In fact, the laser light scattered by one point on the object's surface interferes with the light scattered by other points. In any region of space where these light fields overlap, a random pattern of interference spots is observed. These interference spots are known as

"speckles". The size of the speckles depends on optical
properties of the imaging system and directly influences the
accuracy of measurements:
the finer the speckles the
higher the accuracy.

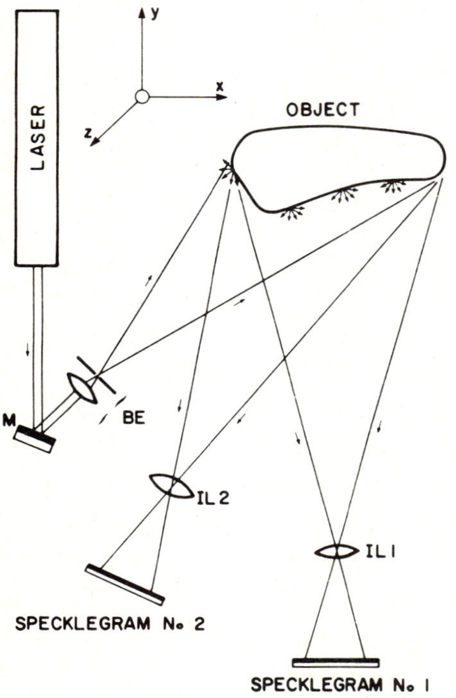

Specklegrams are recorded
by illuminating an object
with a single laser beam;
no reference beam is used
(Fig. 10). The light
scattered by the object
(or transmitting medium in
the case of fluid flow or
gas dynamics applications)
is imaged from one or more
directions onto a high
resolution recording
medium. For
interferometric purposes,
two exposures are made in
the same medium to record
the object's initial and
final configurations,
unless tandem specklegrams
are used where each
configuration is recorded
in separate media which
are later "sandwiched"
together.

Figure 10.
Setup for simultaneous
recording of two specklegrams
from two different directions.

Developed specklegrams are analyzed by sending a narrow laser
beam directly through the specklegram (Fig. 11). Upon passing
through the specklegram, the illuminating beam diffracts and
forms a halo which is modulated by Young's fringes (Fig. 12).
The frequency of Young's fringes is directly proportional to
the magnitude of the displacement recorded by the specklegram,

while their direction is normal to the direction of this
displacement.

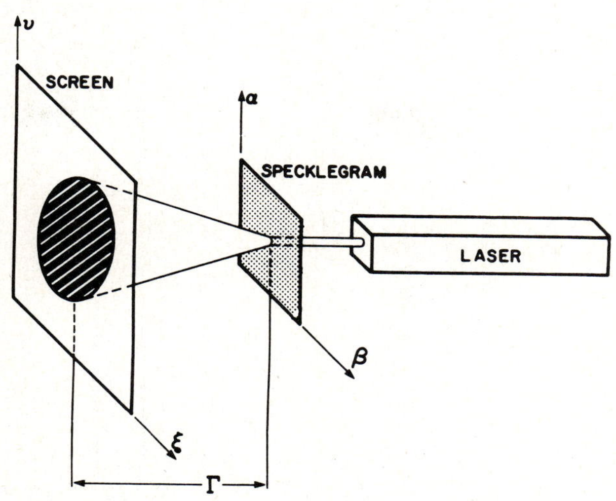

Figure 11.
Setup for reconstruction of specklegrams.

Recent studies (Stetson (1978), Pryputniewicz and Stetson
(1980), Pryputniewicz (1980b))
show that the equations
governing determination of
displacements from specklegrams
are exactly the same as those
used for quantitative
interpretations of holograms.
That is, Eq. 11 applies directly
in quantitative speckle
metrology. This equation
indicates that two specklegrams
recorded from different
directions are sufficient to
compute three-dimensional
displacements of loaded objects.

Figure 12.
Typical Young's fringe
pattern observed
during reconstruction
of a double-exposure
specklegram.

The parameters necessary to interpret specklegrams are
obtained directly from the geometry of the recording and
reconstructing systems (Figs 10 and 11, respectively) and from
the observed Young's fringes (Fig. 12).

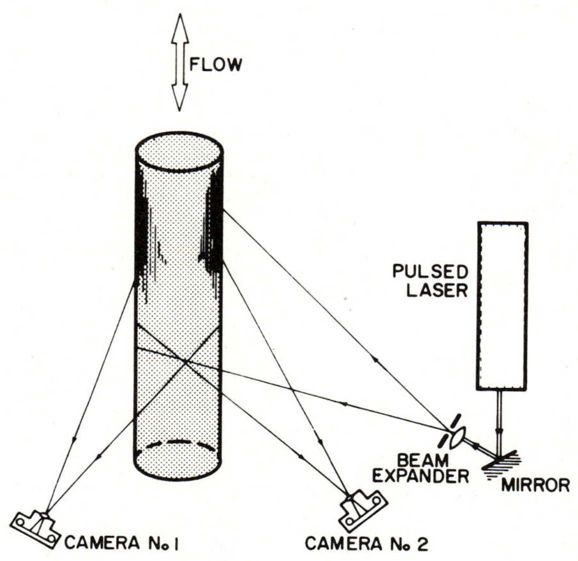

Figure 13.
Setup for simultaneous recording of two
specklegrams in fluid flow analysis.

The speckle metrology finds particular applications in studies
of three-dimensional displacements of solid objects, in
studies of fluid flow (Fig. 13), and in gas dynamics. In
these applications, the speckle methods allow recording of the
displacement and/or deformation pattern over the entire
surface of the object, permit recording of the entire velocity
profiles or the thermal profiles and are particularly suited
to studies of dynamic as well as transient behaviors.

COMPUTER AIDED INTERPRETATION OF LASER IMAGES

In FEM modeling, coordinates of nodal points are known. To specify boundary conditions at these nodes, their position has to be established and reproduced while using the experimental methods. One of such methods involves scanning the holographically reconstructed image (or a diffraction halo obtained during reconstruction of a specklegram) with a computer compatible video digitizer, as shown in Fig. 14. The digitizer, in addition to converting the scene being observed into a composite analog video signal which is viewed on a monitor, produces a digital signal that is transmitted directly to a computer. The computer, in turn, rapidly reads the

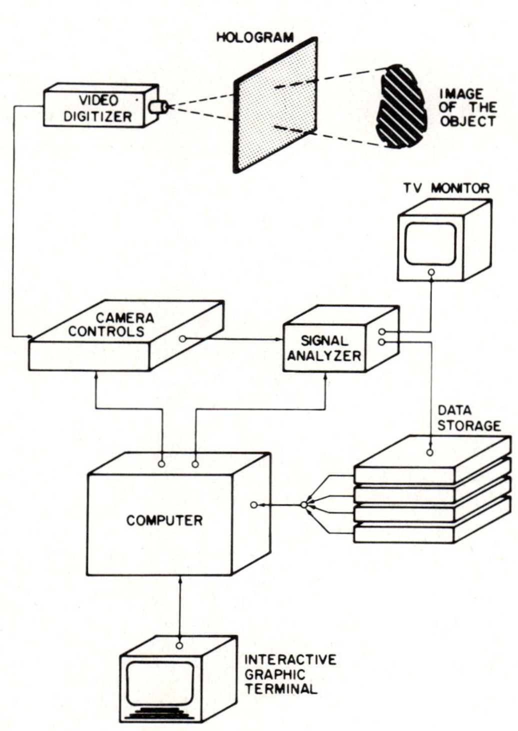

Figure 14.
Schematic of a computer
controlled system for
automated interpretation
of holograms.

electronic signal corresponding to the video image being digitized. It processes the digitized data, producing plots of intensity distribution within the image plane. Data characterizing these intensity distributions, together with other pertinent parameters, are used in quantitative interpretation of laser images. These results can be obtained

for any point within the reconstructed image by simply
instructing the computer to perform calculations for a point,
or a number of points, at specified coordinates.

A system such as that shown in Fig. 14 will provide a unique
capability for unification of finite element methods with
laser experimentation. As such, it will lead to the
development of a fully automated system for quantitative
analysis of structural deformations, which will provide highly
accurate and precise results at any point on the surface of
the studied objects.

REFERENCES

1.   Babuska, I., and Rheinboldt, W. C., Computational aspects
     of the finite element method, in: Mathematical Software,
     Vol. III (Academic Press, New York, 1977).

2.   Babuska, I., and Rheinboldt, W. C., A posteriori error
     estimates for the finite element method, Int. J. Num.
     Meth. Engr., 12 (1978) 1597-1615.

3.   Babuska, I., and Rheinboldt, W. C., Reliable error
     estimation and mesh adaptation for the finite element
     method, in: Oden, J. T. (ed.), Computational Methods in
     Nonlinear Mechanics (1980) 67-108.

4.   Dandliker, R., Marom, E., and Mottier, F. M.,
     Two-reference beam holographic interferometry, J. Opt.
     Soc. Am., 66 (1976) 23-30.

5.   Kardestuncer, H., Tensors in discrete mechanics, Tensor
     Quarterly - TSGB, 20 (1969) 1-9.

6.   Kardestuncer, H., Descrete Mechanics: A Unified Approach
     Springer-Verlag, Vienna, 1975).

7.   Kardestuncer, H., Proceedings of the UFEM Symposium
     Series (University of Connecticut, Storrs, CT, 1978,
     1979, 1980, 1982).

8.   Kardestuncer, H., Tensors versus matrices in discrete
     mechanics, in: Branin, F. H., Jr., and Huseyin, K.
     (eds.), Problem Analysis in Science and Engineering
     (Academic Press, New York, 1977).

9.   Kelly, D. W., de Gago, J. P., Zienkiewicz, O. C., and

Babuska, I., A posteriori error analysis and adaptive
processes in the finite element method: Part I -- Error
analysis, Part II -- Adaptive mesh refinement, Int. J.
Num. Meth. Engr., 19 (1983) 1593-1619.

10. Melosh, R. J., and Utku, S., Efficient finite element
analysis, to appear in: Kardestuncer, H. (ed.), Finite
Element Handbook (McGraw-Hill, New York).

11. Peano, A. G., Pasini, A., Riccioni, R., and Sardella, L.,
Adaptive approximation in finite element structural
analysis, Comp. & Struct., 10 (1979) 332-342.

12. Pryputniewicz, R. J., Laser Holography (Worcester
Polytechnic Institute, Worcester, MA, 1979).

13. Pryputniewicz, R. J., State-of-the-art in hologrammetry
and related fields, Internat. Arch. Photogram., 23
(1980a) 620-629.

14. Pryputniewicz, R. J., Projection matrices in
specklegraphic analysis, SPIE, 243 (1980b) 158-164.

15. Pryputniewicz, R. J., Unification of FEM modeling with
laser experimentation, in: Kardestuncer, H. (ed.), Finite
Elements - Finite Differences and Calculus of Variations,
(University of Connecticut, Storrs, CT, 1982a).

16. Pryputniewicz, R. J., High precision hologrammetry,
Internat. Arch. Photogram., 24 (1982b) 377-386.

17. Pryputniewicz, R. J., Quantitative interpretation of
time-average holograms in vibration analysis, in print.

18. Pryputniewicz, R. J., and Stetson, K. A., Holographic
strain analysis: extension of fringe-vector method to
include perspective, Appl. Opt., 15 (1976) 725-728.

19. Pryputniewicz, R. J., and Stetson, K. A., Fundamentals
and Applications of Laser Speckle and Hologram
Interferometry (Worcester Polytechnic Institute,
Worcester, MA, 1980).

20. Schuman, W., and Dubas, M., Holographic Interferometry
(Springer-Verlag, Berlin, 1979).

21. Smith, H. M., Holographic Recording Materials
(Springer-Verlag, Berlin, 1977).

22. Stetson, K. A., Miscellaneous topics in speckle
metrology, in: Erf, R. K. (ed.), Speckle Metrology
(Academic Press, New York, 1978).

23. Stetson, K. A., The use of projection matrices in
hologram interferometry, J. Opt. Soc. Am., 69 (1979)

     1705-1710.

24. Szabo, B. A., and Mehta, A. U., P-convergence finite
     element approximations in fracture mechanics, Int. J.
     Num. Meth. Engr., 12 (1978) 551-560.

25. Utku, S., and Melosh, R. J., Solution errors in finite
     element analysis, Comp. & Struct., 18 (1984) 379-393.

26. Vest, C. M., Holographic Interferometry (Wiley, New York,
     1978).

27. Zienkiewicz, O. C., Kelly, D. W., and Bettess, P., The
     coupling of the finite element method and boundary
     solution procedures, Int. J. Num. Meth. Engr., 11 (1977)
     355-373.

28. Zienkiewicz, O. C., Kelly, D. W., and Bettess, P.,
     Marriage a la mode -- the best of both worlds (Finite
     elements and boundary integrals) in: Glowinski, R.,
     Rodin, E. Y., and Zienkiewicz, O. C. (eds.), Energy
     Methods in Finite Element Methods, Ch. 5 (John Wiley, New
     York, 1980).

CHAPTER 10

LINEAR CROSSED TRIANGLES FOR INCOMPRESSIBLE MEDIA

D.S. Malkus & E.T. Olsen

This paper examines the error analysis for a rather remarkable type of finite element, which seems to be ideally suited for solving steady flow problems involving fluids with integral constitutive equations. The element is a quadrilateral macroelement of four linear triangles, arranged so that their interior edges form the diagonals of the quadrilateral. The properties which are most useful in such calculations are the constancy of velocity-gradients on the subtriangles and exact incompressibility of the weakly constrained Lagrange multiplier solution. On the other hand, these elements have abundant "spurious pressure modes" and thus fail to satisfy the requirements of the Brezzi – Babuska convergence theory, thought to be necessary to establish convergence of the finite element solutions in simple, Stokesian flows. There is an apparent paradox in this, because without the spurious modes, a simple count of unconstrained degrees of freedom would predict that the element is useless for incompressible media. This paper discusses a new approach to error analysis for finite elements for incompressible media. Though error estimates can only be obtained for a rather restrictive class of problems at present, our results and those in a similar vein by other investigators seem to resolve the apparent paradox of the crossed triangle macroelement: the reason for its success seemed to be the very same as the reason for its expected failure. While these results do not apply rigorously to non-Newtonian flow, they give us reason to expect that the good results so far obtained in such problems are more than fortuitous coincidence.

1. INTRODUCTION

The computation of steady-flow solutions for problems involving fluids with integral constitutive equations has attracted much interest among numerical modellers recently [1–7]. Much of this interest derives from the relevance of such calculations to the modelling of industrial polymer processing, and the potential usefulness of numerical modelling in sorting out which among the many proposed constitutive theories for viscoelastic materials gives the most faithful representation of observed polymer behavior. But there has also been a great deal of interest in such problems because of the peculiar challenges and difficulties encountered in attempts to compute apparently very simple, two-dimensional flows. Part of the challenge is that, for fluids with integral constitutive equations, the stress involves a path integral along the historical path followed by the particle at which the stress is evaluated. The purpose of this paper is to examine the error analysis of a particular finite element proposed for the computation of such solutions. In ref. 2 it is argued that this finite element is ideal for such computations—allowing the determination of exact relative strains (apart from rounding errors) in a finite element trial velocity field. The element is the crossed-triangle macroelement, discovered by Nagtigaal, et al. [8] and analyzed by Mercier [9]. In ref. 8, the element was found to be effective for modelling elasto-plastic materials, and in ref. 10, was sucessfully employed in problems involving large inelastic deformation. We refer to this element as the NRC element (for "Nagtigaal redundant constraint").

This element fails to satisfy the "discrete LBB condition" (the primary requirement of the Brezzi – Babuska theory [2,11–14,21,22,24,25]) in a most dramatic way. Therefore the discussion in this paper centers on the deceptively simple question as to whether the element can be expected to work even in Stokes-flow. This paper builds upon arguments presented in ref. 2, where the usefulness of the crossed triangle element for non-Newtonian flows is argued in some detail. We discuss in more detail the theorems proved in ref. 13, which lead to the establishment of error estimates for the NRC element in Stokes flow on simple meshes. We will show that the failure of the element to satisfy the LBB condition and the reasons for its success are two sides of the same coin, and are explainable in terms of an error analysis which does not require the LBB condition.

In non-Newtonian flows, normal forces are crucially important, which implies that accurate pressures must be obtainable from elements chosen for these flows. Because the NRC element fails the LBB condition, it needs a post-processing of the computed pressures to remove unstable modes [14]. In ref. 2 a practical pressure-smoothing scheme is discussed, which seems to work well. An error analysis for the practical scheme is unknown to us at this time. Here we show that there is at least one smoothing scheme for which error estimates can actually be proved, even though it is not as computationally convenient as the scheme employed in practice.

2. STEADY FLOWS OF MEMORY FLUIDS

2.1 Equations of Motion

We solve the equations of steady flow,

$$\nabla \cdot \underline{\sigma} + \underline{F} = \rho(\underline{u} \cdot \nabla)\underline{u}, \tag{1}$$

for a velocity field, $\underline{u}$. Incompressibility implies $\rho = $ constant and

$$\begin{aligned} \underline{\sigma} &= \underline{\sigma}' - p\underline{I} \\ \nabla \cdot \underline{u} &= 0, \end{aligned} \tag{2}$$

for a suitably chosen hydrostatic pressure function, p.

2.2 The Constitutive Equations

The constitutive equations we employ are of the following form proposed by Curtiss and Bird [15]:

$$\begin{aligned} \underline{\sigma}'' &= \mu_0[T_d^{-2} \int_{-\infty}^0 \underline{A}_0(\tau)m_1(\tau)\,d\tau + \epsilon T_d^{-1} \int_{-\infty}^0 \underline{B}_0(\tau)m_2(\tau)\,d\tau] \\ \underline{\sigma}' &= \underline{\sigma}'' - \tfrac{1}{2}(\sigma''_{11} + \sigma''_{22})\underline{I}. \end{aligned} \tag{3}$$

T_d is the disengagement time. Its magnitude determines the effective memory of the fluid. ϵ is the link-tension coefficient of the Curtiss-Bird model [15]. The kinematic tensors $\underline{A}_0$ and $\underline{B}_0$ are functions of the Cauchy and Finger-strain tensors of the deformation which carries the fluid from its reference state at time $\tau = 0$ to its configuration at historical time τ. $\underline{B}_0$ is also a function of the strain-rate, $\dot{\gamma}(0)$, at the present time $\tau = 0$. The memory functions of eq. (3) are given by (keeping in mind that $\tau \le 0$):

$$\begin{aligned} m_1(\tau) &= \tfrac{96}{\pi^4} \sum_{k=0}^{\infty} exp[(2k+1)^2 \tfrac{\tau}{T_d}] \\ m_2(\tau) &= \tfrac{192}{\pi^4} \sum_{k=0}^{\infty} \frac{exp[(2k+1)^2 \tfrac{\tau}{T_d}]}{(2k+1)^2}. \end{aligned} \tag{4}$$

With this normalization, the parameter μ_0 may be used to adjust the zero-shear viscosity [15]. It should be noted that both $\underline{A}_0$ and $\underline{B}_0$ are functions of the position variable and are, in principle, defined everywhere in the domain of the problem. The time dependence implied by eq. (3) arises by evaluating these tensors along the historical path followed by a particle which is at point $\underline{x}_0$ at time $\tau = 0$. Thus, both of the integrals in eq. (3) are path integrals along a path parameterized by τ. Construction of the path and evaluation of the path integral are numerical procedures crucial to the success of our methods and are discussed in ref. 2. In ref. 2 it is shown that the properties of the NRC element are vital to the efficient evaluation of the path integral in eq.(3). Many other constitutive equations have a form similar to that described here [1] and can be treated by techniques described in ref. 2.

Ref. 2 describes the calculation of the integrand of eq. (3) at various historical times, given a finite element trial velocity field. Quadrature formulas with specified degrees of precision with respect to the weighting functions $m_1(\tau)$ and $m_2(\tau)$ can be generated by classical orthogonal polynomial techniques [3]. The time integral in eq. (3) is replaced by a finite weighted sum:

$$\int_{-\infty}^{0} f(s)m(s)\,ds \approx \sum_i \omega_i f(\tau_i), \tag{5}$$

where $f(s)$ can be either integrand of eq. (3) and $m(s)$ the corresponding memory function. Then ω_i and τ_i are the weights and points computed once and for all for the appropriate memory function.

2.3 The Galerkin Equation

As in ref. 2, we employ a standard Galerkin form of the equations of motion, in which the incompressibility is enforced by a penalty [2,3,13,14]

$$\int_{\Omega} [\underline{\sigma}' \cdot \nabla\underline{v} + 2z(\nabla \cdot \underline{u})(\nabla \cdot \underline{v}) + \rho[(\underline{u} \cdot \nabla)\underline{u}] \cdot \underline{v} - \underline{v} \cdot \underline{F}]\,d\Omega = 0, \tag{6}$$

where z is the penalty parameter and Ω is the spatial domain of the problem. Pressure is computed by

$$p = -2z\nabla \cdot \underline{u} \tag{7}$$

Eq. (6) is restricted to $\underline{u},\underline{v}$ drawn from a finite element trial space, S^h. Given any estimate $\underline{u}^h \in S^h$ for the solution, we need to be able to compute the residual of eq. (6) on which to base an iterative correction scheme. In order to do this, a spatial numerical integration scheme must be employed, which results in a discrete finite sum in place of the space integral in eq. (6).

3. THE NRC MACROELEMENT

As illustrated in Figure 1, the NRC element is a quadrilateral formed by four linear triangles [16] whose interior sides define the diagonals of the quadrilateral. A standard area-coordinate [16] transformation is used for generation of all element quantities from a reference triangle. The geometry of each macro is uniquely specified by specifying the corner coordinates, $\underline{S}_i^e$ (the location of the central node, $\underline{C}^e$, is uniquely determined by $\underline{S}_i^e$). The rectangle pictured in reference configuration in Figure 1 can be mapped to an arbitrary quadrilateral.

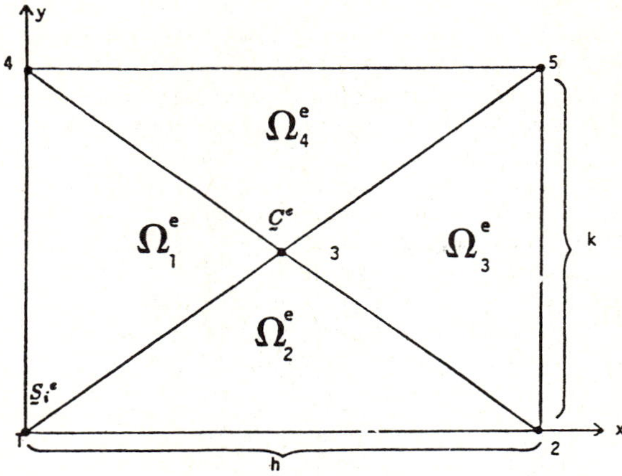

Figure 1

We employ a one-point centroidal integration formula on all terms of eq. (6). Note that for Stokes flow (i.e. let $\rho = 0$ and take the limit of eq. (6) as $T_d \mapsto 0$), this formula is exact. Let T^h denote the trial space of piecewise constant pressures, based on the individual triangles. Consider the following related Galerkin equations:

A. Lagrange Multiplier

$$\int_\Omega [\underline{\sigma}' \cdot \nabla \underline{v}^h - p^h \nabla \cdot \underline{v}^h - q^h \nabla \cdot \underline{u}^h + \rho[(\underline{u}^h \cdot \nabla)\underline{u}^h] \cdot \underline{v}^h$$
$$-\underline{v}^h \cdot \underline{F}]\, d\Omega = 0, \qquad V\, \underline{v}^h \in S^h, \quad q^h \in T^h;$$

B. Perturbed Lagrangian

$$\int_\Omega [\underline{\sigma}' \cdot \nabla \underline{v}^h - p^h \nabla \cdot \underline{v}^h - q^h \nabla \cdot \underline{u}^h + \rho[(\underline{u}^h \cdot \nabla)\underline{u}^h] \cdot \underline{v}^h$$
$$-\tfrac{1}{2z}p^h q^h - \underline{v}^h \cdot \underline{F}]\, d\Omega = 0, \qquad V\, \underline{v}^h \in S^h, \quad q^h \in T^h.$$

The equivalence theorem of ref. 17 which generalizes that ref. 18 to non-self adjoint problems implies that when the one-point formula is applied uniformly to A, B and eq. (6), eq. (6) and B produce the same solutions. We assume that conditions are satisfied that guarantee the convergence of these solutions to the solution of A as $z \mapsto \infty$. Note that the integration is exact for Stokes-flow and for the pressure terms for any T_d and ρ.

One may easily deduce that for A

$$\int_\Omega q^h \nabla \cdot \underline{u}^h \, d\Omega = 0, \qquad V\, q^h \in T^h, \tag{8}$$

and $\nabla \cdot S^h \subseteq T^h$. Thus $\nabla \cdot \underline{u}^h = 0$ pointwise for any solution to A. For B or eq. (6)

$$\int_\Omega q^h \nabla \cdot \underline{u}^h \, d\Omega = -\frac{1}{2z} \int_\Omega q^h p^h \, d\Omega, \qquad V\, q^h \in T^h. \tag{9}$$

With the NRC element, we can set $q^h = \nabla \cdot \underline{u}^h |_e$ and thus deduce from eq. (9) that

$$\nabla \cdot \underline{u}^h = \frac{1}{2z} p^h \tag{10}$$

pointwise in element interiors. It is the exact incompressibility of the Lagrange multiplier solution and the exact incompressibility of the penalty solution in the infinite limit of the penalty that is the key to the element's success in modelling flows of fluids with constitutive equations like eq. (3) [2]. It means that such a solution has a stream-function, and since its curl gives the velocity field, which is demonstrably continuous, this stream-function is evidently continuously differentiable. Furthermore, since the derivatives of the stream-function are linear, the stream-function is a piecewise quadratic (see ref. 2 for an expression for it). Thus the stream-functions associated with weakly incompressible velocity fields of the NRC element are quadratic splines [19]. We should point out that only very special elements like the NRC have discrete Lagrange multiplier solutions which satisfy eq. (8) and discrete penalty solutions which satisfy eq. (10). It is more usual for the right-hand sides of these equations (8) and (10) to involve additional terms which are governed by mesh spacing and do not vanish in the infinite limit of the penalty. The bilinear / constant-p elements of ref. 4 have this feature, and the larger compressibility of the discrete solutions obtained using this element probably contributed to its less than adequate performance [1,4] in non-Newtonian flow calculations. But the NRC fails to satisfy the discrete LBB condition [11–14]. There is then a serious question as to whether the element is stable, even for Stokes-flow. This is the question which is addressed in the remainder of this paper.

4. THE ILL-DISPOSED PRESSURES OF THE NRC ELEMENT

In ref. 20 the term "spurious mode" is applied to the pressures $c^h \in T^h$ which satisfy

$$\int_\Omega c^h \nabla \cdot \underline{v}^h \, d\Omega = 0, \qquad V \, \underline{v}^h \in S^h, \tag{11}$$

and differ from a constant "hydrostatic mode" which occurs in some problems. These c^h have also been referred to as "chessboards" [21] or "checkerboards" owing to their geometric pattern. We prefer to refer to these modes as "ill-disposed" following ref. 22 — without the negative connotation which has become attached to such c^h. In ref. 22 it is argued that ill-disposed modes play no role in determining natural or physical modes; likewise they may lead to inconsistent algebraic systems. But alternatively each ill-disposed mode has a dual velocity mode which is weakly incompressible. A prime example of this simple algebraic consequence of ill-disposed pressures is provided by the NRC element.

A simple count of unconstrained degrees of freedom [18] for the NRC element according to

$$N_{udof} = \dim S^h - \dim T^h, \tag{12}$$

on sequences of regular meshes leads to the conclusion that the NRC is grossly overconstrained — as are other arrangements of linear triangles [18]. Let $U^h \subseteq T^h$ be the subspace satisfying eq. (11) of ill-disposed modes and possibly the hydrostatic mode. What the algebraic argument of ref. 22 is saying is that a careful constraint count gives

$$N_{udof} = \dim S^h - \dim T^h + \dim U^h. \tag{13}$$

Mercier [9] recognized that $\dim U^h$ was significant for the NRC. Given $\underline{v}^h \in S^h$, consider a typical NRC macro in any mesh, and let e_1, e_2, e_3, e_4 denote the four subtriangles, ordered as in Figure 1. Mercier proved that

$$\nabla \cdot \underline{v}^h \mid_{e_1} - \nabla \cdot \underline{v}^h \mid_{e_2} + \nabla \cdot \underline{v}^h \mid_{e_3} - \nabla \cdot \underline{v}^h \mid_{e_4} = 0. \tag{14}$$

Mercier then argues that for T^h as we have chosen, only three of the four incompressibility constraints are independent, since setting $\nabla \cdot \underline{v}^h = 0$ on three triangles forces it to be zero on the fourth.

This argument which shows that the NRC is not overconstrained can be turned around to show that the NRC also has many ill-disposed modes. Consider again our typical macro, labelled "M", and define

$$c^h_M \equiv \begin{cases} \pm 1/a^{e_i}, & \text{on triangle } e_i [+, \, i = 1, 3 \text{ and } -, \, i = 2, 4]; \\ 0 & \text{outside of macro } M. \end{cases} \tag{15}$$

where a^{e_i} is the area of triangle e_i. Now

$$\int_{\Omega} c_M^h \nabla \cdot \underline{v}^h \, d\Omega = \int_M c_M^h \underline{v}^h \, d\Omega = \sum_{i=1}^{4} a^{e_i} \left[c_M^h \nabla \cdot \underline{v}^h \right]_{e_i}, \tag{16}$$

since c_M^h and $\nabla \cdot \underline{v}^h$ are constant on triangles. But substituting eqs. (14) and (15) into eq. (16) shows that c_M^h satisfies eq. (11), thus $c_M^h \in U^h$. Note that there is one such ill-disposed mode for each macro. It is not hard to show that on regular rectangular meshes with boundary conditions which cause the bilinear/ constant- p element [20] to have a checkerboard mode, the NRC has a similar checkerboard mode satisfying eq. (11), constant on each macro and alternating in sign between macros [13]. We refer to this as the "global ill-disposed mode" and those of eq. (15) as "local ill-disposed modes" for obvious reasons.

What we have seen is an instance of a simple algebraic truth [22] — redundant constraints and ill-disposed modes are one in the same. The NRC has a favorable constraint count only because it has "spurious pressure modes". The global mode of the NRC can cause algebraic consistency problems related to inhomogeneous boundary conditions [20]. This must be avoided the way it is with bilinear/ constant-p rectangles. We shall now prove that the local ill-disposed modes cannot lead to inconsistency.

Inhomogeneous boundary conditions are often imposed as described in ref. 23. S^h is taken to be the trial space satisfying homogeneous boundary conditions wherever essential b.c. are specified. $\underline{u}_o^h$ is chosen to be zero at all mesh nodes except those which coincide with boundaries on which inhomogeneous essential b.c. are specified. At those nodes $\underline{u}_o^h$ interpolates to the inhomogeneous data. Thus $\underline{u}^h = \underline{w}^h + \underline{u}_o^h$ is sought with $\underline{w}^h \in S^h$ determined by the Galerkin equation. Constraint equation (8) becomes

$$\int_{\Omega} q^h \nabla \cdot \underline{w}^h \, d\Omega = - \int_{\Omega} q^h \nabla \cdot \underline{u}_o^h \, d\Omega. \tag{17}$$

Note that eq. (8) has the following algebraic solvability condition [20,22]. There is a solution $\underline{w}^h \in S^h$ to eq. (8) only if

$$\int_{\Omega} c^h \nabla \cdot \underline{u}_o^h \, d\Omega = 0 \qquad V \, c^h \in U^h. \tag{18}$$

We have

Theorem 1: Let c_M^h be a local ill-disposed mode of the NRC element, *i.e.* c_M^h is as in eq. (11). Then for $\underline{u}_o^h$ as defined above, the solvability equation (18) is satisfied.

Proof : We note that the only reason eq. (18) does not follow from eq. (11) is that $\underline{u}_o^h \notin S^h$. Consider the larger space $\overline{S}^h \supseteq S^h$ of functions free on all boundaries: $\underline{u}_o^h \in \overline{S}^h$. Now observe that the argument of eq. (16) applies equally well to all $\underline{v}^h \in \overline{S}^h$, since it involves only one macroelement and does not depend on the boundary conditions.   QED

Remark 1.1: The above argument does not apply to the global ill-disposed mode. As with the bilinear/ constant-pressure rectangle, the global mode occurs on some meshes because there is a c^h which is orthogonal to $\nabla \cdot S^h$ but not $\nabla \cdot \overline{S}^h$.

Remark 1.2: The local ill-disposed modes of the NRC are pieced together from ill-disposed modes of the element weak-gradient matrix [22] and are not related to boundary conditions.

Theorem 1 says that inhomogeneous boundary conditions cause no more difficulty for the NRC element than with the bilinear/ constant-p element. Boundary conditions chosen so that $\underline{u}_o^h$ is such that eq. (18) is satisfied for the global ill-disposed mode will avoid the pathologies described in ref. 20. This, however, is a long way from guaranteeing convergence of the NRC approximation, even in Stokes-flow. We now turn our attention to that question.

5. APPROXIMATION ERRORS USING THE NRC IN STOKES-FLOW

The ill-disposed pressures lead to a non-uniqueness of the pressure solution from the Lagrange Multiplier method A for Stokes-flow. This assumes that any algebraic inconsistency with the global mode has been avoided and results from the fact that if p_o^h is a solution, so is $p_o^h + c^h$ for $c^h \in U^h$. The velocity solution is unique [20]. The penalty method gives a unique pressure solution which tends to a representative pressure solution of the Lagrange multiplier method as $z \mapsto \infty$ [24]. The question of convergence as the mesh is refined can be resolved by resolving the question for the Lagrange multiplier solutions $(\underline{u}^h, p_{\underline{u}}^h)$, where

$$\int_\Omega c^h p_{\underline{u}}^h \, d\Omega = 0 \qquad V \, c^h \in U^h \tag{19}$$

which is unique.

To determine whether $(\underline{u}^h, p_{\underline{u}}^h)$ converges to the exact solution $(\underline{u}^*, p^*)$ as the mesh is refined, we first consider a generalized LBB condition [2,11–14,21,22,24,25]:
For q^h orthogonal to U^h in $L_2(\Omega)$:

$$\sup_{\underline{w}^h \in S^h, \, \underline{w}^h \neq 0} \int_\Omega \frac{q^h \nabla \cdot \underline{w}^h}{\| \underline{w}^h \|_1} \geq k_h \| q^h \|_0 \tag{20}$$

where $\| \cdot \|_1$ denotes the energy norm [22] (we assume essential boundary conditions imply that the energy-norm is equivalent to the $W^{1,2}(\Omega) \times W^{1,2}(\Omega)$-norm). $\| \cdot \|_0$ denotes the $L_2(\Omega)$-norm. The desired result for $(\underline{u}^h, p_{\underline{u}}^h)$ could be established if k_h were bounded away from zero independent of h; unfortunately for many NRC meshes, it is not. This has been rigorously established on some simple meshes [13].

5.1 Velocity Estimates without the Discrete LBB Condition

It will be convenient in what follows to use the energy inner product [13,14,22]:

$$a(\underline{u}, \underline{v}) = \int_\Omega \underline{\varepsilon} \cdot \underline{f} \, d\Omega \tag{21}$$

where $\dot{e}_{ij} = \frac{1}{2}(u_{i,j} + u_{j,i})$ and $\dot{f}_{ij} = \frac{1}{2}(v_{i,j} + v_{j,i})$ for $i, j = 1, 2$. We consider the subspace $W^h \subseteq S^h$ of trial functions satisfying the incompressibility constraint equation (8) and define the operator

$$Z_h : H_1 \mapsto W^h \tag{22}$$

which is the projection from the space H_1 onto W^h with respect to the energy norm. H_1 is a subspace of $W^{1,2}(\Omega) \times W^{1,2}(\Omega)$ satisfying homogeneous essential boundary conditions on some or all of the boundary of Ω. For admissible pressures in the Lagrange multiplier method we take $H_o = L_2(\Omega)$. The exact solution $(\underline{u}^*, p^*)$ satisfies the continuous analogue to A above:

$$a(\underline{u}^*, \underline{v}) - (p^*, \nabla \cdot \underline{v})_o - (q, \nabla \cdot \underline{u}^*)_o = (\underline{v} \cdot \underline{F}, 1)_o$$
$$V \, \underline{v} \in H_1 \text{ and } q \in H_o \tag{23}$$

where $(\cdot, \cdot)_o$ denotes the $L_2(\Omega)$ inner product. We assume that the domain, boundary conditions and $\underline{F} \in L_2(\Omega) \times L_2(\Omega)$ are such that $\underline{u}^* \in W^{k,2}(\Omega) \times W^{k,2}(\Omega)$ for $k \geq 2$ with $\underline{u}^*$ unique, and $p^* \in W^{m,2}(\Omega)$ for $m \geq 1$ [12]. p^* may not be unique when there is a hydrostatic mode [20], but hereafter we assume that in such cases p^* is the unique representative pressure solution for which $(p^*, 1)_o = 0$. We now prove

Theorem 2: If $(\underline{u}^*, p^*)$ is the solution to eq. (23) and $(\underline{u}^h, p_{\underline{u}}^h)$ is the finite element solution to the Lagrange multiplier method A in the Stokes- flow case,

$$\| \underline{u}^h - Z_h \underline{u}^* \|_1 \leq c \inf_{q^h \in T^h} \| q^h - p^* \|_0 \tag{24}$$

where $\| \cdot \|_1$ is the norm induced by $a(\cdot, \cdot)$ and c is a constant independent of h.

Proof: Let $q^h = 0$ in problem A and $q = 0$ in eq. (23). Also in eq. (23) let $\underset{\sim}{v} = \underset{\sim}{v}^h \in S^h$,

$$a(\underset{\sim}{u}^{*}, \underset{\sim}{v}^h) - (p^{*}, \nabla \cdot \underset{\sim}{v}^h)_0 = (\underset{\sim}{v}^h \cdot \underset{\sim}{F}, 1)_0$$
$$a(\underset{\sim}{u}^h, \underset{\sim}{v}^h) - (p^h, \nabla \cdot \underset{\sim}{v}^h)_0 = (\underset{\sim}{v}^h \cdot \underset{\sim}{F}, 1)_0$$

Subtracting gives

$$a(\underset{\sim}{u}^h - \underset{\sim}{u}^{*}, \underset{\sim}{v}^h) = (p^{*} - p^h, \nabla \cdot \underset{\sim}{v}^h)_0$$

Now observe that $\underset{\sim}{u}^h \in W^h$ so that $Z^h \underset{\sim}{u}^h = \underset{\sim}{u}^h$ thus letting $\underset{\sim}{v}^h = \underset{\sim}{u}^h - Z_h \underset{\sim}{u}^{*}$ and using the fact that $Z_h^2 = Z_h$ and that Z_h is self-adjoint in $a(\cdot, \cdot)$ implies

$$a(\underset{\sim}{u}^h - \underset{\sim}{u}^{*}, \underset{\sim}{u}^h - Z_h \underset{\sim}{u}^{*}) = a(\underset{\sim}{u}^h - Z_h \underset{\sim}{u}^{*}, \underset{\sim}{u}^h - Z_h \underset{\sim}{u}^{*}) = (p^{*} - p^h, \nabla \cdot \left[\underset{\sim}{u}^h - Z_h \underset{\sim}{u}^{*}\right])_0 \qquad (25)$$

We note that the right-hand side of eq. (25) may not be zero since $p^{*} - p^h \notin T^h$, but it is small since we can let $\bar{p}^h$ = best $L_2(\Omega)$ approximation to p^{*} and add zero to the right-hand side of eq. (25) in the form of $(\bar{p}^h, \nabla \cdot \left[\underset{\sim}{u}^h - Z_h \underset{\sim}{u}^{*}\right])_0$. Also $(p^h, \nabla \cdot \left[\underset{\sim}{u}^h - Z_h \underset{\sim}{u}^{*}\right])_0 = 0$, which gives

$$a(\underset{\sim}{u}^h - Z_h \underset{\sim}{u}^{*}, \underset{\sim}{u}^h - Z_h \underset{\sim}{u}^{*}) = (p^{*} - \bar{p}^h, \nabla \cdot \left[\underset{\sim}{u}^h - Z_h \underset{\sim}{u}^{*}\right])_0 \leq c \| p^{*} - \bar{p}^h \|_0 \| \underset{\sim}{u}^h - Z_h \underset{\sim}{u}^{*} \|_1$$

The inequality follows from the boundedness of the weak divergence, whose operator norm gives c [12]. Division by $\| \underset{\sim}{u}^h - Z_h \underset{\sim}{u}^{*} \|_1$ gives the desired result.   QED

Remark 2.1: This result should be compared to the classical result for unconstrained problems, which shows that the finite element solution is the best energy approximation to the exact solution. Here the FEM solution is as close to the best weakly imcompressible approximation to the exact solution as the pressures are accurate.

Remark 2.2: Theorem 2 shows that the primary role of the Lagrange multiplier is to constrain the FEM solution to be close to $Z_h \underset{\sim}{u}^{*}$.

Remark 2.3: Only accuracy of the Lagrange multiplier space T^h is required, not the stability of the pressure approximation.

Theorem 2 is quite easily turned into an estimate for $\| \underset{\sim}{u}^h - \underset{\sim}{u}^{*} \|_1$ which does not require the LBB condition:

Theorem 3: $\| \underset{\sim}{u}^h - \underset{\sim}{u}^{*} \|_1 \leq \| \underset{\sim}{u}^{*} - Z_h \underset{\sim}{u}^{*} \|_1 + c \inf_{q^h \in T^h} \| p^{*} - q^h \|_0$

Proof: Add and subtract $\underset{\sim}{u}^{*}$ and use the triangle inequality in Theorem 2.   QED

Remark 3.1: The idea of constraint counting [18] was designed to give a heuristic idea of whether $\| \underset{\sim}{u}^{*} - Z_h \underset{\sim}{u}^{*} \|_1$ could be expected to be small based on an estimate of $N_{udof} = dim\, W^h$.

Results similar to Theorem 3 have been established by others. We point out particularly the work of B. Mercier in ref. 24. He proves a result which differs from ours in the present application only in that he uses the standard norm on $W^{1,2}(\Omega) \times W^{1,2}(\Omega)$ rather than the energy norm. This gives different absolute constants in the estimate. Mercier also points out an interesting sharpening of our estimate which applies to the NRC element:

Corollary 3.1 (Mercier): If the functions in W^h are weakly incompressible with respect to multipliers in H_0, then

$$\| \underset{\sim}{u}^h - \underset{\sim}{u}^{*} \|_1 = \| \underset{\sim}{u}^{*} - Z_h \underset{\sim}{u}^{*} \|_1$$

Proof: The right-hand side of eq. (23) *is* zero in this case, implying the right-hand side in Theorem 2 is zero. QED

Remark 3.2: The result applies to the NRC since weakly incompressibility implies pointwise incompressibility implies incompressibility w.r.t. H_o multipliers.

The work of Mercier predates ours but seems to have received less attention than it deserves because emphasis has since been placed on assuring the optimality of $\| \underset{\tilde{}}{u}^{\, *} - Z_h \underset{\tilde{}}{u}^{\, *} \|_1$ by choice of elements which satisfy the LBB conditon. Satisfaction of the LBB condition guarantees that $\| \underset{\tilde{}}{u}^{\, *} - Z_h \underset{\tilde{}}{u}^{\, *} \|_1$ is on the order of optimal approximation of strain-rates by S^h. This follows directly from the work of Brezzi [11]. There are several ways which $\| \underset{\tilde{}}{u}^{\, *} - Z_h \underset{\tilde{}}{u}^{\, *} \|_1$ can be shown to be small when the LBB condition is not satisfied. The interested reader is referred to ref. 14.

In ref. 24 a construction is sketched which is intended to show that there is a $\underset{\tilde{}}{w}^h \in W^h$ with

$$\| \underset{\tilde{}}{w}^h - \underset{\tilde{}}{u}^{\, *} \|_1 \le Ch \| \underset{\tilde{}}{u}^{\, *} \|_2$$

on rectangular domain discretized by square elements. Since the transformation of one rectangular domain to another is infinitely differentiable, a simple change of variable would lead to the same conclusion for a rectangular domain with rectangular elements. The construction in ref. 24 is very brief and contains a confusing misprint, and most important does not show how the construction can be carried out near no-slip boundaries [2]. Mercier's construction appears to be correct if no b.c. are imposed. It can be modified to make $\underset{\tilde{}}{w}^h$ incompressible quite easily, but this seems to require the sacrifice of approximation accuracy. The most desireable result would be to be able to extend the construction of Mercier to other boundary conditions. This would be desireable because the construction requires only that $\| \underset{\tilde{}}{u}^{\, *} \|_2 < \infty$, but such an extension of Mercier's work does not seem obvious to us at this time. Instead, we turn to techniques inspired by Johnson and Pitkäranta [25] for the bilinear/constant-p element. We will refer to this element by the acronym, "BCP," in what follows. The arguments of ref. 25 for the BCP element rely on a superconvergence result; our arguments for the NRC will directly use the results for the BCP element, and both estimates will require $\| \underset{\tilde{}}{u}^{\, *} \|_3 < \infty$. This is not optimal in the smoothness required of the exact solution in a linear element method, but is the best we have been able to argue rigorously, at present. Numerical experiments seem to suggest that the extra smoothness required is only an artifact of proof technique [2,13,14].

5.2 The NRC—BCP Correspondence

We continue to consider the Lagrange multiplier method A. In this subsection we present several results from ref. 13 which establish that there exists a $\underset{\tilde{}}{w}^h \in S^h$ such that

$$\| \underset{\tilde{}}{w}^h - \underset{\tilde{}}{u}^{\, *} \|_1 \le Ch \| \underset{\tilde{}}{u}^{\, *} \|_3$$
$$\nabla \cdot \underset{\tilde{}}{w}^h = 0 \tag{26a}$$

This, of course, establishes that

$$\| \underset{\tilde{}}{u}^{\, *} - Z_h \underset{\tilde{}}{u}^{\, *} \|_1 \le Ch \| \underset{\tilde{}}{u}^{\, *} \|_3 \tag{26b}$$

We refer to eq. (26b) as *satisfaction of the constrained approximation condition* ("CAC") [14]. We will deal with "triangulations" of domains into rectangles which have the usual restrictions [24], the most important of which is that the rectangles meet vertex to vertex. We shall say that a domain Ω, has the *rectangular regularity property* if:

1. Ω is the interior of its closure.
2. There exists a bounded rectangle, R, and a uniform triangulation, T^h, of R by rectangles of dimension $h \times k$ such that $\Omega \subseteq R$ and $\Omega^h \bigcap \Omega$ is either empty or equal Ω^e for all $\Omega^e \in T^h$. It then follows that $[\![\Omega^e \bigcap \Omega]\!]$ is a uniform triangulation of Ω.

The rectangular regularity property assures that the domain, Ω has no cracks and that it can be triangulated by triangulations which can be extended to triangulations of the rectangle, R, by extending the lines which form interelement boundaries. Some of the results below require the rectangular regularity property, and while it is somewhat restrictive, it does apply to triangulations of T-shaped and L-shaped domains and other unions of rectangles.

We first state a result for the BCP element which extends a similar result of Johnson and Pitkäranta [25] to domains with the rectangular regularity property. It is the fundamental building block of our arguments for the NRC element. As with all of the remaining theorems we state here, this theorem is given without proof. The interested reader is referred to ref. 13 for the proofs and further discussion of the results. The numbers in parentheses which immediately follow the theorem numbering of this paper give the theorem(s) and page numbers of the corresponding theorem(s) in ref. 13.

Theorem 4(4.12, p. 74)(The CAC for the BCP element): Let Ω be a connected domain with the rectangular regularity property, and let $[\![T^h]\!]$ be a collection of uniform triangulations of Ω. Let $S^h \subseteq W_0^{1,2}(\Omega) \times W_0^{1,2}(\Omega) = H_1$ be the velocity trial space constructed on T^h using the BCP element, and let $W^h \subseteq S^h$ be the subspace consisting of weakly incompressible velocity fields. Then given an incompressible $\underset{\sim}{u}^{\bullet} \in H_1 \bigcap [W^{3,2}(\Omega) \times W^{3,2}(\Omega)]$, there exists $\underset{\sim}{w}^h \in W^h$ satisfying eq. (26a) with C independent of h and $\underset{\sim}{u}^{\bullet}$.

This theorem is of use in obtaining estimates for the NRC element because of the following two results which make a correspondence between the NRC and BCP elements:

Theorem 5(5.1, p. 80): Let S^h be the FEM velocity trial space constructed using the NRC element for some domain, Ω, and some uniform triangulation, $T^h = [\![\Omega^e]\!]$, of Ω by rectangles. A $\underset{\sim}{u}^h \in S^h$ is incompressible iff the nodal values, (u_i^e, v_i^e), satisfy

$$\frac{1}{k}(v_1^e + v_2^e - v_4^e - v_5^e) = \frac{1}{h}(-u_1^e + u_2^e - u_4^e + u_5^e)$$
$$u_3^e = \frac{1}{4}(u_1^e + u_2^e + u_4^e + u_5^e) + \frac{1}{4}\frac{h}{k}(v_1^e - v_2^e - v_4^e + v_5^e) \tag{27}$$
$$v_3^e = \frac{1}{4}(v_1^e + v_2^e + v_4^e + v_5^e) + \frac{1}{4}\frac{k}{h}(u_1^e - u_2^e - u_4^e + u_5^e)$$

for each Ω^e in T^h, where Ω^e has dimensions $h \times k$ and the nodes of $\Omega^{e^{\bullet}}$ are labelled as in Figure 1.

Theorem 6(5.3, p. 83): Let Ω be any domain which can be triangulated by rectangles, and let $T^h = [\![\Omega^e]\!]$ be a triangulation of Ω by rectangles (T^h need not be uniform, but we still require that elements in the triangulation join vertex to vertex). Let $S_1^h \times T_1^h$ be the FEM trial space constructed on T^h using the BCP element, and let $S_2^h \times T_2^h$ be the FEM trial space constructed on T^h using the NRC element. Denote by R_h the $L_2(\Omega)$ projection onto T_1^h. Then:
(1) If $\underset{\sim}{u}^h$ is any weakly incompressible element of S_1^h, i. e. $R_h \nabla \cdot \underset{\sim}{u}^h = 0$, there is an element, $\underset{\sim}{v}^h \in S_2^h$, which is incompressible and agrees with $\underset{\sim}{u}^h$ on $\partial\Omega^e$ for all $\Omega^e \in T^h$.
(2) If $\underset{\sim}{v}^h$ is any incompressible element of S_2^h, the element, $\underset{\sim}{u}^h$ of S_1^h which agrees with $\underset{\sim}{v}^h$ on $\partial\Omega^e$ for all $\Omega^e \in T^h$ is weakly incompressible.

Remark 6.1: The implication of Theorems 5 and 6 which is crucial to obtaining estimates for the NRC is that, given a BCP weakly incompressible velocity field (which is *not* necessarily exactly incompressible), there is a corresponding *exaclty* incompressible NRC field. The NRC field is obtained by interpolating to the BCP at the corner nodes of the NRC and assigning the central nodal values according to eq. (27).

A technical result not cited here (Theorem 5.4, p. 85 [13]) implies that if the BCP field is a good approximation to $\underset{\sim}{u}^*$, then the assignment of the central nodal values according to eq. (27) carries that property over to the corresponding NRC field. This leads to

Theorem 7(5.5, p. 87 and 5.6, p. 89) (The CAC for the NRC element): The NRC satisfies the CAC to the same order as the BCP element, with the same requirements of smoothness on $\underset{\sim}{u}^*$, on appropriate domains and meshes. In particular, let Ω be a connected domain with the rectangular regularity property, and let $[\![\mathcal{T}^h]\!]$ be a collection of uniform triangulations of Ω. Let $S^h \subseteq W_0^{1,2}(\Omega) \times W_0^{1,2}(\Omega) = H_1$ be the velocity trial space constructed on $\mathcal{T}^h$ using the NRC element, and let $W^h \subseteq S^h$ be the subspace consisting of incompressible velocity fields. Then given a divergenceless $\underset{\sim}{u}^* \in [H_1 \bigcap W^{3,2}(\Omega) \times W^{3,2}(\Omega)]$, there exists a $\underset{\sim}{w}^h \in W^h$ satisfying eq. (26a), with a C independent of h.

In the light of Theorem 3, Theorem 7 establishes that the velocity estimate in Stokes flow on a domain satisfying the recatangular regularity property is $O(h)$, when uniform triangulations of NRC rectangles are employed, and the exact solution has at least three L_2 derivatives. This has been done without obtaining any estimate for the pressures. We turn our attention to the question of the pressures in the remainder of this paper.

5.3 Pressure Estimates without the Discrete LBB Condition

Let us assume that velocity estimates along the lines of the previous subsection have been obtained. Subtracting discrete and continuous weak equations as in the proof of Theorem 2 gives for all $\underset{\sim}{v}^h \in S^h$

$$a(\underset{\sim}{u}^h - \underset{\sim}{u}^*, \underset{\sim}{v}^h) = (p^* - p^h_{\underset{\sim}{w}}, \nabla \cdot \underset{\sim}{v}^h)_\circ = (p^* - \bar{p}^h + \bar{p}^h - p^h_{\underset{\sim}{w}}, \nabla \cdot \underset{\sim}{v}^h)_\circ$$

where, as before, $\bar{p}^h$ is the best $L_2(\Omega)$ approximation to p^* from $\mathcal{T}^h$. Thus

$$(p^h_{\underset{\sim}{w}} - \bar{p}^h, \nabla \cdot \underset{\sim}{v}^h)_\circ = (\bar{p}^h - p^*, \nabla \cdot \underset{\sim}{v}^h)_\circ - a(\underset{\sim}{u}^h - \underset{\sim}{u}^*, \underset{\sim}{v}^h) \tag{28}$$

Under usual assumptions of approximation accuracy, it suffices to show that $\| p^h_{\underset{\sim}{w}} - \bar{p}^h \|_0$ is small. This does not immediately follow from eq. (28) even though the right-hand side is small when $\| \underset{\sim}{v}^h \|_1 = 1$ [The right-hand side is indeed small because $\| \underset{\sim}{u}^h - \underset{\sim}{u}^* \|_1$ and $\| \bar{p}^h - p^* \|_0$ are small.] The problem is two-fold: First, $p^h_{\underset{\sim}{w}} - \bar{p}^h$ may have a component in U^h. The smallness of the right-hand side might then only show that this component is substantial, not that $\| p^h_{\underset{\sim}{w}} - \bar{p}^h \|_0$ is small. In ref. 14 this problem is addressed. There it is shown that with some reasonable assumptions there is an approximation $\bar{\bar{p}}^h \in [\![U^h]\!]^\perp \bigcap \mathcal{T}^h$ such that $\| p^* - \bar{\bar{p}}^h \|_0$ is as small as the maximum of $\| \underset{\sim}{u}^h_I - \underset{\sim}{u}^* \|_1$ and $\| \bar{p}^h - p^* \|_0$, where $\underset{\sim}{u}^h_I$ is the nodal interpolate to $\underset{\sim}{u}^*$. Thus proving that $\| p^h - \bar{\bar{p}}^h \|_0$ is small will suffice to produce a pressure estimate. Furthermore, since $(p^h - \bar{\bar{p}}^h, \nabla \cdot \underset{\sim}{v}^h)_\circ = 0$ for all $\underset{\sim}{v}^h \in S^h$, $\bar{\bar{p}}^h$ can be substituted for $\bar{p}^h$ in eq. (28).

Second, since the constant k_h of eq. (20) can evidently only be bounded by $k_h \geq C_1 h$ on many meshes, the best that follows from eq. (28) by taking the supremum of both sides with $\| \underset{\sim}{v}^h \|_1 = 1$ is

$$\| p^h_{\underset{\sim}{w}} - \bar{\bar{p}}^h \|_0 \leq \frac{1}{C_1 h}\left[C_2 \| \underset{\sim}{u}^h - \underset{\sim}{u}^* \|_1 + C_3 \| \bar{p}^h - p^* \|_0\right] \tag{29}$$

$\| \bar{p}^h - p^* \|_0$ is usually $\leq Ch\| p^* \|_1$. We believe that $\| \underset{\sim}{u}^h - \underset{\sim}{u}^* \|_1 \leq Ch\| \underset{\sim}{u}^* \|_2$ as discussed earlier. The best that eq. (29) gives then is that $\| p^h_{\underset{\sim}{w}} - \bar{\bar{p}}^h \|_0$ is bounded.

Fortunately, we have had good success with smoothing the pressure field by projection of the raw pressures computed on each subtriangle onto an auxiliary trial space [2,13,14,20,23,25]. In ref. 2 a simple and easy to implement projection method is described which uses L_2 projection of the raw pressures onto the conforming bilinear space based on NRC macro corner nodes. Unfortunately, we do not know any error estimate for that scheme to assure us that the possible lack of convergence pointed out by eq. (29) does not occur in practice. We have computational experience which seems to indicate that good results can be obtained using the projection method of ref. 2 [2,14]. We present the final theorem to illustrate that the idea of pressure projection can be well founded in theory, even though we prefer to use the scheme of ref. 2 for reasons of practicality. The final theorem is a composite of results from Chapter 7 of ref. 13.

Theorem 8: Let Ω be a rectangle and $[\![T^h]\!]$ be a sequence of uniform triangulations of Ω with 2^m rectangles per side for $m > 2$. Denote these rectangles by $[\![\Omega^e]\!]$. Let $S^h \times T^h$ be the FEM trial space constructed on T^h, using the crossed triangle. Let r_h denote the L_2 projection from T^h to the space of piecewise constant pressures based on 2×2 rectangular patches of Ω^e's. If p^h is the raw pressure solution to the Lagrange multiplier problem A from T^h, and ϵ is an arbitrary positive number, then the "filtered" pressure, $r_h p^h$, satisfies

$$\| r_h p^h - p^* \|_0 \leq C h^{1-\epsilon}(\| \underline{u}^* \|_3 + \| p^* \|_1) \tag{30}$$

6. CONCLUSIONS

We believe that the NRC crossed triangle element is an ideal element for steady flows of memory fluids. We believe that the theoretical results for linear problems given here provide a logical scenario, explaining why such elements can give convergent approximations when employed with care. The advantage provided by exact incompressibility of Lagrange multiplier FEM solutions is of fundamental importance to implementation of numerical procedures for memory fluids. These procedures also take much advantage of the fact that the velocity gradients are constant on each subtriangle. The several advantages just mentioned make the NRC element completely unique among finite elements for plane and axisymmetric incompressible flows. We believe that this uniqueness is ample justification for putting up with the added computational cost of the internal macro node. Looking beyond the application to non-Newtonian memory fluid problems, we believe that the NRC element will prove useful in many situations where efficient analytic construction of smooth streamlines is important. In fact, because of the BCP — NRC correspondence, an exactly incompressible NRC velocity field can easily be computed from every weakly incompressible BCP field, by assigning central nodal values as we have described and using the corner nodal values of the BCP field. This seems to show promise as a method by which to obtain smooth streamlines from results computed using the BCP element.

Acknowledgement: The research described in this paper was performed while both authors were at Illinois Institute of Technology, and was partially supported by N. S. F. Grant MCS-81-02089. The manuscript was prepared at the Mathematics Research Center, University of Wisconsin – Madison, where the first author holds a visiting position, and is sponsored by the United States Army under Contract No. DAAG29-80-C-0041.

7. REFERENCES

[1] M. Crochet and K. Walters, Numerical methods in non-Newtonian fluid mechanics, Ann. Rev. Fluid Mech. 15 (1983) 241.

[2] B. Bernstein, D. S. Malkus, and E. T. Olsen, A finite element for incompressible plane flows of fluids with memory, Int. J. Numer. Meths. Fluids, to appear.

[3] D. S. Malkus and B. Bernstein, Flow of a Curtiss-Bird fluid over a transverse slot using the finite element drift-function method, J. Non-Newtonian Fluid Mechs., to appear.

[4] B. Bernstein, M. K. Kadivar, and D. S. Malkus, Steady flows of memory fluids with finite elements: Two test problems, Comp. Maths. Appl. Mech. Eng. 27 (1981) 279–302.

[5] M Viriyayuthakorn and B. Caswell, Finite element simulation of viscoelastic flows, J. Nc Newtonian Fluid Mechs. 8 (1981) 245–267.

[6] H. Court, K. Walters and R. Davies, Long-range memory effects in flows involving abrupt changes in geometry. Part 4: Numerical simulation using integral rheological models, J. Non-Newtonian Fluid Mechs. 8 (1981) 95.

[7] O. Hassager, A Lagrangian finite element method for the simulation of flow of non-Newtonian liquids, J. Non-Newtonian Fluid Mechs. 12 (1983) 153.

[8] J. Nagtigaal, D. M. Parks, and J. R. Rice, On numerically accurate finite element solutions in the fully plastic range, Comp. Meths. Appl. Mech. Eng. 4 (1974) 153–178.

[9] B. Mericier, A conforming finite element method for two dimensional, incompressible elasticity, Int. J. Numer. Meths. Eng. 14 (1979) 942–945.

[10] J. H. Argyris, J. St. Doltsinis, W. C. Knudson, J. Szimmat, H. Wüstenberg, and K. Willam, Eulerian and Lagrangian techniques for elastic and inelastic large degormation processes, I.S.D. Report No. 256 (1976).

[11] F. Brezzi, On the existence, uniqueness and approximation of saddle point problems arising from Lagrangian multipliers, R.A.I.R.O Analyse Numerique 8 (1974) 129–151.

[12] I. Babuska and A. K. Aziz, Mathematical Foundations of the Finite Element Method (Academic Press, New York, 1972).

[13] E. T. Olsen, Stable finite elements for non-Newtonian flows: First order elements which fail the LBB condition, Ph. D. Thesis, Department of Mathematics, Illinois Institute of Technology, Chicago (1983).

[14] D. S. Malkus and E. T. Olsen, Obtaining error estimates for optimally constrained incompressible finite elements, Comp. Meths. Appl. Mech. Eng. 42 (1984).

[15] C. Curtiss and R. B. Bird, Kinetic theory for polymer melts. Parts I and II, J. Chem. Phys. 74 (1981) 2016–2033.

[16] P. Ciarlet, The Finite Element Method for Elliptic Problems (North-Holland, Amsterdam, 1978).

[17] D. S. Malkus, Penalty methods in finite element analysis of fluids and structures, Nuc. Eng. Design. 57 (1980) 441–448.

[18] D. S. Malkus and T. J. R. Hughes, Mixed finite element methods — reduced and selective integration techniques: A unification of concepts, Comp. Meths. Appl. Mech. Eng. 15 (1978) 63–81.

[19] A. Ralston and P. Rabinowitz, A First Course in Numerical Analysis (McGraw-Hill, New York, 1978).

[20] R. L. Sani, P. M. Gresho, R. L. Lee, D. F. Griffiths, and M. Engelman, The cause and cure (?) of spurious pressures generated by certain FEM solutions to the incompressible Navier-Stokes equations. Parts I and II, Int. J. Numer. Meths. Fluids 1 (1981) 17–43 (part I), 171–204 (part II).

[21] M. Fortin, An analysis of convergence of mixed finite element methods, R.A.I.R.O. Analyse Numerique 8 (1977) 341–254.

[22] D. S. Malkus, Eigenproblems associated with the discrete LBB condition for incompressible finite elements, Int. J. Eng. Sci. 19 (1981) 1299–1310.

[23] T. J. R. Hughes, W. K. Liu, and A. Brooks, Finite element analysis of incompressible viscous flows by the penalty function formulation, J. Comp. Phys. 30 (1779) 1–60.

[24] B. Mercier, Topics on finite element solution of elliptic problems (Tata Institute of Fundamental Research in Bombay Lecture Series, Springer-Verlag, Berlin, 1979).

[25] C. Johnson and J. Pitkäranta, Analysis of some mixed finite element methods related to reduced integration, Math. Comp. 38 (1982) 375–400.

Unification of Finite Element Methods
H. Kardestuncer (Editor)
© Elsevier Science Publishers B.V. (North-Holland), 1984

CHAPTER 11

THE NUMERICAL ANALYSIS OF NECKING INSTABILITIES

A. Needleman

Some issues arising in the finite element analysis
of necking instabilities are addressed within the
context of specific problems.   In addition to
analyses based on classical Mises type constitutive
laws, we discuss numerical solutions using
constitutive relations that more accurately model
plastic slip processes and progressive rupture on
the microscale. We illustrate the ability of
finite element solutions based on these
nonclassical constitutive relations to reproduce
essential features of the development of failure in
the necked down region.

1. INTRODUCTION

When a ductile metal is deformed past the maximum load point in tension,
the deformations localize into a neck-like region. Failure ultimately
occurs within this "neck," although the mode of failure is both material
and geometry dependent. "Necking" has also come to be used as a generic
term denoting any tensile instability that leads to localized thinning.
Necking instabilities play an important role in setting deformation and
stress patterns in material testing situations, in limiting ductility in
sheet forming operations and in setting macroscopic conditions for
ductile rupture. Hence, there is strong motivation for developing a
quantatative description of necking phenomena. Much progress toward this
end has been made in the past decade or so and numerical solutions have
played a major role in this development.

Here, no attempt is made to give an extensive review of the subject.
Instead, attention is directed toward a particular topic in order to
illustrate the progress that has been made and some of the unresolved
issues that remain. The topic that will serve as the focus for this
discussion is the difference in behavior exhibited by structural metals
in axisymmetric and plane strain tensile testing. Typical
phenomenologies that develop in these two tests are illustrated in
Fig. 1, from Speich and Spitzig [1]. In both tests, the deformations
remain essentially homogeneous up to the maximum load point after which a
diffuse neck develops. In the round bar tension test, Fig. 1a, diffuse
necking leads to a cup and cone type fracture. On the other hand, in the
plane strain tensile test of the same material, Fig. 1b, the deformation

mode shifts to one involving localized shearing while the diffuse neck is rather shallow. It is worth emphasizing that the specimens shown in Figs. 1a and 1b are made of the same material; the variation in geometry accounts for the difference in behavior.

Analyses of necking date back, at least, to the one dimensional model for the onset of necking due to Considere [2]. This analysis predicts that necking initiates at the maximum load point, in both axisymmetric and plane strain tension. An attractive feature of this one dimensional theory is that the only material property required as input is the uniaxial stress-strain curve. However, the development of the neck and the nonhomogeneous stress and deformation states that occur within it cannot be analyzed within this framework. These stress and strain distributions are needed in order to relate measured quantities to material properties. In the 1940's, Bridgman [3] developed his now classical approximate analysis which gives expressions for field quantities of interest across the plane of the neck. Bridgman's [3] analysis is for an ideally plastic solid and requires as input the neck curvature as a function of imposed strain. Also, the question of how the neck initiates is not addressed.

Since necking in tensile specimens is such a common and striking manifestation of finite strain plastic behavior, there has been considerable interest in predicting the observed phenomenology from basic principles of plasticity theory. This is not entirely straightforward, since it was demonstrated [4] that a necking type bifurcation is not possible in axisymmetric tension for a classical rigid-plastic solid. However, a necking bifurcation does occur in plane strain tension, Cowper and Onat [5]. Even though elastic strains are very small compared with plastic strains, it turns out that the rigid plastic framework is too restrictive for analyzing necking instabilities.

A general framework for analyzing finite strain elastic-plastic problems, including variational principles and uniqueness and bifurcation criteria is largely due to Hill [6,7]. This framework permits a unified treatment of plastic instabilities, including tensile necking as well as the more thoroughly explored problems of structural buckling. For both the axisymmetric and plane strain specimen geometries, the onset of a necking bifurcation is found to be delayed beyond the maximum load point, with the delay being greater for stubbier specimens, Cheng et al. [8], Needleman [9], Hutchinson and Miles [10] and Hill and Hutchinson [11]. For axisymmetric specimens the magnitude of the delay increases with the elastic shear modulus and becomes infinite in the rigid-plastic limit, [10].

The variational structure of the governing equations permits convenient application of finite element methods. In fact, one of the first applications of the finite element method to a finite strain plasticity problem was an analysis of neck development in an axisymmetric tensile bar [9]. This solution and other numerical solutions [12-14] for the stress and strain distributions in the neck make it possible to address issues such as the effect of strain hardening on neck development. Such numerical solutions also play a useful role in assessing conditions governing fracture initiation [15-18]. However, analyses which employ

classical elastic-plastic flow theory to characterize the material
behavior are incapable of modelling the tensile fracture shown in Fig. la.
Nevertheless, the successes of these numerical analyses played a useful
role in clarifying the finite strain (as opposed to material modelling)
issues arising in formulating constitutive relations for elastic-plastic
Mises-type solids. In fact, the formulation and use of numerical methods
for finite strain plasticity have advanced to the point where the
generation of solutions for neck development, based on classical flow
laws, is virtually routine.

Numerical solutions for neck development, using isotropically hardening
Mises-type plasticity theory, have also been carried out for the plane
strain tension test [19,20]. These plane strain calculations have found
continued growth of the diffuse neck without any tendency for the flow
pattern to shift to one involving the localized shearing shown in Fig. 1b.
This is a consequence of the constitutive assumption, since "material
instability" analyses reveal that the classical smooth yield surface
elastic-plastic solid is quite resistant to localized shearing, Rice [21].
Deviations from the classical smooth yield surface idealization do permit
shear bands to emerge at achievable strain levels. Yield surface vertex
effects, which arise from the discrete nature of crystalline slip, and the
dilational plastic flow induced by the nucleation and growth of micro-
voids are particularly significant in this regard, [21,22]. When finite
element analyses of neck development are carried out using constitutive
relations incorporating these effects, the numerical solutions reproduce
the observed phenomena in considerable detail [23-25].

We begin by outlining a Lagrangian convected coordinate formulation of the
governing equations that has been extensively used in analyzing finite
deformation plasticity problems. This formulation is readily adapted for
a variety of plastic constitutive relations. Next, plastic flow rules
that incorporate models of yield surface vertex effects and progressive
cavitation are discussed in addition to a finite deformation version of
the classical isotropic hardening Mises solid. A brief resume is then
given of some analytical results on necking and shear localization. The
analyses give information needed for an effective mesh design as well as
provide a perspective for interpreting the numerical results. Finally,
numerical studies of necking are reviewed with a focus on recent
calculations of the development of failure in the necked down region.

2. FIELD EQUATIONS

Relative to a fixed Cartesian frame, the position of a material point in
the initial configuration is denoted by $\underset{\sim}{x}$. In the current configuration
the material point initially at $\underset{\sim}{x}$ is at $\bar{\underset{\sim}{x}}$. The displacement vector $\underset{\sim}{u}$ and
deformation gradient $\underset{\sim}{F}$ are defined by

$$\underset{\sim}{u} = \bar{\underset{\sim}{x}} - \underset{\sim}{x} \qquad \underset{\sim}{F} = \frac{\partial \bar{\underset{\sim}{x}}}{\partial \underset{\sim}{x}} \qquad\qquad (2.1)$$

To express equilibrium, we first introduce the force transmitted across a
material element, $d\underset{\sim}{P}$, which can be related to either the symmetric Cauchy
stress tensor $\underset{\sim}{\sigma}$ or the unsymmetric nominal stress tensor $\underset{\sim}{n}$

$$d\underset{\sim}{P} = \overline{\underset{\sim}{\nu}}\cdot\underset{\sim}{\sigma}d\overline{S} = \underset{\sim}{\nu}\cdot\underset{\sim}{n}dS \tag{2.2}$$

where $d\overline{S}$ and $\overline{\underset{\sim}{\nu}}$ are the area and orientation in the current configuration
of a material element that had area, dS, and orientation, $\underset{\sim}{\nu}$, in the
initial configuration. It is convenient to introduce the Kirchhoff stress
defined by

$$\underset{\sim}{\tau} = \det(\underset{\sim}{F})\underset{\sim}{\sigma} \tag{2.3}$$

where $\det(\ )$ denotes the determinant.

Although general theoretical issues are conveniently addressed using a
coordinate free notation, actually carrying out a computation requires
dealing with the component form of the equations. A Lagrangian, convected
coordinate formulation of the governing equations, as described in Green
and Zerna [26] and Budiansky [27], has been extensively employed in finite
deformation plasticity analyses. Discussions of this formulation, with
particular focus on the finite element implementation are given by
Needleman [28] and Needleman and Tvergaard [29].

Convected coordinates, y^i, are introduced which serve as particle labels
and all field quantities are considered as functions of y^i and a time like
parameter t which is identified with a monotonically increasing quantity
that parameterizes the loading history. Covariant base vectors in the
initial configuration, $\underset{\sim}{g}_i$, and in the current configuration, $\overline{\underset{\sim}{g}}_i$, are given
by

$$\underset{\sim}{g}_i = \frac{\partial\underset{\sim}{x}}{\partial y^i} \qquad \overline{\underset{\sim}{g}}_i = \frac{\partial\overline{\underset{\sim}{x}}}{\partial y^i} \tag{2.4}$$

so that from (2.1)

$$\overline{\underset{\sim}{g}}_i = \underset{\sim}{F}\cdot\underset{\sim}{g}_i \tag{2.5}$$

The metric tensor in the initial configuration is $g_{ij} = \underset{\sim}{g}_i\cdot\underset{\sim}{g}_j$ and in the
current configuration is $\overline{g}_{ij} = \overline{\underset{\sim}{g}}_i\cdot\overline{\underset{\sim}{g}}_j$, with inverses g^{ij} and $\overline{g}^{ij}$,
respectively. The reciprocal (or contravariant) base vectors are given by

$$\underset{\sim}{g}^i = g^{ij}\underset{\sim}{g}_j \qquad \overline{\underset{\sim}{g}}^i = \overline{g}^{ij}\overline{\underset{\sim}{g}}_j \tag{2.6}$$

Components of vectors and tensors on the convected coordinates are
obtained simply via "dot" products with the appropriate base vectors, for
example the components of $\underset{\sim}{n}$ on the initial base vectors are $n^{ij} = \underset{\sim}{g}^i\cdot\underset{\sim}{n}\cdot\underset{\sim}{g}^j$

while the components of $\underset{\sim}{\tau}$ on the current base vectors are $\tau^{ij} = \underset{\sim}{g}_i \cdot \underset{\sim}{\tau} \cdot \underset{\sim}{g}_j$. These components are related by

$$n^{ij} = \tau^{ik} F^{j}_{.k} \tag{2.7}$$

With body forces neglected and given that the body is in equilibrium at t, equilibrium at t+dt requires the incremental principle of virtual work to be satisfied.

$$\int_V \dot{n}^{ij} \delta \dot{F}_{ji} \, dV = \int_S \dot{T}^i \delta \dot{u}_i \, dS \tag{2.8}$$

where $\underset{\sim}{T} = \underset{\sim}{\nu} \cdot \underset{\sim}{n}$ and $(\dot{}) = d()/dt$ at fixed y^i.

For the rate independent elastic-plastic solids considered here, the constitutive relations are specified as a piecewise linear relation between the Jaumann derivative of Kirchhoff stress, $\overset{\Delta}{\underset{\sim}{\tau}}$, and the rate of deformation tensor, $\underset{\sim}{d}$ (= the symmetric part of $\dot{\underset{\sim}{F}} \cdot \underset{\sim}{F}^{-1}$), so that we write

$$\overset{\Delta}{\underset{\sim}{\tau}} = \underset{\sim}{C} : \underset{\sim}{d} \tag{2.9}$$

here : denotes the dyadic product, i.e. the component form of $\underset{\sim}{C} : \underset{\sim}{d}$ is $c^{ijk\ell} d_{\ell k}$.

Since $\dot{\eta}_{ij} = \overline{\underset{\sim}{g}}_i \cdot \underset{\sim}{d} \cdot \overline{\underset{\sim}{g}}_j$, where $\dot{\eta}_{ij}$ are the components of Lagrangian strain rate on the undeformed base vectors, the component form of (2.9) on the current base vectors is

$$\begin{rcases} \overset{\Delta}{\tau}{}^{ij} = C^{ijk\ell} \dot{\eta}_{k\ell} \\[2ex] \dot{\eta}_{k\ell} = \frac{1}{2}\left(\dot{u}_{k,\ell} + \dot{u}_{\ell,k} + u^{m}_{,k} \dot{u}_{m,\ell} + u^{m}_{,k} \dot{u}_{m,\ell} \right) \end{rcases} \tag{2.10}$$

For use in the incremental principle of virtual work (2.8) we need the relation between $\dot{n}^{ij}$ (which are the contravariant components of the nominal stress rate on the initial undeformed base vectors) and the contravariant components of $\overset{\Delta}{\underset{\sim}{\tau}}$ on the current deformed base vectors. This relation is conveniently specified in two steps. First,

$$\overset{\Delta}{\tau}{}^{ij} = \dot{\tau}^{ij} + \overline{g}^{ik} \tau^{j\ell} \dot{\eta}_{k\ell} + \overline{g}^{j\ell} \tau^{ik} \dot{\eta}_{k\ell} \tag{2.11}$$

then from the incremental form of (2.7)

$$\dot{n}^{ij} = \dot{\tau}^{ik} F^{j}_{\cdot k} + \tau^{ik} \dot{F}^{j}_{\cdot k} \tag{2.12}$$

Now we can write

$$\dot{n}^{ij} = K^{ijk\ell} \dot{F}_{\ell k} \tag{2.13}$$

where

$$K^{ijk\ell} = L^{mink} F^{\ell}_{\cdot n} F^{j}_{\cdot m} + \tau^{ik} g^{j\ell} \tag{2.14}$$

with

$$L^{ijk\ell} = C^{ijk\ell} - \frac{1}{2}\left[\overline{g}^{ik} \tau^{j\ell} + \overline{g}^{jk} \tau^{i\ell} + \overline{g}^{i\ell} \tau^{jk} + \overline{g}^{j\ell} \tau^{ik} \right] \tag{2.15}$$

The only relation that changes with material model is (2.9). The
relations (2.14) and (2.15) are model independent and can be incorporated
into a finite element code in a modular fashion. We also note that the
components of nominal stress and deformation gradient appearing in (2.8)
are defined on the undeformed coordinate net which is chosen for
convenience, e.g. as a Cartesian system or as a cylindrical coordinate
system for axisymmetric problems.

If the incremental moduli (2.9) posess the symmetries $C^{ijk\ell} = C^{k\ell ij}$, then
for a standard boundary value problem, prescribed traction rates $\dot{T}$ on S_T
and displacement rates $\dot{u}$ on S_u, the following variational principle holds
[6,7]: Among all displacement rate fields that satisfy the displacement
rate boundary conditions, for any actual field

$$\left.\begin{aligned}
\delta I &= 0 \\[1em]
I &= \frac{1}{2} \int_V K^{ijk\ell} \dot{F}_{ji} \dot{F}_{\ell k} dV - \int_{S_T} \dot{T}^{i} \dot{u}_{i} dS
\end{aligned}\right\} \tag{2.16}$$

3.   CONSTITUTIVE RELATIONS

Here, we discuss a finite strain formulation of the constitutive relations
for an isotropically hardening Mises solid and generalizations of this
relation that permit investigation of (i) yield surface vertex effects as
predicted by physical models of the crystalline slip process and (ii) the
dilational and pressure sensitive plastic flow due to the nucleation and
growth of micro-voids. We confine attention to isothermal, rate
insensitive and isotropic material response. We also assume that elastic
strains remain small. For structural metals this is quite an approprite
simplification, except possibly when the hydrostatic stress is very large.

We aim here only at pointing out features of these material models of main significance for the analysis of necking instabilities. More complete descriptions can be found in the references cited. Also, Tvergaard [30] has reviewed issues arising in the implementation of these constitutive relations in finite element analyses.

3.1 Classical Smooth Yield Surface Plasticity Theory

The elastic-plastic constitutive relation most commonly used in finite element analyses is the isotropically hardening Mises solid, also called J_2 flow theory or Prandtl-Reuss theory. For infinitesimal strains the flow rule takes the form

$$\dot{\underset{\sim}{\varepsilon}}^{P} = \frac{3}{2}\left[\frac{1}{E_t} - \frac{1}{E}\right]\frac{\dot{\sigma}_e}{\sigma_e}\underset{\sim}{s} \tag{3.1}$$

where $\underset{\sim}{s}$ is the stress deviator, σ_e is the Mises effective stress, E is Young's modulus and E_t is the tangent modulus, the slope of the uniaxial effective stress-effective strain curve. Of course, (3.1) is understood to apply only when the current state is at yield and the imposed deformations are such as to enforce continued plastic deformation. The total strain rate is then written as the sum of the plastic strain rate rate from (3.1) and the elastic strain rate from the incremental version of Hooke's law.

Several issues arose in formulating this constitutive relation for finite deformations, including:

(i) the choice of appropriate stress and strain measures to use in (3.1). Note that the question of an invariant stress rate does not arise in relation to the plastic flow rule since the only rate that appears is the "time" derivative of a scalar. For finite deformations, the plastic strain rate in (3.1) is identified with the plastic part of the rate of deformation tensor d and the stress deviator and effective stress can be based on either the Cauchy stress or the Kirchhoff stress, Budiansky [31], Hutchinson [32]. The formulations are physically equivalent for small elastic strains and the one based on Kirchhoff stress is preferred since it gives a symmetric tensor of moduli $\underset{\sim}{C}$ in (2.9) and hence leads to the variational structure (2.16). This in turn has the computational advantage of a symmetric finite element stiffness matrix. With this choice of variables the flow rule takes the form

$$\left.\begin{aligned}
\underset{\sim}{d}^{P} &= \frac{3}{2}\left[\frac{1}{E_t} - \frac{1}{E}\right]\frac{\dot{\sigma}_e}{\sigma_e}\underset{\sim}{s} \\[2em]
\underset{\sim}{s} &= \underset{\sim}{\tau} - \frac{1}{3}(\underset{\sim}{\tau}:\underset{\sim}{I})\underset{\sim}{I} \\[2em]
\sigma_e^2 &= \frac{3}{2}\underset{\sim}{s}:\underset{\sim}{s}
\end{aligned}\right\} \tag{3.2}$$

where $\underset{\sim}{I}:\underset{\sim}{\tau}$ is τ^k_k.

(ii) the appropriate finite strain generalization of the incremental
statement of Hooke's law. While it is possible to use a fully consistent
finite elasticity formulation for the elastic part of the strain rate, it
turns out that this leads to an awkward formulation of the incremental
relations. A convenient approximation involves using the linear hypo-
elastic relation [31,32]

$$\underset{\sim}{d}^e = \frac{1+\nu}{E} \overset{\Delta}{\underset{\sim}{\tau}} - \frac{\nu}{E} (\overset{\Delta}{\underset{\sim}{\tau}} : \underset{\sim}{I}) \underset{\sim}{I} \qquad\qquad (3.3)$$

As long the stresses remain small compared to Young's modulus, which is
equivalent to small elastic strains, (3.3) is an acceptable approximation.
It is (3.3) that involves a choice of stress rate, other choices than the
Jaumann rate could be used but these would lead to a less convenient
formulation. Again we emphasize that there is no question concerning a
choice of stress rate with regard to the plastic flow rule (3.2). On the
other hand for plastic constitutive relations that have a tensor internal
variable, as for example kinematic hardening (the tensor internal variable
being the yield surface center), there is a choice of rate in writing the
evolution equation for the internal variable. In this context, the
question of choice of stress rate is part of a larger issue concerning
appropriate evolution laws for tensor internal variables.

(iii) at various times the appropriateness of summing the elastic and
plastic strain rates has been questioned. A key distinction here is
between the rate of accumulation of plastic strain, where by plastic
strain is meant the residual strain remaining on unloading, and the
plastic part of the rate of deformation, i.e. the plastic part of the
strain increment from the current loaded state. Writing a flow rule for
the former quantity does not, in general, lead to an additative relation,
expressing the flow rule in terms of the latter, as in (3.2), does.

Writing the rate of deformation, $\underset{\sim}{d}$, as the sum of (3.2) and (3.3) and
inverting gives a relation in the form (2.10) for which the tensor of
moduli have the symmetries $C^{ijk\ell} = C^{k\ell ij}$, required for the variational
formulation.

Basic physical assumptions underlying this constitutive relation are that
the plastic strain rate that has a direction normal to the yield surface
in stress space and, furthermore, that this surface is smooth, with a
unique normal at the current stress point. Also, the plastic strain rate
is volume preserving and pressure insensitive. These features have a
strong influence on the predicted course of neck development. Therefore,
we consider constitutive relations which incorporate, in a physically
appropriate way, deviations from these attributes.

3.2 J_2 Corner Theory

A consequence of deriving the flow rule (3.2) from a smooth yield surface
is that the plastic strain rate direction is not influenced by the stress
rate. This is at variance with predictions of physical plasticity models
for polycrystalline aggregates based the concept of single crystal
slip. Such models lead inevitably to the prediction of a yield surface

corner at the current loading point [33,34], with the plastic strain rate
direction depending, within limits, on the stress rate. Direct
experimental evidence for corners is conflicting, although in some cases
there is evidence for a region of high curvature at the current loading
point [35]. A recent analysis by Pan and Rice [36] shows that slight
material rate sensitivity can account for the experimental ambiguity.

The main significance of a yield surface corner for bifurcation phenomena
lies in the softer response to an abrupt change of loading path that
occurs when the plastic strain rate can follow the stress rate.
Structural buckling predictions based on J_2 deformation theory of
plasticity often give much better agreement with experimental results than
predictions based on J_2 flow theory [37]. Very early, Batdorf [38] noted
that bifurcation predictions of deformation theory from pre-bifurcation
states arrived at via proportional loading could be justified by appealing
to the presence of a corner on the yield surface. In the post-bifurcation
regime strongly non-proportional loading takes place and use of J_2
deformation theory, which is a path independent non-linear elastic
constitutive relation, as a plasticity theory is not acceptable.

The J_2 corner theory of Christoffersen and Hutchinson [39] is an
analytically tractable phenomenological theory of plasticity that
incorporates key features exhibited by physical models of polycrystalline
aggregates. In J_2 corner theory the moduli for nearly proportional
loading are taken to be those of a finite strain deformation theory solid.
For increasing deviation from proportional loading the moduli stiffen
monotonically until they coincide with the linear elastic moduli for
stress rates directed along or within the yield surface.

The yield surface in the neighborhood of the loading point is taken to be
a cone in stress deviator space with the cone axis in the direction

$$\lambda = \frac{s}{[s:M:s]^{1/2}} \tag{3.4}$$

Here, M are the J_2 deformation theory plastic compliances, $d^P = M:\overset{\Delta}{\tau}$ and s
is the Kirchhoff stress deviator from (3.2). The positive angular measure
of the stress rate relative to the cone axis is defined by

$$\cos\theta = \frac{\lambda:M:\overset{\Delta}{s}}{[\overset{\Delta}{s}:M:\overset{\Delta}{s}]^{1/2}} \tag{3.5}$$

At the corner the plastic potential is given by

$$w^P = \frac{1}{2}\, f(\theta)\, \overset{\Delta}{s}:M:\overset{\Delta}{s} \tag{3.6}$$

and the plastic strain rate is obtained from the gradient of the potential
as

$$\underset{\sim}{d}^P = \frac{\partial^2 W^P}{\partial \underset{\sim}{\overset{\Delta}{s}} \partial \underset{\sim}{\overset{\Delta}{s}}} : \underset{\sim}{\overset{\Delta}{s}} \tag{3.7}$$

Combining (3.3) and (3.7) gives the total incremental compliance tensor. Then, inverting gives the incremental moduli $\underset{\sim}{C}$ to be used in (2.9). The incremental stiffnesses depend on the angle θ defined in (3.5), with $f(\theta) = 1$ in the total loading regime ("softest response") and decreasing monotonically to zero for elastic unloading ("stiffest response"). In the finite element calculations to be reviewed subsequently the transition function that has been used is the one found in [39] to give behavior in good agreement with predictions based on the physical model [34]. The interested reader is referred to [40], as well as to the original paper of Christoffersen and Hutchinson [39], for a complete specification of J_2 corner theory. This theory is more involved to implement in finite element program than is the constitutive relation for a Mises solid, but a relatively straightforward procedures has proved effective in practice [30,40].

The features of this constitutive relation of greatest significance for neck development are that within the total loading regime the response to a change in loading path is much softer than for the Mises solid and that as the loading path leaves the total loading regime the response becomes stiffer.

3.3 Continuum Model of a Ductile Porous Solid

Based on an approximate rigid-plastic analysis of a single spherical void, Gurson [41,42] proposed a yield function for a solid with a randomly distributed volume fraction, f, of the form

$$\phi = \frac{\sigma_e^2}{\bar{\sigma}^2} + 2fq_1\cosh\left(\frac{\underset{\sim}{\sigma} : \underset{\sim}{I}}{2\bar{\sigma}}\right) - \left(1+q_1^2 f^2\right) = 0 \tag{3.8}$$

where $\underset{\sim}{s}$ is the current matrix flow strength, σ_e is the macroscopic effective stress, defined in terms of the Cauchy stress and q_1 is an additional parameter introduced by Tvergaard [43,44] from comparisons of predictions of this model with full field numerical results. A further modification of the yield function by Tvergaard [45] and Tvergaard and Needleman [25] accounts for the loss of load carrying capacity associated with void coalescence at void spacings of the order of the void radius. This modification introduces additional parameters to specify the coalescence rate and is necessary to account for complete loss of stress carrying capacity at realistic void volume fractions. However, even with this modification the flow rule maintains the structure described here.

The matrix material is characterized as a Mises solid (when f = 0 (3.8) reduces to the Mises yield surface) and the matrix effective plastic strain rate and effective stress rate are related by a uniaxial stress-strain relation. These quantities are related to macroscopic quantities via the equivalence of macroscopic and microscopic plastic work

$$\dot{\bar{\varepsilon}} = \left(\frac{1}{E_t} - \frac{1}{E}\right)\dot{\bar{\sigma}} \qquad \underset{\sim}{\sigma}:\underset{\sim}{d}^P = (1-f)\bar{\sigma}\,\dot{\bar{\varepsilon}} \qquad\qquad (3.9)$$

The increase in the void volume fraction, f, arises partly from the growth of existing voids and partly from the nucleation of new voids, so that

$$\dot{f} = \dot{f}_{growth} + \dot{f}_{nucleation} \qquad\qquad (3.10)$$

Since the matrix material is plastically incompressible

$$\dot{f}_{growth} = (1-f)\underset{\sim}{I}:\underset{\sim}{d}^P \qquad\qquad (3.11)$$

One nucleation criterion, suggested by Gurson's [41,42] analysis of data obtained by Gurland [47], is

$$\dot{f}_{nucleation} = A\dot{\bar{\varepsilon}} \qquad\qquad (3.12)$$

using $\underset{\sim}{d}^P = \dot{\Lambda}\partial\phi/\partial\underset{\sim}{\sigma}$ together with (3.8) through (3.12) and the consistency condition $\dot{\phi} = 0$ during plastic loading, gives a flow rule of the form

$$\underset{\sim}{d}^P = \frac{1}{H}\,\underset{\sim}{P}(\underset{\sim}{P}:\overset{\triangle}{\underset{\sim}{\sigma}}) \qquad\qquad (3.13)$$

where explicit expressions for the terms in (3.13) are given in the references cited and in [46,48]. Here, we note that plastic flow is dilational, with possibly large plastic volume changes, and pressure sensitive due to the presence of the micro-voids. Furthermore, although the matrix material continues to harden, the macroscopic hardening H in (3.13) can become negative due to the weakening effects of void nucleation and growth.

The flow rule (3.13) is necessarily phrased in terms of Cauchy stress and not Kirchhoff stress (the Cauchy stress appears from averaging over the plastically incompressible matrix). In this case an elastic strain rate expression of the form (3.3) is used but is expressed in terms of Cauchy stress quantities. Although the plastic strain rate in (3.13) has a direction normal to the yield surface in Cauchy stress space it is not normal to the yield surface in Kirchhoff stress space. Hence, the tensor of incremental moduli, $\underset{\sim}{C}$ in (2.9), do not satisfy the symmetry conditions necessary for the variational formulation to hold and the resulting finite element stiffness matrix is unsymmetric.

4. BIFURCATION ANALYSES

We consider a uniform tensile specimen subject to a prescribed axial displacement at ends that remain shear free. Also, the lateral faces of the specimen are taken to be traction free, except, of course, for the normal tractions required to enforce the plane strain constraint in the plane strain tension specimen. For these boundary conditions one possible solution is a homogeneous one. When the elastic-plastic constitutive relations admit the variational formulation (2.16), bifurcation from this homogeneous state of deformation can be addressed within the framework laid out by Hill [6,7]. A central role is played in Hill's theory by the concept of a linear comparison solid. For the loading histories envisioned here, the linear comparison solid has, at each material point, moduli equal to the plastic loading moduli. The onset of bifurcation for the linear comparison solid is governed by a conventional eigenvalue problem and then this eigenmode can be related to the bifurcation mode of the actual elastic-plastic solid [37]. When the variational structure is absent, as is the case for the porous plastic solid, there is no analogous framework for addressing questions of bifurcation and uniqueness, although steps toward developing such a framework have been taken [49].

Here, we first present a one-dimensional Considere type argument for the onset of necking. Next, we summarize results on diffuse necking in plane strain and axisymmetric tension which exhibit the delay in the onset of necking due to three dimensional effects. Finally, we mention some results on shear localization for the material models discussed in Section 3. Shear localizations play a central role in determining the failure behaviors shown in Fig. 1.

4.1   One-Dimensional Analysis

The Considere treatment of necking can be viewed as a one dimensional bifurcation (and imperfection) analysis. To do this, focus attention on two cross-sections, A and B, of a tensile bar. Cross section B is the minimum section and cross section A is far from the neck. Equilibrium requires

$$\sigma_A A_A = \sigma_B A_B \tag{4.1}$$

where σ is the average true stress, A is the cross-sectional area and quantities associated with each cross-section are denoted by the appropriate subscript.

Writing (4.1) in incremental form and neglecting elastic volume changes (so that cross-section area changes can be simply related to length changes), gives

$$A_B(K_B^t - \sigma_B)\dot{\varepsilon}_B = A_A(K_A^t - \sigma_A)\dot{\varepsilon}_A \tag{4.2}$$

with ε the logarithimic tensile strain and K^t the slope of the σ-ε curve; $K^t = E_t$ for axisymmetric tension and $K^t = 4E_t/3$ for plane strain tension.

For a homogeneous bar (4.2) reduces to

$$A_A(K_A^t - \sigma_A)(\dot{\varepsilon}_B - \dot{\varepsilon}_A) = 0 \tag{4.3}$$

Hence, a bifurcation is possible when

$$\sigma_A = K_A^t \tag{4.4}$$

This is, of course, the classical Considere result which can be phrased as stating that the necking bifurcation point and the maximum load point coincide.

If A_A and A_B differ initially or if grip conditions give rise to inhomogeneous straining, then an incremental solution of (4.2) is needed to calculate the evolution of the neck. The solution continues until

$$\sigma_B = K_B^t \tag{4.5}$$

at which point $d\varepsilon_B/d\varepsilon_A \to \infty$; unloading begins at cross-section A and this marks the onset of necking. The rate at which the neck subsequently develops in the neck is not determined by this analysis. Also, the spatial extent of the neck is determined by factors not incorporated into the one dimensional model.

4.2 Diffuse Necking

Accounting for the actual three dimensional necking deformations delays the onset of bifurcation to a point subsequent to the attainment of the maximum load. In order to carry out a three dimensional analysis a choice of plasticity theory has to be made and the delay between the maximum load point and the bifurcation point can be constitutive relation as well as geometry dependent.

Analytical bifurcation analyses have been carried out for the special case of incompressible elastic as well as plastic material behavior. The small effect of elastic compressibility is completely negligible. Although these bifurcation results do not directly pertain to the porous plastic solid, for structural metals, void volume fractions prior to the onset of necking are generally small, and the onset of necking in these circumstances is well represented by the analysis based on incompressible material behavior.

Necking type bifurcation modes can be written in the form

$$\dot{u}_i = g_i \cos(m\pi z/L + \psi_i) \tag{4.6}$$

where z is the direction of the tensile axis, m is a wave number, ψ_i is 0 or $\pi/2$, L is the current specimen length and the functions g_i have as independent variables coordinates in the cross-sectional area. The critical mode is the longest wavelength mode consistent with the boundary conditions [10,11]. When symmetry about the mid-plane of the specimen is required this gives the familiar hour glass shaped necking mode [10,11], m = 2 in (4.6).

Hutchinson and Miles [10] analyzed the onset of bifurcation for a class of incompressible elastic-plastic solids with smooth yield surfaces. For a Mises solid, the critical bifurcation stress is

$$\frac{\sigma}{E_t} = 1 + \frac{(2\pi)^2}{8} \left(\frac{R}{L}\right)^2 + \frac{(2\pi)^4}{192} \left(\frac{R}{L}\right)^4 \frac{G}{E_t} + \ldots \qquad (4.7)$$

asymptotically as $R/L \to 0$. Here, R is the current tensile bar radius and G is the elastic shear modulus.

As $G/E_t \to \infty$, the onset of necking is delayed indefinitely. This is the rigid plastic limit. For finite G/E_t, bifurcation is delayed only slightly beyond the maximum load point if the bar is sufficiently slender. For representative material properties and reasonably slender specimens, say $R/L = 4$, the necking bifurcation occurs at an imposed axial strain 10-20 percent greater than the imposed axial strain at maximum load, as found from finite element bifurcation results [9]. For stubby specimens the delay can be considerably greater [9,10]. The appearance of the elastic shear modulus in (4.7) is a consequence of using a smooth yield surface plasticity theory. As discussed by Hutchinson and Miles [10], a corner theory of plasticity leads to a vertex reduced shear modulus appearing in (4.7); for J_2 corner theory G is replaced by the secant shear modulus. Except in limiting cases, stubby specimens or nearly rigid-plastic behavior, the predicted onset of necking is not very much affected.

A similar eigenvalue problem can be formulated for a plane strain rectangular slab. The asymptotic expression analogous to (4.7) is, [11],

$$\frac{\sigma}{(4E_t/3)} = 1 + \frac{(2\pi)^2}{3} \left(\frac{w}{L}\right)^2 + \frac{7(2\pi)^4}{45} \left(\frac{w}{L}\right)^4 + \ldots \qquad (4.8)$$

where w is the current specimen width in the plane of deformation.

The elastic shear modulus does not appear in (4.7) and there is a finite bifurcation stress for rigid-plastic solids [5,11]. However, in addition to the long wavelength necking mode, short wavelength modes (larger values of m in (4.6)) are available. The critical conditions for bifurcation into these modes are quite constitutive model sensitive, with short wavelength modes essentially ruled out for the Mises type solid [11].

For elastic-plastic solids, the one dimensional result emerges from both (4.7) and (4.8) as the limit for infinitely long specimens. For finite length specimens, the onset of the long wavelength diffuse necking mode is delayed beyond the maximum load point and, for representative specimen geometries and material properties, the magnitude of this delay is not very constitutive model sensitive.

4.3 Localized Necking

Localized necking in the form of a shear band can be analyzed within a framework that regards localization as a "material instability," where a material element is considered to be subject to prescribed all around

displacements that are consistent with a homogeneous deformation.
Deformations in a localized band are permitted provided the velocity field
remains continuous and continuing equilibrium at the band interface is
satisfied, Hill [50], Rice [21]. The effects of initial imperfections on
localization can also be analyzed within this material instability
framework, Rice [21].

An elastic-plastic solid with a smooth yield surface, such as a Mises
solid, is quite resistant to localization. In plane strain a nearly
ideally plastic state is required, exactly ideally plastic in the rigid-
plastic limit, while in axisymmetric tension the strain hardening rate
must be strongly negative for a localization bifurcation to be possible
[21]. Yield surface vertex effects, as modeled by J_2 corner theory [39]
and the dilational and pressure sensitive plastic flow incorporated in
Gurson's [41,42] model (when small initial imperfections are accounted
for, [48]), lead to shear localization at strain levels achieved in the
neck of tensile specimens. However, the greater resistance to shear
localization under axisymmetric conditions than under plane strain
conditions remains [21,22]. This is illustrated by some simple formulas
of Needleman and Rice [22] for shear band bifurcations in a finite strain
J_2 deformation theory solid. These shear band bifurcation results pertain
to a J_2 corner theory solid, since the J_2 deformation theory moduli are
those of the linear comparison solid for a J_2 corner theory solid.

For a material with a power hardening effective stress versus effective
logarithmic strain relation of the form

$$\sigma_e = K\varepsilon_e^N \tag{4.9}$$

the critical axial tensile strains for the onset of shear bands are given
by

$$\varepsilon_{crit} = \sqrt{N(1-N)} \qquad \text{plane strain tension } N \leq 1/2$$

$$\varepsilon_{crit} = \sqrt{(1+3N)(1-N)/3} \quad \text{axisymmetric tension } N \leq 1/3 \tag{4.10}$$

With $N = 0.1$ the critical strain in plane strain tension is 0.30 and the
critical strain in axisymmetric tension is 0.62. The finite strain
generalization of J_2 deformation theory used in [22] is not unique; an
alternative finite strain J_2 deformation theory can be constructed [51].
This alternative deformation theory has the advantage of being a finite
strain hyper-elastic solid but its bifurcation formulas do not take on as
simple a form as (4.10). In any case, for $N = 0.1$ the shear band
bifurcation predictions obtained from these two formulations differ
little.

The vertex theory model attributes the onset of localization to an
inherent feature of the plastic flow process while the porous plastic
solid model ties shear localization to the factors responsible for the
initiation of ductile rupture. Clearly, the ductile failures in Fig. 1
involve both intense shearing and void nucleation and growth within the
shear band. What is not clear from looking at Fig. 1b, however, is
whether localized shearing occurred first, leading to large plastic

strains which then induce void nucleation and growth or whether the micro-
rupture process itself precipitated the observed localization. While not
as strikingly evident from the figure, localized shearing also plays a
prominent role in the cup-cone fracture process depicted in Fig. 1a.

When shear bands develop from nonhomogeneous deformation states, as in the
neck of a tension specimen, the analytical procedures referred to above
are not directly applicable. The onset of localization and its initial
direction of propagation can be determined by a material instability
analysis [52]; but the determination of shear band propagation and of the
varying shear intensity within the band involves a full boundary problem
solution. This generally requires a large scale numerical calculation and
accurate numerical analyses of localized shearing require proper mesh
design. As described by Tvergaard et al. [23], knowledge of the
preferred shear band orientation can be used to design a mesh that is
capable of resolving narrow bands.

5. NUMERICAL ANALYSES OF NECKING

Chen [53] published the first analysis of neck development in an
axisymmetric tensile specimen. He initiated neck development by
specifying an initial reduced cross sectional area and used a Kantorovich
method in conjunction with the principle of virtual work (2.8) to solve
for field quantities at each stage of the loading history. The material
was characterized by the finite strain Mises solid described in
Section 3.1. Needleman [9] used the same formulation of the field and
constitutive equations as Chen [53] but employed the finite element method
for the spatial discretization. Also the effect of grip conditions on
neck development was studied in [9]. Subsequently, Norris et al. [13] and
Saje [14] have used various finite difference formulations to study neck
development.

Before discussing the results of numerical analyses of necking, a few
comments will be made on finite element methods for this class of
problems. Although diffuse necking can be analyzed using a Mises
constitutive description, analyzing the failure phenomena shown in Fig. 1
requires incorporating a material description into the analysis that
models microstructural features, such as yield surface vertices and/or
void nucleation and growth. The algorithims employed need to be flexible
enough to accommodate models of these phenomena. Also, since the systems
being analyzed are marginally stable physically, the numerical methods
should neither introduce artificial numerical instabilities nor mask any
actual physical instabilities that develop. Hence, it should not be
surprising that relatively simple methods have had the greatest success in
this class of problems; a spatial discretization using the displacement
finite element method with a grid of linear displacement triangular
elements and simple incremental "time" integration procedures. It is
important for the triangular elements to be arranged in quadrilaterals
consisting of four triangles formed by the quadrilateral diagonals. Such
a grid can accommodate the nearly isochoric deformations characteristic of
plastic deformation (when void nucleation and growth are absent),
Nagtegaal et al. [54]. Also, such a mesh, when designed in light of
critical shear band orientations, is well suited for resolving localized
deformation in a narrow band [23-25].

Some of the results from [9] will now be described. Two different types
of boundary condition were used. An axial displacement is prescribed at
the ends of the specimen. In one case, shear free end conditions are
prescribed so that the deformations remain homogeneous until a bifurcation
is encountered. In the other case, the ends are taken to be cemented to
rigid grips, prohibiting tangential displacements at the ends and giving
inhomogeneous deformations from the outset. In both cases axisymmetric
deformations are assumed and furthermore symmetry about the mid-plane of
the specimen is required.

The uniaxial response is specified as a power law of the form (4.9) for
stresses in excess of the yield strength. Computations were carried out
for N = 0.125 and for an initial specimen length to radius ratio of 4.
The load versus end-displacement curve for the bar with shear free ends is
shown in Fig. 2. Bifurcation takes place at an end-displacement about
10 percent greater than the end-displacement at maximum load. The onset
of bifurcation was determined by a numerical implementation of Hill's
[6,7] bifurcation theory [9].

The development of the neck is shown in Fig. 3 for both sets of end
conditions. This development is measured either by the radial
displacement at the neck or by the ratio of current to initial neck cross-
sectional area. Such a curve clearly reveals the course of neck
development and the same type of plot has subsequently been used by Burke
and Nix [20] and Argyris et al. [55]. When considered as a function of
imposed displacement, as in Fig. 3, the area of the neck decreases more
rapidly for the rigid grip end condition. However, the area reduction
versus total load relation is nearly the same for both sets of end
conditions [9]. In fact, at a given value of area reduction, the stress
and strain distributions at the neck are nearly independent of end
condition, once necking is well underway [9].

The strain distributions across the neck that emerge from the numerical
calculations [9,13,14,53] are in good agreement. The stress distribution
obtained by Chen [53] develops a peak stress off the neck axis and is not
reliable in the latter stages of necking. The distribution obtained by
Needleman [9] has a somewhat sharper peak on the axis than found by
subsequent investigators, but otherwise the level and distribution of
stresses in [9,13,14] are in good agreement. Of particular interest is
the fact that the calculations all indicate that the Bridgman [3] solution
underestimates the hydrostatic stress in the neck. This is of particular
interest in relating conditions in the neck of a tensile specimen to the
onset of ductile rupture.

Once ductile hole growth begins, however, the assumptions underlying an
analysis based on a Mises type solid are no longer applicable. Quite
recently, Tvergaard and Needleman [25] have carried out an analysis of the
axisymmetric tensile test based on Gurson's [41,42] model of a porous
plastic solid, modified so as to account for a complete loss of stress
carrying capacity at a realistic void volume fraction. The matrix
material was taken to follow a power hardening relation of the form (4.9)
with N = 0.10 and void nucleation was plastic strain controlled as in
(3.12).

The results in [25] reproduce the essential features of the cup-cone
fracture shown in Fig. 1a. A crack forms in the center of the neck and
propagates across the specimen as shown in Fig. 4, where the shaded region
corresponds to material that has undergone a complete loss of stress
carrying capacity. It can be seen in this figure that there is a tendency
for the crack to zig-zag. As discussed by Tvergaard and Needleman [25]
this is a consequence of shear localization being inhibited by the
additional plastic work associated with the hoop strains that accompany
shearing in the axisymmetric geometry. As the free surface is approached
this axisymmetric constraint is relaxed, permitting the cone of the cup-
cone fracture to form [25]. This analysis shows in detail how the
interaction of the tendency to localization in a material weakened by void
nucleation and growth together with a constraining geometrical effect lead
to the cup-cone fracture in Fig. 1a.

There is no corresponding geometrical constraint in plane strain tension
so that, once initiated, a shear band can propagate across the entire
specimen. However, whether or not a shear band occurs in a calculation
depends on the constitutive relation employed. Calculations based on an
isotropically hardening Mises solid give continued growth of a diffuse
neck with no tendency for localized shearing to develop. This is, of
course, expected based on the analytical flow localization results
discussed in Section 4.3. Numerical analyses of the plane strain tensile
test using an isotropically hardening Mises solid to characterize the
material cannot reproduce the behavior shown in Fig. 1b.

A finite element analysis of the plane strain tensile test was carried out
by Tvergaard, Needleman and Lo [23] using J_2 corner theory to characterize
the material behavior. The uniaxial stress strain curve is a power law
(4.9), with N = 0.10. Various small initial thickness inhomogeneities are
prescribed in the form of a linear combination of the long wavelength
diffuse necking mode shape (m = 2 in (4.6)) and various short wavelength
mode shapes. Tvergaard et al. [23] display various shear band patterns,
where the particular pattern that occurs depends on the initial
imperfection. However, in each case the orientation of the band, or
bands, is in good agreement with that predicted from a shear band
bifurcation analysis. Fig. 5 shows deformed finite element meshes for one
of their calculations. The actual computations were carried out for one
quadrant and symmetry boundary conditions were imposed. Shear bands
develop naturally during the course of the calculation and, eventually, a
second shear band pattern develops in the interior of the specimen,
unconnected to the original one. This is a consequence of the vertex
stiffening incorporated into the J_2 corner theory model. As the
deformation pattern shifts, the moduli in the shear band stiffen, while
the moduli outside the shear band remain the less stiff total loading
moduli. The increasing strain in the neck interior then induces the
internal shear bands as illustrated in Fig. 5. The extent to which the
separation is affected by the discretization of the problem is not fully
understood. This is one of several issues concerning mesh effects in the
numerical analysis of localized shearing that merit investigation [23,29].

Localization in plane strain can also occur as a consequence of the
weakening induced by void nucleation and growth. This is demonstrated in
Tvergaard's [24] calculation on shear band formation at a free surface in

a solid subject to plane strain tension. Shear banding in a plane strain
tensile specimen is depicted in Fig. 6, where contours of constant void
volume fraction are shown at two stages of loading. The material is
characterized by the porous plastic constitutive relation described in
Section 3.3, without the additional failure terms used by Tvergaard [24]
and Tvergaard and Needleman [25]. The matrix material follows the power
law relation (4.9) with $N = 0.10$. At the earlier stage of deformation
shown, few voids have nucleated (none are present initially) and the
contour line follows the distribution expected from the necking behavior
of a Mises solid. At a somewhat later stage of deformation the void
volume fraction distribution clearly reflects the presence of shear bands.

Both yield surface vertex effects and the weakening induced by micro-
rupture phenomena can lead to the localized shear fracture depicted in
Fig. 1b. Which of these mechanisms is operative in a particular
circumstance is material (and, posibly, stress state) dependent.
Advances, within the past decade or so, in finite element methods for
analyzing finite strain plasticity problems, in the theoretical
understanding of plastic instability phenomena and in materials modelling
capability together have made possible a detailed description of the
phenomena shown in Fig. 1. Computer simulation studies of the type
described here can now be used to investigate aspects of ductile failure
mechanics that are difficult, if not impossible, to investigate by other
means.

Although here only analyses of tension tests using constitutive
descriptions for polycrystalline metals and qualitative comparisons with
observation have been mentioned, a variety of localization phenomena have
been analysed [29] and, furthermore, when quantitative comparisons can be
made, Larsson et al. [56], Peirce et al., [57,58], the agreement between
the numerical results and experiment is remarkably good. The agreement
arises because the constitutive descriptions employed in the analyses
model relevant material characteristics. For polycrystalline metals there
is a need for constitutive relations that incorporate the anisotropy that
arises due to crystallographic texture. In fact, it turns out that, on a
more microscopic level than discussed here, shear localization can play a
significant role in texture development [59].

ACKNOWLEDGMENT

The support of the Solid Mechanics Program of the U.S. National Science
Foundation through grant MEA-8101948 is very much appreciated.

REFERENCES

1. Speich, G.R. and Spitzig, W.A., Metall. Trans., 13A, 2239 (1982).

2. Considere, A., Ann. Ponts Chaussees, 9 Ser. 6, 574 (1885).

3. Bridgman, P.W., Studies in Large Plastic Flow and Fracture, McGraw-
   Hill (1952).

4. Miles, J.P., J. Mech. Phys. Solids, 19, 89 (1971).

5. Cowper, G.R. and Onat, E.T., in Proc. 4th U.S. Nat. Congr. Appl. Mech. (ed. Rosenberg, R.M.), 1023 (1962).

6. Hill, R., J. Mech. Phys. Solids, 6, 236 (1958).

7. Hill, R., J. Mech. Phys. Solids, 7, 209 (1959).

8. Cheng, S.Y., Ariaratnam, S.T. and Dubey, R.N., Q. Appl. Math., 29, 41 (1971).

9. Needleman, A., J. Mech. Phys. Solids, 20, 111 (1972).

10. Hutchinson, J.W. and Miles, J.P., J. Mech. Phys. Solids, 22, 61 (1974).

11. Hill, R. and Hutchinson, J.W., J. Mech. Phys. Solids, 23, 239 (1975).

12. Argon, A.S., Im, J. and Needleman, A., Metall. Trans., 6A, 815 (1975).

13. Norris, D.M., Moran, B., Scudder, J.K. and Quinones, D.F., J. Mech. Phys. Solids, 26, 1 (1978).

14. Saje, M., Int. J. Solids Struct., 15, 731 (1979).

15. Argon, A.S., Im, J. and Safoglu, R., Metall. Trans., 6A, 825 (1975).

16. Argon, A.S. and Im, J., Metall. Trans., 6A, 839 (1975).

17. Hancock, J.W. and MacKenzie, A.C., J. Mech. Phys. Solids, 24, 147 (1976).

18. Hancock, J.W. and Brown, D.K., J. Mech. Phys. Solids, 31, 1 (1983).

19. Osias, J.R., Finite Deformation of Elasto-Plastic Solids, Ph.D. Thesis, Carnegie-Mellon Univ. (1972).

20. Burke, M.A., and Nix, W.D., Int. J. Solids Struct., 15, 379 (1979).

21. Rice, J.R., in Proc. 14th Int. Congr. Theoret. Appl. Mech. (ed. Koiter, W.T.), 207 (1977).

22. Needleman, A. and Rice, J.R., in Mechanics of Sheet Metal Forming, (ed. Koistinen, D.P. and Wang, N.-M.), 237 (1978).

23. Tvergaard, V., Needleman, A. and Lo, K.K., J. Mech. Phys. Solids, 29, 115 (1981).

24. Tvergaard, V., J. Mech. Phys. Solids, 30, 399 (1982).

25. Tvergaard, V. and Needleman, A., Acta. Metall., in press.

26. Green, A.E. and Zerna, W. Theoretical Elasticity, Oxford (1968).

27. Budiansky, B., in Problems of Hydrodynamics and Continuum Mechanics, (ed. Lavrent'ev, M.A. et al.), 77 (1969).

28. Needleman, A., in Plasticity of Metals at Finite Strain (ed. Lee, E.H. and Mallett, R.L.) 387 (1982).

29. Needleman, A. and Tvergaard, V., in Finite Elements - Special Problems in Solid Mechanics, Vol. 5, (ed. Oden, J.T. and Carey, G.F.), 94 (1983).

30. Tvergaard, V., in Plasticity of Metals at Finite Strain (ed. Lee, E.H. and Mallett, R.L.) 480 (1982).

31. Budiansky, B., unpublished work.

32. Hutchinson, J.W., in Numerical Solution of Nonlinear Structural Problems, (ed. Hartung, R.F.), 17 (1973).

33. Hill, R., J. Mech. Phys. Solids, 15, 79 (1967).

34. Hutchinson, J.W., Proc. R. Soc. Lond., A314, 457 (1970).

35. Hecker, S.S., in Constitutive Equations in Viscoplasticity, ASME AMD 20, 1 (1976).

36. Pan, J. and Rice, J.R., Int. J. Solids Struct., 19, 973 (1983).

37. Hutchinson, J.W., Adv. Appl. Mech., 14, 67 (1974).

38. Batdorf, S.B., J. Aeronaut. Sci. 16, 405 (1949).

39. Christoffersen, J. and Hutchinson, J.W., J. Mech. Phys. Solids, 27, 465 (1979).

40. Hutchinson, J. W. and Tvergaard, V., Int. J. Mech. Sci., 22, 339 (1980).

41. Gurson, A.L., Plastic Flow and Fracture Behavior of Ductile Materials Incorp. Void Nucl., Growth and Interac. Ph.D. Thesis Brown Univ. (1975).

42. Gurson, A.L., J. Engr. Mat. Tech., 99, 2 (1977).

43. Tvergaard, V., Int. J. Fract., 17, 389 (1981).

44. Tvergaard, V., Int. J. Fract., 18, 237 (1982).

45. Tvergaard, V., Int. J. Solids Struct., 18, 659 (1982).

46. Chu, C.C. and Needleman, A., J. Engr. Mat. Tech., 99, 2 (1980).

47. Gurland, J., Acta Metall., 20, 735 (1972).

48. Saje, M., Pan, J. and Needleman, A., Int. J. Fract., 19, 163 (1982).

49. Raniecki, B. and Bruhns, O.T., J. Mech. Phys. Solids, 29, 153 (1981).

50. Hill, R., J. Mech. Phys. Solids, 10, 1 (1962).

51. Hutchinson, J.W. and Tvergaard, V., Int. J. Solids Struct. 17, 451 (1981).

52. Iwakuma, T. and Nemat-Nasser, S., Int J. Solids Struct., 18, 69 (1982).

53. Chen, W. Int J. Solids Struct., 7, 685 (1971).

54. Nagtegaal, J.C., Parks, D.M., and Rice, J.R., Comp. Meths. Appl. Mech. Eng., 4, 153 (1974).

55. Argyris, J.H., Doltsinis, J.St., Straub, K., Pimenta, P.M., Symeonidis, Sp. and Wustenberg, H., Comp. Meths. Appl. Mech. Eng., 32, 2 (1982).

56. Larsson, M., Needleman, A., Tvergaard, V., and Storakers, B., J. Mech. Phys. Solids, 30, 121-154 (1982).

57. Peirce, D., Asaro, R.J. and Needleman, A., Acta Metall., 30, 1087 (1982).

58. Peirce, D., Asaro, R.J. and Needleman, A., Acta Metall., 31, 1951 (1983).

59. Asaro, R.J. and Needleman, A. Scripta Metall., to appear.

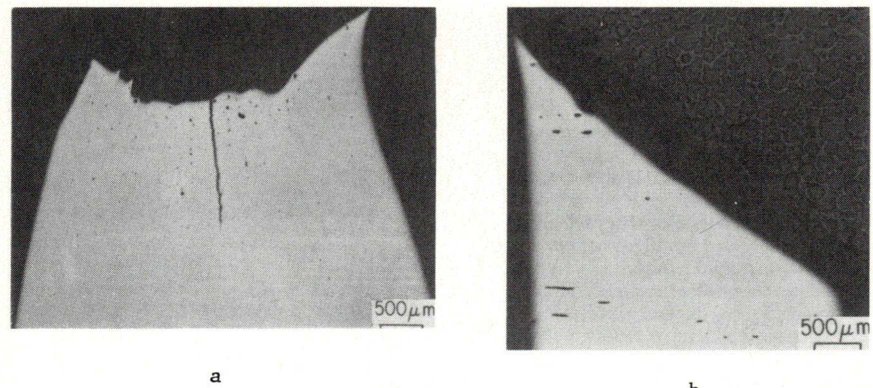

Fig. 1   Effect of stress state on failure of steel tensile specimens
         (a) axisymmetric tension; (b) plane strain tension. From [1].

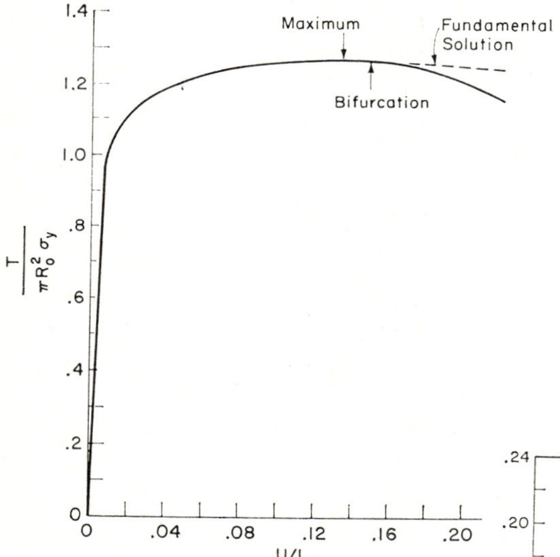

Fig. 2   Load versus end-displacement for an axisymmetric tensile specimen with shear free ends. From [9].

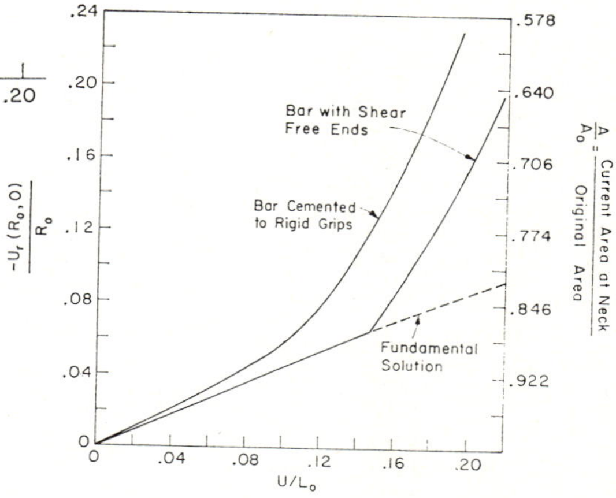

Fig. 3   Neck development versus end-displacement for axisymmetric tensile specimens. From [9].

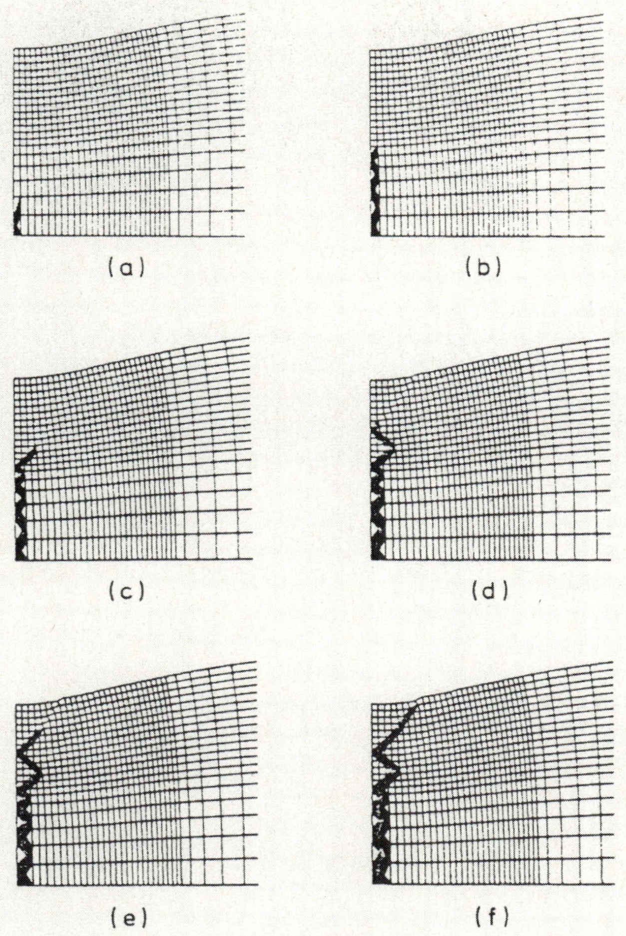

Fig. 4 Crack growth in the neck of an axisymmetric tensile specimen
with the material characterized by a constitutive relation
for a porous plastic solid that permits a loss of stress
carrying capacity at realistic void volume fractions.
From [25].

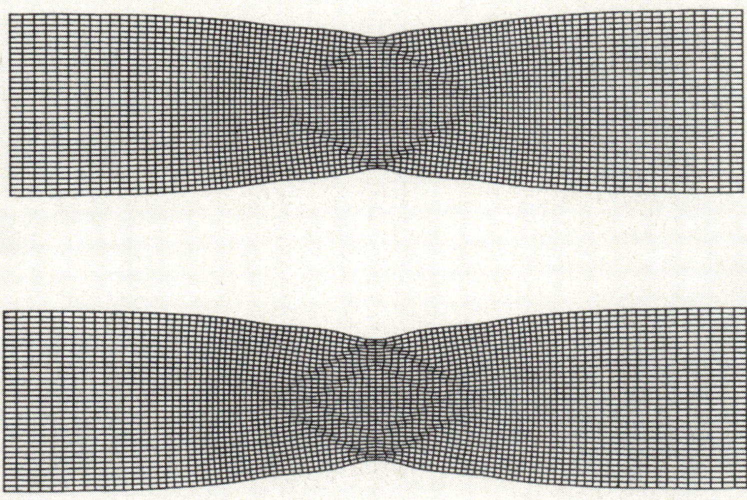

Fig. 5 Two stages of shear band development in a plane strain tensile specimen with the material characterized by J_2 corner theory. From [23].

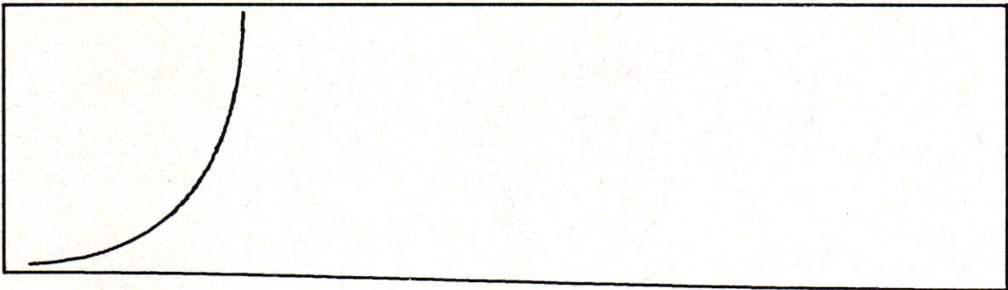

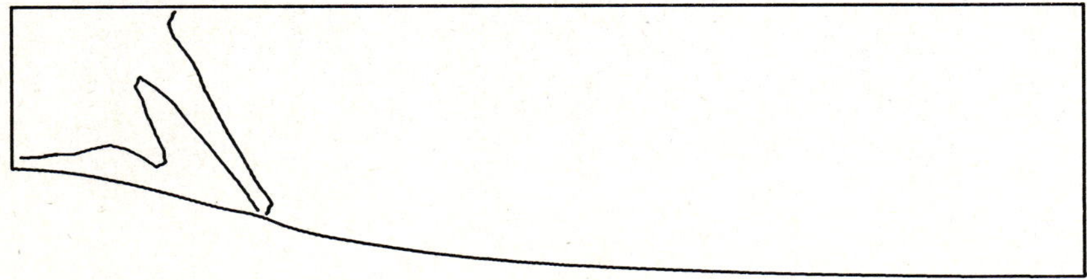

Fig. 6 Contours of constant void volume fraction in a plane strain tensile specimen with the material characterized by a constitutive relation for a porous plastic solid. The contour line corresponds to a void volume fraction of 0.005 in (a) and 0.055 in (b).

CHAPTER 12

RECENT ADVANCES IN THE APPLICATION OF VARIATIONAL METHODS TO NONLINEAR PROBLEMS

A.K. Noor

Two recent advances in the application of direct variational methods to nonlinear steady-state problems are discussed. The first is a hybrid analysis technique based on the combined use of regular perturbation expansion and the classical direct variational techniques for predicting the nonlinear response of the system. The second is a two-stage direct variational technique which allows a substantial reduction in the number of degrees of freedom to be made. The advantages of each of the two techniques are discussed and their effectiveness for the solution of nonlinear steady-state thermal and structural problems are demonstrated by means of four numerical examples.

NOMENCLATURE

A	cross-sectional area of fin
A, B, C	undetermined coefficients used in the expressions of the approximation functions
a	radius of circular plate
c	circumference of fin cross section
E	elastic modulus of isotropic material
E_L, E_T	elastic moduli of the individual layers in the direction of fibers and normal to it, respectively
G_{LT}	shear modulus in the plane of fibers
g, h	*a priori* chosen approximation functions (defined in Eqs. 8, 9 and 13)
h	total thickness of plate (or shell)
h	convective heat transfer coefficient
k	thermal conductivity coefficient
L	length of fin (also side length of plate)
m, m_1	ranges of indices (see Eqs. 4 and 5)
n	total number of coordinate functions
n_1	range of indices (see Eqs. 3 and 11)

P	concentrated load
p_0	intensity of uniform distributed loading
R	radius of curvature of the shell
r	radial coordinate (for circular plates)
T	temperature
T_b	fin base temperature
U	total strain energy of the structure
u, v, w	displacement components of the plate (or shell) middle plane (or middle surface), with w normal to the plate (or shell) middle plane (or middle surface)
x	fundamental unknown (or vector of fundamental unknowns)
$x_i, x_{i,j}$	coordinate functions (see Eqs. 2 and 3)
x_1, x_2, x_3	Cartesian coordinate system with x_3 normal to the middle plane of the plate
α_0	edge angle of spherical cap
$\bar{\beta}$	condition number of the Gram matrix of the basis vectors
β, q	parameters
$[\Gamma]$	matrix of basis vectors defined in Eqs. 16 and 17
$\bar{\gamma}$	constant
δ	symbol of first variation
θ	nondimensional temperature
$\lambda, \lambda_1, \lambda_2$	control (or perturbation) parameters
ν	Poisson's ratio of isotropic materials
ν_{LT}	major Poisson's ratio of the individual layers of laminated plate
$\xi = x_1/L$	dimensionless coordinate
Π	functional
$\Pi_i, \Pi_{i,j}$	perturbation functionals defined in Eqs. 4 and 5
$\{\Phi\}$	vector of undetermined coefficients
ϕ	rotation component of the middle surface of the shell
$\psi_i, \psi_{i,j}$	amplitudes of coordinate functions

INTRODUCTION

Variational methods such as Rayleigh-Ritz and Bubnov-Galerkin techniques
have been, and continue to be, popular tools for nonlinear analysis. To
date there are three practical implementations of variational methods. The
first is the direct (or global) variational approach in which the field
variables (or fundamental unknowns) are sought in the form of a series of
a priori chosen coordinate functions (or modes) with unknown coefficients.
The coordinate functions are chosen to cover the entire domain of the field
variables. The second approach is the finite element method (or local

variational approach) based on dividing the domain into finite elements and using piecewise approximation for the field variables within each element. The third approach is the global (or macro) element method which is a compromise between the first two. In this technique the region under consideration is divided into a small number of subregions and a suitable approximation is made within each subregion. The continuity of the field variables across the interfaces is imposed implicitly in the variational functional [1].

The direct (or global) variational approach was the standard tool for nonlinear analysis in the precomputer era. However, the widespread availability of computers and the fascination with the finite element method, caused by its versatility in handling complex domains and simplicity of computer implementation, has resulted in a relative stagnation in the development of effective direct variational methods.

When contrasted with finite element methods, direct variational methods combine the following two advantages: a) they provide physical insight into the nature of the solution of the problem; and b) they generally have a higher rate of (asymptotic) convergence and result in a much smaller system of (nonsparse) equations (see [2]). However, their major drawback, from a practical point of view, is the difficulty of selecting good approximation functions for complicated domains and/or complex system response. The global (or macro) element method alleviates this drawback, but it does not realize the full potential of variational methods.

In the past few years considerable progress has been made in combining the finite element method with direct variational techniques into effective hybrid numerical procedures (see, for example, [3]-[6]). These hybrid numerical procedures are referred to as *reduction methods* and were successfully used for predicting the nonlinear thermal and structural responses of solids. Also, application of reduction methods to the prediction of nonlinear response of structures subjected to multiple independent loads, and to tracing post-limit-point and post-bifurcation point paths have been reported in [7]-[9]. The successful experience with the hybrid numerical procedures prompted new attempts to be made for the realization of the full potential of variational methods in the solution of nonlinear problems through: 1) the combined use of variational methods in conjunction with other analytical techniques [10]; and 2) innovative ways of applying direct variational methods [11]. The present paper summarizes some of the recent work in this area. The topics discussed herein include: a) the combined use of regular perturbation technique and direct variational methods; and b) the two-stage application of direct variational methods.

Numerical examples are presented to demonstrate the effectiveness of the proposed two techniques. Also, research areas which have high potential for novel application of variational methods are identified.

2. HYBRID PERTURBATION/DIRECT VARIATIONAL TECHNIQUE

Perturbation methods share with direct variational techniques the advantage, over finite element methods, of providing physical insight into the nature of the solution of the problem. However, in contrast to direct variational techniques, the regular perturbation method consists of the development of the solution in terms of unknown functions with *preassigned coefficients*.

The unknown functions are obtained by solving a recursive set of differential equations, or equivalently, by minimizing a recursive set of functionals which characterize the response of the physical system. The recursive set of differential equations (and functionals) are, in general, simpler than the original governing differential equations (and original functional) of the problem. Despite their usefulness in solving nonlinear problems, regular perturbation techniques have two major drawbacks. The first stems from the fact that as the number of terms in the perturbation series increases, the mathematical complexity of the differential equations (and the functional) describing the response of the system builds up rapidly. Therefore, for practical applications, the perturbation series has to be restricted to a few terms. The second drawback is the requirement of restricting the perturbation parameter to small values in order to obtain solutions of acceptable accuracy.

The aforementioned drawbacks of the regular perturbation technique have been recognized and a number of remedial actions were proposed. These included the use of a small number of terms (e.g., two or three) in the perturbation expansion and either: 1) generating "mimic functions" which give accurate numerical estimates of the solution over the entire physical domain (see [12]); or 2) applying a nonlinear transformation (e.g., Shanks transformation [13] and [14]) to estimate the solution as the number of terms goes to infinity. However, as shown in [15], the success of these methods cannot be guaranteed, in general, and the remedial actions may fail to produce satisfactory results. In this section a hybrid perturbation/ direct variational technique is described which alleviates the major drawbacks of the two parent techniques while retaining their advantages.

The hybrid technique is used herein in conjunction with the variational formulation of the problem. Application of the hybrid technique in conjunction with the differential equation formulation, to the solution of nonlinear thermal problems, is described in [10].

Many situations can be cited wherein hybridization provided an innovative way of overcoming problems and proved to be effective. In the case of computational algorithms, examples are provided by the hybrid explicit/ implicit temporal integration schemes for transient problems, and the combined use of direct and iterative techniques for solution of linear algebraic equations (particularly those associated with hierarchical finite elements or multigrid finite differences).

2.1 Basic Idea of Hybrid Technique

The steady-state response of the system is described by the stationary condition of a functional Π which characterizes the system, i.e.

$$\delta\Pi(\mathbf{X}) = 0 \tag{1}$$

where $\mathbf{X}$ is the field variable (or fundamental unknown) and δ is the symbol of first variation. For more than one fundamental unknown $\mathbf{X}$ is a vector. For nonlinear problems the functional $\Pi(\mathbf{X})$ include quadratic as well as higher-order terms in $\mathbf{X}$ (and/or its spatial derivatives). Equation 1 is equivalent to the governing differential equations and the boundary conditions of the system.

The application of the hybrid perturbation/direct variational technique to the solution of Eq. 1 can be conveniently divided into the following two distinct steps: 1) generation of coordinate functions (or modes) using the standard regular perturbation method, and 2) computation of the amplitudes of the coordinate functions via direct variational technique. The procedure is described in detail subsequently.

2.2 Generation of Coordinate Functions

For the purpose of generating the required coordinate functions (or modes), the functional $\pi(\mathbf{x})$ is embedded in a single- or multiple-parameter family of functionals of the form: $\pi(\mathbf{x},\lambda)$ for the single-parameter case; and $\pi(\mathbf{x}, \lambda_1, \lambda_2)$ for the two-parameter case, where λ, λ_1, and λ_2 are normalizing parameters which are also referred to as control or perturbation parameters. Extension to more than two parameters is straightforward and is not discussed herein.

The field variable (or vector of field variables) $\mathbf{x}$ is represented by the regular perturbation expansion:

$$\mathbf{x} = \sum_{i=0}^{n-1} \lambda^i \, \mathbf{x}_i \tag{2}$$

for the single-parameter case, and

$$\mathbf{x} = \sum_{j=0}^{n_1} \sum_{i=0}^{j} \lambda_1^i \, \lambda_2^{j-i} \; \mathbf{x}_{i,j-i} \tag{3}$$

for the two-parameter case, where $\mathbf{x}_i$ and $\mathbf{x}_{i,j}$ are perturbation functions which represent modes; and n is the total number of terms in the expansion. For the two-parameter case $n = 1/2 \, (n_1 + 1)(n_1 + 2)$.

If the perturbation expansions, Eqs. 2 or 3, are substituted into the functional π and the terms having the same powers of the perturbation parameters are grouped together, then π can be written in the following form:

$$\pi = \sum_{i=0}^{m-1} \lambda^i \, \pi_i \tag{4}$$

for the single-parameter case, and

$$\pi = \sum_{j=0}^{m_1} \sum_{i=0}^{j} \lambda_1^i \, \lambda_2^{j-i} \; \pi_{i,j-i} \tag{5}$$

for the two-parameter case, where π_i and $\pi_{i,j}$ are the perturbation functionals, and m is the total number of terms in the expansion of π. For the two-parameter case $m = 1/2 \, (m_1 + 1)(m_1 + 2)$.

The stationary condition of π, Eq. 1, can now be replaced by the following recursive set of stationary conditions:

$$\delta\pi_i = 0 \qquad\qquad (i = 0 \text{ to } m\text{-}1) \tag{6}$$

for the single-parameter case, and

$$\delta\Pi_{i,j-i} = 0 \qquad (j = 0 \text{ to } m_1, \; i = 0 \text{ to } j) \qquad (7)$$

for the two-parameter case.

Note that whereas the original stationary condition, Eq. 1, leads to a *nonlinear set of differential equations*, the recursive set of stationary conditions, Eqs. 6 and 7, lead to a *recursive set of linear differential equations in* X_ℓ *and* $X_{\ell,k-\ell}$, respectively, where i = 2ℓ and j = 2k. Only even values of the indices i,j in Eqs. 6 and 7 need to be considered.

In most practical problems, an exact solution for the differential equations resulting from the stationary conditions, Eqs. 6 and 7, cannot be obtained. Approximate solutions are sought for X_ℓ and $X_{\ell,k}$ through the application of Rayleigh-Ritz technique, i.e.,

$$X_\ell = \sum_\alpha A_\ell^{(\alpha)} \, g_\ell^{(\alpha)} \qquad (8)$$

for the one-parameter case, and

$$X_{\ell,k} = \sum_\alpha B_{\ell,k}^{(\alpha)} \, h_{\ell,k}^{(\alpha)} \qquad (9)$$

for the two-parameter case, where $g_\ell^{(\alpha)}$ and $h_{\ell,k}^{(\alpha)}$ are *a priori* chosen functions of the spatial coordinates (approximation functions) which satisfy the essential boundary conditions as well as the symmetry conditions on X_ℓ and $X_{\ell,k}$, and $A_\ell^{(\alpha)}$, $B_{\ell,k}^{(\alpha)}$ are undetermined coefficients. In the case of more than one field variable, different sets of approximation functions g and h can be assumed for the different field variables (i.e., for the different components of X). The A and B coefficients are obtained by substituting Eqs. 8 or 9 into the recursive set of stationary conditions, Eqs. 6 or 7, and solving the resulting sets of recursive *linear algebraic equations*. Note that the left-hand sides of the recursive sets of algebraic equations, for all values of ℓ (or for all pairs of ℓ and k), are the same.

2.3   Computation of Amplitudes of Coordinate Functions

The perturbation functions X_ℓ (or $X_{\ell,k}$) are now chosen as coordinate functions and the field variable X is expressed as a linear combination of these functions as follows:

$$X = \sum_{i=0}^{n-1} \psi_i \, X_i \qquad (10)$$

for the single-parameter case, and

$$X = \sum_{j=0}^{n_1} \sum_{i=0}^{j} \psi_{i,j-i} \, X_{i,j-i} \qquad (11)$$

for the two-parameter case where ψ_i and $\psi_{i,j}$ are unknown parameters which represent amplitudes of the coordinate functions (or modes) X_i and $X_{i,j}$; and n equals the total number of modes. In the two-parameter case $n = 1/2 \ (n_1 + 1)(n_1 + 2)$.

The parameters ψ_i (or $\psi_{i,j}$) are obtained by substituting the expansion of X in the original functional Π, applying the stationary condition, Eq. 1, and solving the resulting small set of *nonlinear* equations.

2.4 Comments on Selection of Coordinate Functions

The chosen set of coordinate functions has the following three properties:

1. They are linearly independent and span the space of solutions in the neighborhood of the point of their generation. Therefore, they fully characterize the nonlinear solution in that neighborhood.

2. Their generation, using the regular perturbation method in conjunction with the Rayleigh-Ritz technique, requires the solution of a recursive set of linear algebraic equations. The left-hand sides of these equations are the same.

3. They provide a direct measure of the sensitivity of the nonlinear response of the system to changes in the control or perturbation parameters.

The first property is necessary for the convergence of the direct variational technique. The second property enhances the effectiveness of the proposed hybrid technique for solving nonlinear problems. The implication of the third property is that by appropriate choice of the perturbation parameters, sensitivity of the response to changes in the characteristics of the system can be obtained. Note that the sensitivity information is obtained at zero value(s) of the perturbation parameter(s). If the sensitivity is required at other values of the parameter(s), approximate Taylor series expansions may be used.

The mathematical complexity of the right-hand sides of the recursive set of functionals used in generating the coordinate functions builds up rapidly with the increase in the number of these functions. Therefore, for practical applications, only a few functions are generated.

3. TWO-STAGE DIRECT VARIATIONAL TECHNIQUE

As an attempt to improve the efficiency of the classical direct variational techniques in nonlinear analysis, a two-stage technique was proposed for considerably reducing the number of degrees of freedom of the discretized system. The technique is particularly useful for predicting the response of nonlinear systems with simple geometries but complex construction. Examples from the structures and solid mechanics area are provided by ring and stringer stiffened closed cylindrical shells and shell panels with discrete stiffeners and rectangular or circular planform. The technique is outlined in this section. To sharpen the focus of the study, discussion is limited to steady-state nonlinear problems of conservative systems, with two independent path parameters λ_1 and λ_2. In structural and solid mechanics problems, the path parameters can be identified with load, displacement or arc-length parameters in the solution space.

As in the preceding section, the steady-state response of the system is characterized by the stationary condition of a functional Π, i.e.

$$\delta\Pi(\mathbf{X}, \lambda_1, \lambda_2) = 0 \tag{12}$$

The two stages of the solution process are discussed subsequently.

3.1 Stage I - Spatial Discretization

In the first stage of the proposed technique, the system is discretized by using the following approximate expression for the fundamental unknown (or vector of fundamental unknowns):

$$\mathbf{X} = \sum_\alpha \Phi^{(\alpha)} \, \mathfrak{g}^{(\alpha)} \tag{13}$$

where $\mathfrak{g}^{(\alpha)}$ are *a priori* chosen functions of the spatial coordinates (coordinate functions) which satisfy the essential boundary conditions as well as the symmetry conditions on $\mathbf{X}$; and $\Phi^{(\alpha)}$ are undetermined coefficients which are functions of the path parameters λ_1 and λ_2. Note that for the case of more than one field variable different sets of approximation functions $\mathfrak{g}$ can be assumed for the different field variables (i.e., for the different components of $\mathbf{X}$).

The governing equations of the discretized system are obtained by first replacing $\mathbf{X}$ by its expression in terms of the undetermined coefficients $\Phi^{(\alpha)}$ and then applying the stationary condition of the functional, Eq. (12).

If the coefficients $\Phi^{(\alpha)}$ are varied independently and simultaneously, one obtains the following set of nonlinear equations:

$$\{f(\{\Phi\}, \lambda_1, \lambda_2)\} = 0 \tag{14}$$

$$\text{where} \quad \{\Phi\} = \begin{Bmatrix} \Phi^{(1)} \\ \Phi^{(2)} \\ \vdots \end{Bmatrix} \tag{15}$$

3.2 Stage II - Reduction of the Number of Degrees of Freedom

In the second stage of the proposed technique, the vector of undetermined coefficients $\{\Phi\}$ of the discretized system is approximated over a range of values of λ_1 and λ_2, by a linear combination of $\{\Phi\}$ corresponding to a particular pair of values of λ_1, λ_2 (typically $\lambda_1 = \lambda_2 = 0$) and a number of their path derivatives (derivatives of $\{\Phi\}$ with respect to λ_1 and λ_2) evaluated at the same values of λ_1 and λ_2; i.e.

$$\{\Phi\} = [\Gamma]\{\psi\} \tag{16}$$

where $[\Gamma]$ is a transformation matrix which is given by:

$$[\Gamma] = \left[\{\Phi\} \; \left\{\frac{\partial\Phi}{\partial\lambda_1}\right\} \; \left\{\frac{\partial\Phi}{\partial\lambda_2}\right\} \; \left\{\frac{\partial^2\Phi}{\partial\lambda_1^2}\right\} \; \left\{\frac{\partial^2\Phi}{\partial\lambda_1\partial\lambda_2}\right\} \; \left\{\frac{\partial^2\Phi}{\partial\lambda_2^2}\right\} \cdots \right] \tag{17}$$

and $\{\psi\}$ is a vector of reduced unknowns (with $\hbar$ components only) which are functions of the path parameters λ_1 and λ_2. Note that $\hbar$ is considerably smaller than the total number of degrees of freedom of the system (which is equal to the number of components of $\{\Phi\}$). The columns of the matrix $[\Gamma]$ are usually referred to as the basis vectors.

The direct variational technique is now applied a second time to replace the governing equations obtained in the first stage, Eqs. 14, by the following reduced system of $\hbar$ nonlinear equations in the reduced unknowns $\{\psi\}$

$$\{\tilde{f}(\{\psi\}, \lambda_1, \lambda_2)\} = 0 \tag{18}$$

where

$$\{\tilde{f}\} = [\Gamma]^t \{f([\Gamma]\{\psi\}, \lambda_1, \lambda_2)\} \tag{19}$$

In Eqs. 19 superscript t denotes transposition.

The equations used in evaluating the basis vectors are obtained by successive differentiation of the governing equations of the discretized system, Eqs. 14, with respect to λ_1 and λ_2. The left-hand sides of the resulting recursive set of linear algebraic equations are the same. A criterion for selecting the number of basis vectors was proposed in [11]. The criterion is based on monitoring the condition number $\bar{\beta}$ of the Gram matrix of the basis vectors, and the generation of these vectors is terminated when $\bar{\beta}$ exceeds a prescribed value. Also, upper and lower limits for the number of basis vectors are prescribed.

The ranges of λ_1 and λ_2 for which Eqs. 18 provide an acceptable approximation for the original discrete system, Eqs. 14, depends on the degree of nonlinearity of the problem. The ranges can be identified by selecting an error measure (e.g., norm of the residual vector); and monitoring the accuracy of the solutions of the reduced system, Eqs. 18. When the error measure exceeds a prescribed tolerance, the vector $\{\Phi\}$ generated by the reduced system of equations is used as a predictor and the Newton-Raphson iterative technique is used in conjunction with the original equations, Eqs. 14, to obtain a corrected (improved) solution. Then a new (updated) set of basis vectors is generated. The computational procedure is described in [11] and will not be repeated herein.

3.3 Comments on the Proposed Two-Stage Technique

1. The essence of the two-stage direct variational technique for the steady-state nonlinear analysis is to separate the spatial distributions of the fundamental unknowns for any pair of values of λ_1 and λ_2 from their variations with λ_1 and λ_2. The form of the spatial distribution of the fundamental unknowns is given by the assumed coordinate functions in the first stage. The use of the path derivatives as basis vectors in the second stage permits the known information about the nonlinear response in the neighborhood of a point to be brought into the analysis. In this manner

substantially fewer degrees of freedom will be required to achieve a desired
overall accuracy in comparison with that based on the classical direct
variational technique. Note that the time required to solve the reduced
system of equations is relatively small, and the total analysis time to a
first approximation, equals the time required to evaluate the basis vectors
and generate the reduced equations.

2. The effectiveness of the proposed two-stage technique may be attributed to the fact that the spatial distribution of the field variables
varies slowly with the variation of the path parameters λ_1 and λ_2. A large
number of numerical experiments have shown that large changes in the path
parameters are frequently associated with small changes in the spatial
distribution of the response quantities. Gross changes in the spatial
distribution require updating of the basis vectors in the second stage.

3. If the proposed two-stage technique is contrasted with the static
perturbation technique (see [16]), the following can be noted. In both
techniques the vector of undetermined coefficients $\{\Phi\}$ is approximated by
a linear combination of a small number of path derivatives, Eqs. 16 and 17.
However, the coefficients of the linear combination $\{\psi\}$ in the static perturbation technique are fixed and are equal to: 1, $\Delta\lambda_1$, $\Delta\lambda_2$,
$\dfrac{(\Delta\lambda_1)^2}{2}$, $\Delta\lambda_1\Delta\lambda_2$, $\dfrac{(\Delta\lambda_2)^2}{2}$, ... By contrast, the coefficients $\{\psi\}$ in the
proposed approach are left as free parameters, and are determined by applying the direct variational technique a second time. Numerical experiments
indicate that the use of free parameters leads to accurate solutions not
only within the radius of convergence of the Taylor series but also well
beyond it.

4. For the prediction of the nonlinear response of practical systems
with complex domains, the two-stage procedure can be applied in conjunction
with the global (or macro) element method, or even with the finite element
method. In the latter case, the two-stage procedure reduces to the reduction
method described in [9].

4. NUMERICAL STUDIES

To evaluate the effectiveness of both the hybrid perturbation/direct variational technique and the two-stage direct variational technique, several
nonlinear steady-state thermal and structural problems were solved. For
each problem, the solutions obtained by the proposed techniques were compared with converged finite element solutions and with other numerical
approximations, whenever available. Herein the results of four typical
steady-state thermal and structural problems are discussed. The four problems are: 1) steady-state thermal response of one-dimensional conducting-convecting fin with variable heat transfer coefficient; 2) large deflection
analysis of laminated anisotropic plate subjected to uniform transverse
loading; 3) nonlinear axisymmetric response of an isotropic circular plate
subjected to combined uniform and concentrated loadings; and 4) nonlinear
axisymmetric response of a clamped shallow spherical cap subjected to
uniform normal pressure. The first three problems were analyzed by using
the regular perturbation technique and were used to assess the effectiveness
of the hybrid perturbation/direct variational technique. The fourth problem
is used to evaluate the two-stage direct variational technique.

In all the problems considered, all the analytical work, namely, generation
of perturbation functionals, generation of various-order perturbation
equations; evaluation of coordinate functions and of basis vectors, was
done by using the computerized symbolic manipulation system MACSYMA [17].

4.1 Steady-State Thermal Response of a Conducting-Convecting Fin with Variable Heat-Transfer Coefficient

The first problem considered is that of a straight conducting-convecting
fin of length L, cross-sectional area A, and perimeter c, exposed on both
sides to a free convective environment of temperature $T_a = 0$. The boundary
conditions are a constant base temperature and an adiabatic tip. The thermal conductivity κ is assumed to be independent of the temperature. The
convective heat-transfer coefficient is taken to be of the form:

$$h = h_b \, \theta^\beta \qquad\qquad (20)$$

where $\theta = T/T_b$ is a normalized temperature defined in terms of the fin base
temperature T_b; β is a small parameter ($\beta = 0.25$ and 0.33 for laminar and
turbulent conditions, respectively [18]); and $h_b = \bar{\gamma} \, T_b^\beta$, where $\bar{\gamma}$ is a
constant. The variational functional and the governing differential
equation for this problem are given in Appendix I.

Two different choices are made for the perturbation parameter λ. The first
choice is the same as that of [18], namely $\lambda = \beta$. The second choice is

$$\lambda = q^2 = \frac{h_b \, cL^2}{\kappa A} \qquad\qquad (21)$$

where q^2 is a convection-conduction fin parameter. For each of the two
choices of the perturbation parameter, the coordinate functions were obtained by solving the various-order perturbation equations. In each case,
closed-form solutions for the recursive set of differential equations were
obtained. The latter choice of the perturbation parameter, $\lambda = q^2$, resulted
in significantly simplified expressions of the coordinate functions. For
case 1, $\lambda = \beta$, two coordinate functions were used, and for case 2, $\lambda = q^2$,
four coordinate functions were generated. The expressions of the coordinate
functions for the two cases are given in [10] and are not repeated here.
The amplitudes of the coordinate functions were obtained by applying the
Bubnov-Galerkin technique to the original governing differential equation,
Eq. A.2. A single free parameter was used in case 1 and three free parameters were used in case 2 (since one of the free parameters, ψ_0, is used
to satisfy the prescribed nonzero boundary condition).

An indication of the accuracy of the solutions obtained by the hybrid technique and the regular perturbation method are given in Figure 1 for $q = 1.0$
and 2.0 and $\beta = 0.33$ and 1.0. The latter value of β has no physical significance and was selected in order to amplify the effect of the magnitude of
the perturbation parameters on the quality of the solutions. The standard
for comparison was taken to be the finite element solution obtained by using
uniform grid of 15 three-noded finite elements with quadratic Lagrangian
interpolation functions for the temperature. As can be seen from Figure 1,
the accuracy of the perturbation solution is very sensitive to both the

choice and the magnitude of the perturbation parameter. For $\lambda = \beta$, the two-term perturbation expansion is accurate for $\beta \leq 0.33$ but becomes quite inaccurate for $\beta = 1.0$. On the other hand, for $\lambda = q^2$, the four-term perturbation expansion is grossly in error for all $q \geq 1.0$. The perturbation solutions for $q = 2$ could not be shown in Figure 1. By contrast, the accuracy of the solutions obtained by the hybrid technique were found to be insensitive to the choice of the perturbation parameter. The solutions obtained by using $\lambda = \beta$ and $\lambda = q^2$ were equally accurate and were almost indistinguishable from the finite element solutions.

4.2 Laminated Anisotropic Plate Subjected to Uniform Transverse Loading

The second problem considered is that of the nonlinear response of a clamped, square, symmetrically laminated 16-ply graphite-epoxy plate subjected to uniform transverse loading. The material and geometric characteristics of the plate are shown in Figure 2. The perturbation parameter was selected to be the load parameter $\lambda = p_o L^4/(E_T h^4)$. A von Karman type nonlinear plate theory is used and the problem is formulated in terms of the three displacement components of the middle plane of the plate; namely, the in-plane displacements u, v, and the transverse displacement w. The explicit form of the governing differential equations and the functional for this problem are given in [19]. The displacements are expanded in perturbation series as follows:

$$w = w_1 \lambda + w_3 \lambda^3 + w_5 \lambda^5 + \ldots \tag{22}$$

$$u = u_2 \lambda^2 + u_4 \lambda^4 + \ldots \tag{23}$$

$$v = v_2 \lambda^2 + v_4 \lambda^4 + \ldots \tag{24}$$

The perturbation functions w_i (i = 1,3,5, ...), and u_j, v_j (j = 2,4, ...) were sought in the form of modified Fourier series as follows:

$$w_i = \sum_{m=2,4} \sum_{n=2,4} A_{mn}^{(i)} \left[\cos \frac{m\pi x_1}{L} - (-1)^{m/2}\right]\left[\cos \frac{n\pi x_2}{L} - (-1)^{n/2}\right] \tag{25}$$

$$u_j = \sum_{m=2,4} \sum_{n=1,3} B_{mn}^{(j)} \sin \frac{m\pi x_1}{L} \cos \frac{n\pi x_2}{L} \tag{26}$$

$$v_j = \sum_{m=1,3} \sum_{n=2,4} C_{mn}^{(j)} \cos \frac{m\pi x_1}{L} \sin \frac{n\pi x_2}{L} \tag{27}$$

where $A_{mn}^{(i)}$, $B_{mn}^{(j)}$, and $C_{mn}^{(j)}$ are undetermined coefficients obtained by applying the Rayleigh-Ritz technique to the perturbation functionals Π_{2i} and Π_{2j}, and solving the resulting system of linear algebraic equations in $A_{mn}^{(i)}$, $B_{mn}^{(j)}$ and $C_{mn}^{(j)}$. Note that the left-hand sides of these algebraic equations are the same, regardless of the values of i and j; and the equations for $A_{mn}^{(i)}$ are uncoupled from those for $B_{mn}^{(j)}$ and $C_{mn}^{(j)}$.

Three perturbation functions were generated for w and two coordinate functions were generated for each of u and v. The perturbation functions were then used as coordinate functions and the amplitudes of these functions were obtained by applying the Rayleigh-Ritz technique to the original functional (total potential energy of the plate).

An indication of the accuracy of the transverse displacements w_c and the strain energies obtained by the hybrid technique and the regular perturbation method, for different values of loading, is shown in Figure 3. The standard of comparison was taken to be the finite element solution using a uniform 6 x 6 grid of bicubic interpolation functions for the displacements and rotations. As can be seen from Figure 3, the perturbation solutions are grossly in error for $\lambda > 200$ (corresponding to $w_c/h > 0.5$). By contrast, the solutions obtained with the hybrid technique are in close agreement with the finite element solution for all the range of loading considered.

4.3 Nonlinear Axisymmetric Response of an Isotropic Circular Plate Subjected to Combined Uniform and Concentrated Loading

The next problem considered is that of the nonlinear axisymmetric response of a circular plate subjected to combined uniform distributed loading p_o and a concentrated load P applied at the center of the plate. The plate is freely supported with the transverse displacement $w = 0$ and the radial displacement u unrestrained at the edge. The problem was considered in [20] as an application of the two-parameter perturbation technique.

A von-Karman type plate theory is used and the problem is formulated in terms of the transverse displacement w and the radial normal force N_r. The governing differential equations are given in Appendix II. As in [20], the perturbation parameters are selected to be the two normalized loadings $\lambda_1 = \dfrac{p_o a^4}{Eh^4}$ and $\lambda_2 = \dfrac{Pa^2}{Eh^4}$. Two-parameter perturbation series are used for both w and N_r as follows:

$$w = w_{10}\,\lambda_1 + w_{01}\,\lambda_2 + w_{30}\,\lambda_1^3 + w_{21}\,\lambda_1^2\,\lambda_2 + w_{12}\,\lambda_1\,\lambda_2^2 + w_{03}\,\lambda_2^3 + \ldots \quad (28)$$

$$N_r = N_{20}\,\lambda_1^2 + N_{11}\,\lambda_1\,\lambda_2 + N_{02}\,\lambda_2^2 + \ldots \quad (29)$$

The perturbation functions w_{10}, w_{01}, N_{20}, N_{11}, N_{02}, w_{30}, w_{21}, w_{12} and w_{03} are obtained by solving the recursive set of linear ordinary differential equations. The explicit form of these differential equations is given in Appendix II. Closed-form solutions were obtained for all these equations. Six perturbation functions were generated for w and three perturbation functions were generated for N.

The perturbation functions were then used as coordinate functions and the amplitudes of these functions were obtained by applying the Bubnov-Galerkin technique to the original differential equations, Eqs. B.1 and B.2.

An indication of the accuracy of the transverse displacements and the strain energies obtained by the hybrid technique and the two-parameter perturbation method is shown in Figure 4 for various combinations of λ_1 and λ_2. The standard of comparison is taken to be the finite element solution using a uniform grid of 15 elements with first-order Hermitian polynomials for both the radial and transverse displacements. As can be seen from Figure 4, the solutions obtained with the perturbation method are grossly in error for $\lambda_2 > 2$ and $\lambda_1 \geq 0$. By contrast, the solutions obtained with the hybrid technique are in close agreement with the finite element solution for all the range of loadings considered.

4.4 Nonlinear Axisymmetric Response of Shallow Spherical Caps

The last problem considered is that of clamped shallow spherical caps subjected to uniform normal loading p_0. The material and geometric characteristics of the shell are given in Figure 5. Two shallow caps with different thicknesses are considered; namely, $h = 0.0127$ and 0.005 m. The behavior of the two caps is highly nonlinear with an initial softening and subsequent stiffening. Moreover, the response of the thinner cap exhibits limit points. The problem is used to assess the accuracy and effectiveness of the two-stage Rayleigh-Ritz technique. Analytic solutions for the thicker cap, $h = 0.0127$ m., are given in [21]. Finite element solutions for the same problem are given in [22] and [23]. In the present study a shear-deformation Sanders-Budiansky type shell theory is used and the problem is formulated in terms of the three generalized displacements of the middle surface of the shell, u, w, and ϕ. The approximation functions used in the first stage of the analysis are as follows:

$$u = \sum A_n \sin \frac{n\pi\alpha}{\alpha_0} \tag{30}$$

$$w = \sum B_n \cos \frac{(2n-1)\pi\alpha}{2\alpha_0} \tag{31}$$

$$\phi = \sum C_n \sin \frac{n\pi\alpha}{\alpha_0} \tag{32}$$

The control parameter was chosen to be the generalized arc-length in the solution space (see [7]). The basis vectors were generated for the unloaded cap ($q = \frac{p_0 R}{Eh} = 0$, $\lambda = 0$, $\{\Phi\} = 0$), and were thus obtained by solving a linear set of finite element equations. Nine basis vectors (path derivatives with respect to the generalized arc-length) were generated. The basis vectors were orthonormalized using the Gram-Schmidt procedure.

An indication of the accuracy of the solutions obtained by the two-stage Rayleigh-Ritz technique is given in Figures 6 and 8 for the two spherical caps. For the thicker cap the same set of basis vectors were used throughout the range of loading considered. The solutions obtained using nine basis vectors were highly accurate. At $q = 7.95 \times 10^{-3}$, the errors in the transverse displacement w_c and total strain energy U were 0.466% and 0.722%, respectively.

The complex nature of the thin cap response is depicted in Figure 8. The basis vectors were updated two times in this case, corresponding to $q = 9.88 \times 10^{-4}$ and $q = 7.13 \times 10^{-4}$.

It is worth mentioning that the use of generalized arc-length as the control parameter resulted in considerably improving the performance of the solution technique for the nonlinear reduced equations. The number of Newton-Raphson iterations required for the convergence of the reduced equations at each increment was equal to (or less than) 3. In order to demonstrate the advantage of the two-stage Rayleigh-Ritz technique over the static pertur-bation technique, the transverse displacement for the thicker cap obtained by using six, seven, eight and nine nonzero terms in the Taylor series expansion about the solution at $q = 0$ are shown in Figure 7. As can be seen from Figure 7, the solution drifts from the true equilibrium path. The drift is more pronounced when the number of terms in the Taylor series is small.

Note that for conservative systems the efficiency of the two-stage Rayleigh-Ritz technique can be increased if the error tolerance is increased and the accuracy of the reduced system of equations is maintained by backtracking the solution path each time a new (updated) set of basis vectors is generated.

5. POTENTIAL OF THE PROPOSED HYBRID AND TWO-STAGE TECHNIQUES

The two techniques described in the preceding sections have high potential for solution of nonlinear steady-state problems, especially for systems with complex construction but simple geometries. Examples from the struc-tures area are provided by ring and stringer stiffened closed cylindrical shells, and shell panels with discrete stiffeners and rectangular or circular planform. The numerical studies conducted clearly demonstrated the accuracy and effectiveness of the two techniques. In particular, the following two points are worth mentioning:

1. The hybrid perturbation/direct variational technique can be thought of as either of the following:

a) A generalized perturbation method in which 1) the perturbation expansions of the field variables contain free parameters rather than fixed coefficients; and 2) the perturbation parameters need not be small. Since the accuracy of the solutions obtained with the hybrid technique appears to be insensitive to the choice of perturbation parameters, they may be intro-duced artificially to simplify the form of the recursive set of differential equations (or the recursive set of functionals) used in evaluating the various-order perturbation solutions.

b) An extended direct variational technique with the coordinate functions generated by using the standard regular perturbation technique rather than chosen *a priori*.

2. The foregoing hybrid technique is the analytic counterpart of the two-stage direct variational technique presented herein. The primary ob-jective of using the two-stage technique is to reduce considerably the

number of degrees of freedom in the initial discretization, and hence,
reduce the computational effort involved in the solution of the problem.
By contrast, the objectives of the foregoing hybrid technique are: a) to
extend the range of validity of the regular perturbation method by removing
the restriction of a small perturbation parameter; and b) to enhance the
effectiveness of the direct variational technique by removing (or reducing)
the arbitrariness in the selection of the coordinate functions.

6. FUTURE DIRECTIONS FOR RESEARCH ON DIRECT VARIATIONAL METHODS

Among the different research areas which have high potential for application
of direct variational methods are the following:

   a) Development of effective approximation functions for nonlinear
problems. This includes an assessment of the relative merits of using
nonlinear forms of coordinate functions instead of linear combinations of
these functions.

   b) Use of the hybrid analytical technique and the two-stage technique
in conjunction with the weighted residual-least squares procedure. Also, a
systematic comparison between the merits of using least squares procedure
versus Bubnov-Galerkin or Rayleigh-Ritz technique.

   c) Application of the two-stage variational methodology to the analysis
of one- and two-dimensional structural components wherein the fundamental
unknowns are obtained from a one- or two-dimensional theory and are used as
coordinate functions with undetermined parameters in the three dimensional
theory. The undetermined parameters are obtained by applying the direct
variational technique.

7. CONCLUDING REMARKS

Two recent advances in the application of direct variational methods to
nonlinear steady-state problems are discussed. The first is a hybrid analysis technique based on the combined use of regular perturbation expansion
and the classical direct variational techniques for predicting the nonlinear
steady-state response of the system. The second is a two-stage direct
variational technique.

The application of each of the two techniques to the solution of nonlinear
problems can be divided into two stages. For the hybrid technique the first
stage consists of generating the coordinate functions (or modes) using the
standard regular perturbation method and approximating the field variables
by a linear combination of these modes. The classical direct variational
technique is then used to compute the coefficients of the linear combination
(amplitudes of the modes). In the two-stage direct variational technique,
the first stage consists of discretizing the system by using coordinate
functions which cover the entire domain. In the second stage a substantial
reduction in the number of degrees of freedom is achieved by expressing the
vector of unknown parameters as a linear combination of a small number of
basis vectors. The direct variational technique is applied a second time
to approximate the nonlinear equations of the discretized system by a reduced system of nonlinear equations. The basis vectors used in the second
stage are chosen to be those commonly used in static perturbation technique,
namely, a nonlinear solution and a number of its path derivatives.

Four numerical examples are presented to demonstrate the effectiveness of the two techniques for the solution of nonlinear steady-state problems.

Several conclusions can be made regarding the two techniques. These conclusions are as follows:

1. The hybrid technique exploits the best elements of the regular perturbation method and the direct variational technique as follows:

a) The regular perturbation method is used as a systematic and general approach for generating coordinate functions.

b) The direct variational technique is used as an efficient procedure for minimizing and distributing the error, in the field variables, throughout the domain.

2. The hybrid technique extends the range of applicability of the perturbation method and enhances the effectiveness of the direct variational technique. It also alleviates the following major drawbacks of the classical techniques:

a) The requirement of using a small parameter in the regular perturbation expansion is avoided.

b) The method provides a systematic selection of the coordinate functions (or modes) needed in the direct variational technique.

3. The accuracy of the solutions obtained by the hybrid technique is relatively insensitive to the choice of the perturbation parameter(s). Therefore, the parameter(s) may be introduced artificially to simplify the form of the recursive set of differential equations (or recursive functionals) used in evaluating the various-order perturbation solutions (viz, the coordinate functions).

4. The two-stage direct variational technique greatly alleviates one of the major drawbacks of the classical direct variational technique, namely, the repeated solution of a large system of nonsparse nonlinear equations.

APPENDIX I - FUNCTIONAL AND GOVERNING DIFFERENTIAL EQUATION FOR THE ONE-
            DIMENSIONAL CONDUCTING-CONVECTING FIN

The functional characterizing the steady-state thermal response of a one-dimensional conducting-convecting fin with variable heat-transfer coefficient is given by:

$$\Pi(\theta) = \int_0^1 [\tfrac{1}{2}(\tfrac{d\theta}{d\xi})^2 + \tfrac{q^2}{2+\beta}\,\theta^{2+\beta}]\,d\xi \qquad (A.1)$$

where $\theta = T/T_b$ is a normalized temperature defined in terms of the fin base temperature T_b; β is a small parameter; $\xi = x_1/L$; k is conductivity coefficient; $q^2 = \dfrac{h_b\,cL^2}{k\,A}$ is a convection-conduction fin parameter; $h_b = \bar\gamma\,T_b^\beta$; and $\bar\gamma$ is a constant.

The governing differential equation is given by:

$$\frac{d^2\theta}{d\xi^2} - q^2\,\theta^{1+\beta} = 0 \qquad (A.2)$$

The boundary conditions used in the present study are:

$$\text{At } \xi = 0 \ , \quad \theta = 1 \tag{A.3}$$

$$\text{At } \xi = 1 \ , \quad \frac{d\theta}{d\xi} = 0 \tag{A.4}$$

APPENDIX II - GOVERNING DIFFERENTIAL EQUATIONS AND PERTURBATION EQUATIONS FOR THE CIRCULAR ISOTROPIC PLATE

The von-Karman plate theory is used to describe the nonlinear axisymmetric response of the isotropic circular plate used in the present study. The problem is formulated in terms of the transverse displacement w and the radial normal force N_r. The two governing differential equations are given by:

$$D \frac{d}{dr} \left[\frac{1}{r} \frac{d}{dr} \left(r \frac{dw}{dr} \right) \right] - N_r \frac{dw}{dr} = \frac{1}{2} p_o + \frac{P}{2\pi r} \tag{B.1}$$

$$r \frac{d}{dr} \left[\frac{1}{r} \frac{d}{dr} \left(r^2 N_r \right) \right] + \frac{Eh}{2} \left(\frac{dw}{dr} \right)^2 = 0 \tag{B.2}$$

where $D = \dfrac{Eh^3}{12(1-\nu^2)}$ is flexural rigidity of the plate; p_o is the intensity of uniform distributed loading; P is a concentrated load at the center; and h is the thickness of the plate.

The following recursive sets of perturbation equations are obtained by using the two-parameter perturbation expansions for w and N_r, Eqs. 28 and 29.

$$\mathcal{L}(w_{ij}) = \mathcal{R}_{ij} \left(w_{\ell k}, \ N_{\ell k} \right) \tag{B.3}$$

$$\bar{\mathcal{L}}(N_{ij}) = \bar{\mathcal{R}}_{ij} \left(w_{\ell k} \right) \ , \tag{B.4}$$

$$\ell \le i, \quad k < j \quad \text{or} \quad \ell < i, \quad k \le j$$

where $\mathcal{L}, \bar{\mathcal{L}}$ are linear differential operators given by:

$$\mathcal{L} = \frac{d^3}{dr^3} + \frac{1}{r} \frac{d^2}{dr^2} - \frac{1}{r^2} \frac{d}{dr} \tag{B.5}$$

$$\bar{\mathcal{L}} = 2r^2 \frac{d^2}{dr^2} + 6r \frac{d}{dr} \tag{B.6}$$

The explicit form of the first six nonzero components of $\mathcal{R}_{ij}$ and the first three nonzero components of $\bar{\mathcal{R}}_{ij}$ are given in Table 1.

Table 1 - Explicit form of R and $\bar{R}$ functions

i	j	R_{ij}	i	j	$\bar{R}_{ij}$
1	0	$6(1 - \nu^2)\, \dfrac{rh}{a^4}$	2	0	$- Eh \left(\dfrac{dw_{10}}{dr}\right)^2$
0	1	$\dfrac{6(1 - \nu^2)}{\pi r}\, \dfrac{h}{a^2}$			
3	0	$\dfrac{1}{D}\, N_{20}\, \dfrac{dw_{10}}{dr}$	1	1	$- Eh\, \dfrac{dw_{01}}{dr}\, \dfrac{dw_{10}}{dr}$
2	1	$\dfrac{1}{D}\left(N_{20}\, \dfrac{dw_{01}}{dr} + N_{11}\, \dfrac{dw_{10}}{dr}\right)$			

The expressions for R_{12}, R_{03}, and $\bar{R}_{02}$ are obtained from those of R_{21}, R_{30}, and $\bar{R}_{20}$ by interchanging the two subscripts of each of w and N.

REFERENCES

[1] Delves, L.M. and Phillips, C., A fast implementation of the global element method, Journal of the Institute of Mathematics and Its Applications 25 (1980) 177-197.

[2] Delves, L.M. and Mead, K.O., On the convergence rates of variational methods. I. Asymptotically diagonal systems, Mathematics of Computation 25 (1971) 699-716.

[3] Almroth, B.O., Brogan, F.A. and Stern, P., Automatic choice of global shape functions in structural analysis, AIAA Journal 16 (1978) 525-528.

[4] Noor, A.K. and Peters, J.M., Reduced basis technique for nonlinear analysis of structures, AIAA Journal 18 (1980) 455-462.

[5] Noor, A.K., Balch, C.D. and Shibut, M.A., Reduction methods for nonlinear steady-state thermal analysis, NASA TP-2098 (March 1983).

[6] Noor, A.K., Recent advances in reduction methods for nonlinear problems, Computers and Structures 13 (1981) 31-44.

[7] Noor, A.K. and Peters, J.M., Tracing post-limit-point paths with reduced basis technique, Computer Methods in Applied Mechanics and Engineering 28 (1981) 217-240.

[8] Noor, A.K. and Peters, J.M., Bifurcation and postbuckling analysis of laminated composite plates via reduced basis technique, Computer Methods in Applied Mechanics and Engineering 29 (1981) 271-295.

[9] Noor, A.K. and Peters, J.M., Recent advances in reduction methods for instability analysis of structures, Computers and Structures 16 (1983) 67-80.

[10] Noor, A.K. and Balch, C.D., Hybrid perturbation/Bubnov-Galerkin technique for nonlinear thermal analysis, NASA TP-2145 (June 1983).

[11] Noor, A.K., Peters, J.M. and Andersen, C.M., Two-stage Rayleigh-Ritz technique for nonlinear analysis of structures, in: Proceedings of the Second International Symposium on Innovative Numerical Analysis in Applied Engineering Science, June 16-20, 1980, Montreal, Canada.

[12] Guttmann, A.J., Derivation of 'mimic functions' from regular perturbation expansions in fluid mechanics, Journal of the Institute of Mathematics and Its Applications 15 (1975) 307-317.

[13] Van Dyke, M., Analysis and improvement of perturbation series, Quarterly Journal of Mechanics and Applied Mathematics 27 (1974) 423-450.

[14] Watson, L.G. and Chudobiak, J.M., Solution of von Karman's plate equations with perturbation and series summation, in: Research in Structural and Solid Mechanics, NASA CP-2245 (1982).

[15] Aziz, A. and Hamad, G., Regular perturbation expansions in heat transfer, International Journal of Mechanical Engineering Education 5 (1977) 167-182.

[16] Thompson, J.M.T. and Walker, A.C., The nonlinear perturbation analysis of discrete structural systems, International Journal of Solids and Structures 4 (1968) 757-768.

[17] Mathlab Group, MACSYMA Reference Manual (version Ten, first printing, Massachusetts Institute of Technology, January 1983).

[18] Aziz, A. and Benzies, J.Y., Application of perturbation techniques to heat transfer problems with variable thermal properties, International Journal of Heat and Mass Transfer 19 (1976) 271-276.

[19] Chia, C.Y., Nonlinear Analysis of Plates (McGraw-Hill Book Company, 1980).

[20] Nowinski, J.L. and Ismail, I.A., Application of a multi-parameter perturbation method to elastostatics, in: Shaw, W.A. (ed.), Developments in Theoretical and Applied Mechanics, Vol. 2, Proceedings of the Second Southeastern Conference on Theoretical and Applied Mechanics, sponsored by Georgia Institute of Technology, Atlanta, March 5-6, 1974 (Pergamon Press, New York, 1965) 35-45.

[21] Kornishin, M.S. and Isanbaeva, F.S., Flexible Plates and Panels (Nauka, Moscow, 1968; in Russian).

[22] Yaghmai, S., Incremental analysis of large deformations in mechanics of solids with applications to axisymmetric shells of revolution, NASA CR-1350 (1969).

[23] Bathe, K.J., Ramm, E. and Wilson, E.L., Finite element formulations for large deformation dynamic analysis, International Journal for Numerical Methods in Engineering 9 (1975) 353-386.

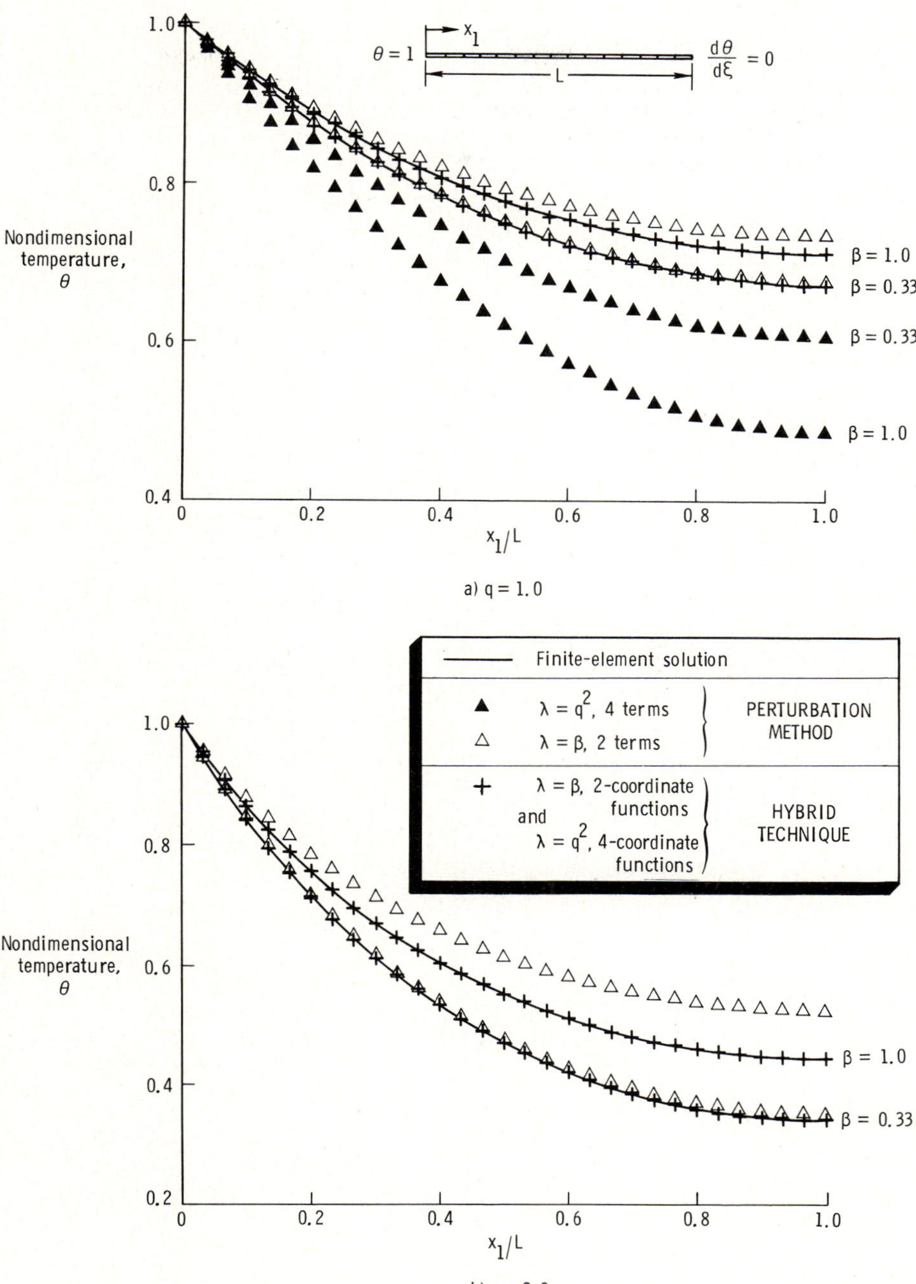

Figure 1. – Comparison of solutions obtained by Perturbation method and Hybrid technique for one-dimensional conducting-convecting fin with variable heat-transfer coefficient.

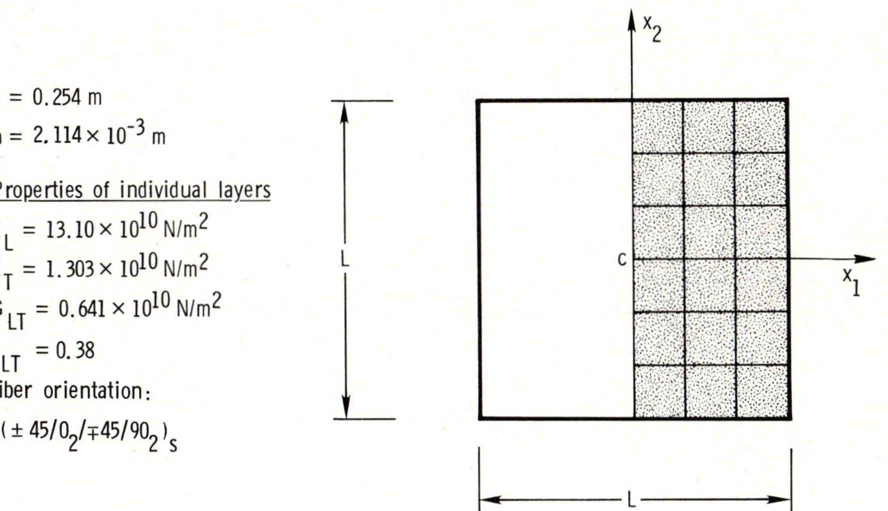

$L = 0.254$ m

$h = 2.114 \times 10^{-3}$ m

<u>Properties of individual layers</u>

$E_L = 13.10 \times 10^{10}$ N/m^2

$E_T = 1.303 \times 10^{10}$ N/m^2

$G_{LT} = 0.641 \times 10^{10}$ N/m^2

$\nu_{LT} = 0.38$

Fiber orientation:

$(\pm 45/0_2/\mp 45/90_2)_s$

<u>Boundary conditions</u>

All edges clamped $u = v = w = \dfrac{\partial w}{\partial x_1} = \dfrac{\partial w}{\partial x_2} = 0$

Figure 2. - Sixteen-ply graphite-epoxy plate used in the present study.

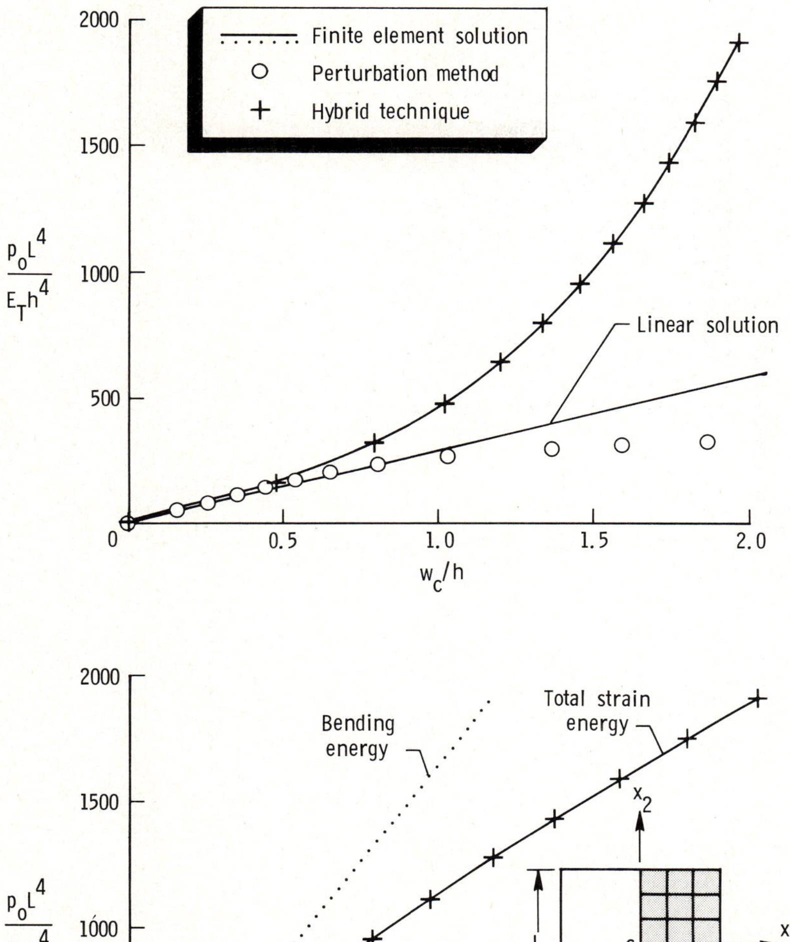

Figure 3. - Accuracy of solutions obtained by Perturbation method and
Hybrid technique at various load levels. Clamped, square
sixteen-ply graphite-epoxy plate subjected to uniform
transverse loading (see Fig. 2).

A.K. Noor

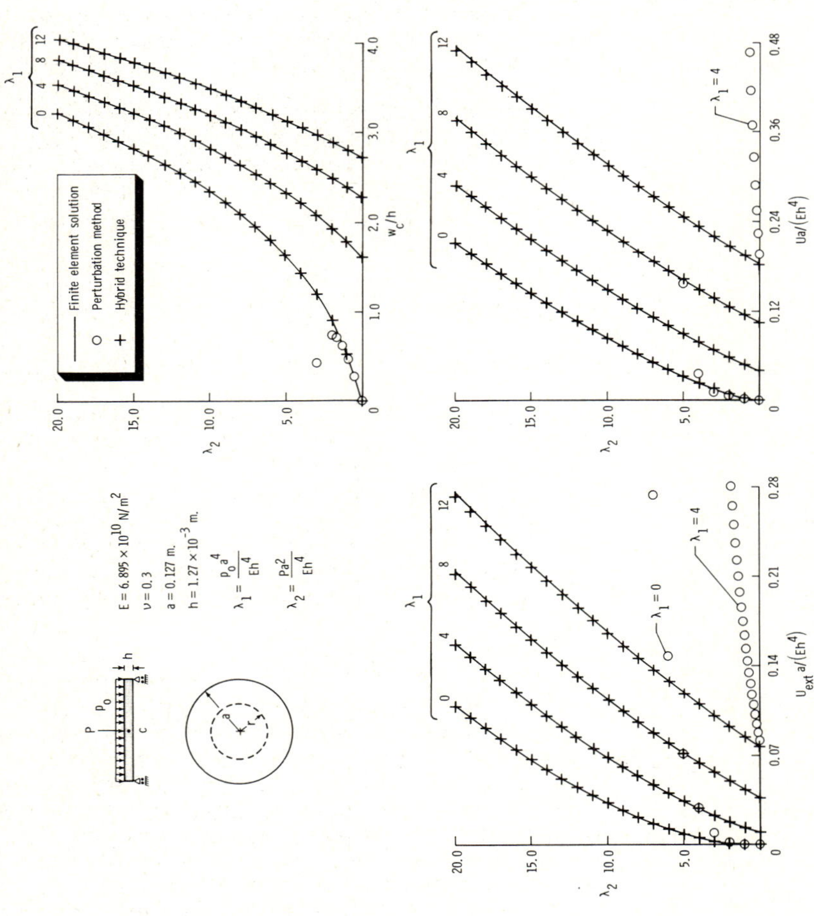

Figure 4. — Accuracy of solutions obtained by two-parameter Perturbation method and Hybrid technique. Isotropic circular plate subjected to combined uniform and concentrated loading.

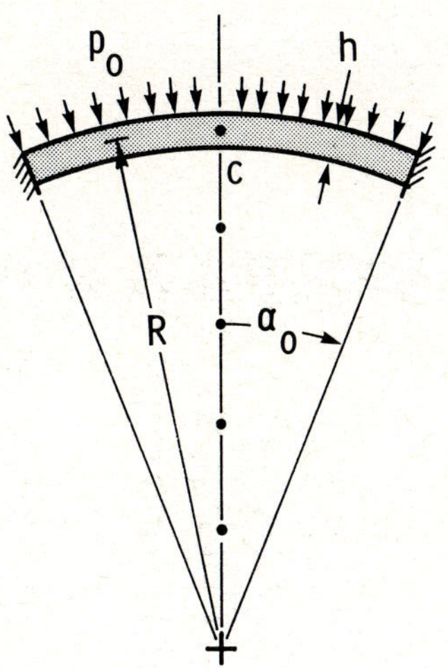

Figure 5. - Shallow spherical caps used in the present study.

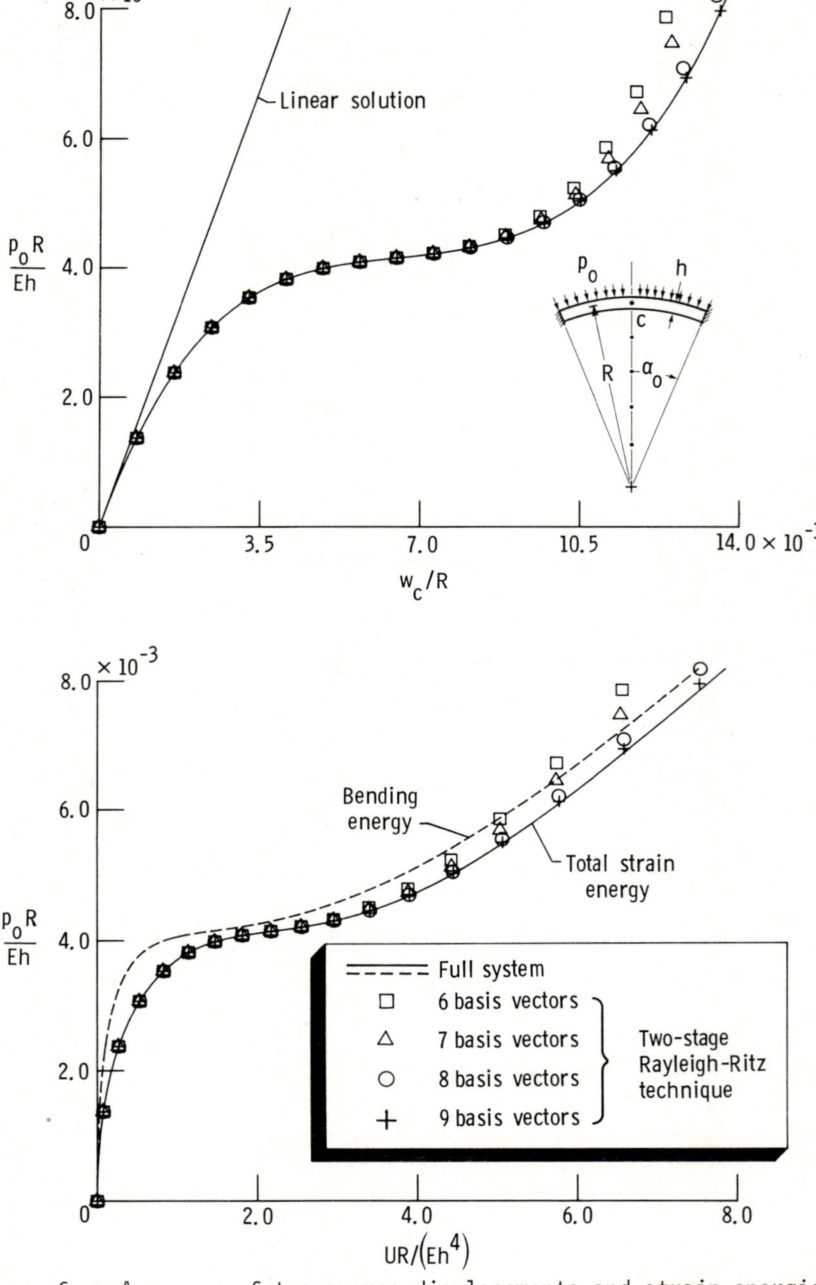

Figure 6. – Accuracy of transverse displacements and strain energies obtained by the two-stage Rayleigh-Ritz technique at various load levels. Shallow spherical cap shown in Fig. 5, $h = 0.0127$.

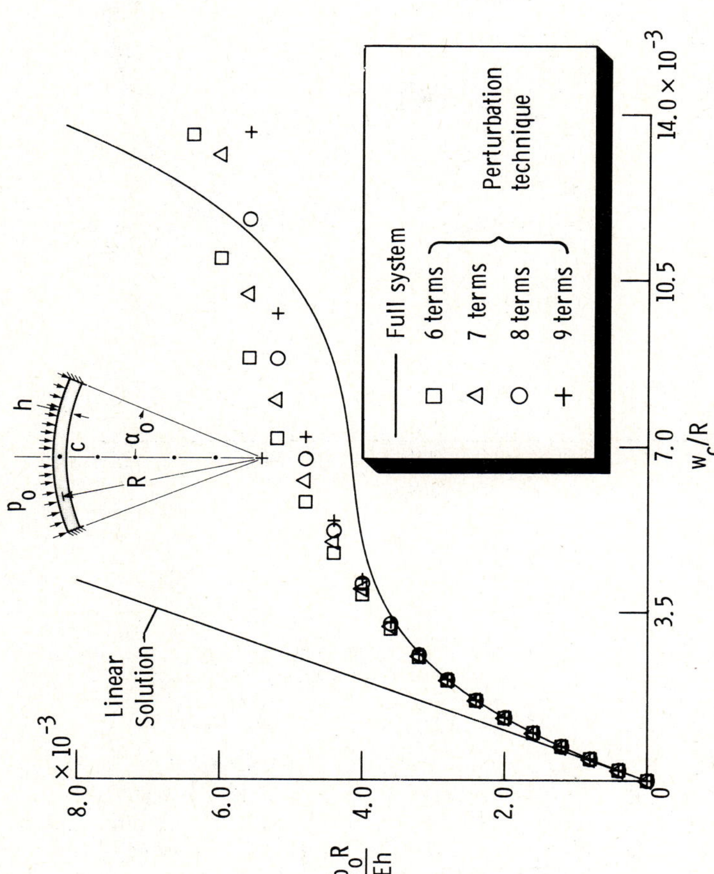

Figure 7. – Accuracy of transverse displacements obtained by the static Perturbation technique at various load levels. Shallow spherical cap shown in Fig. 5, h = 0.0127 m.

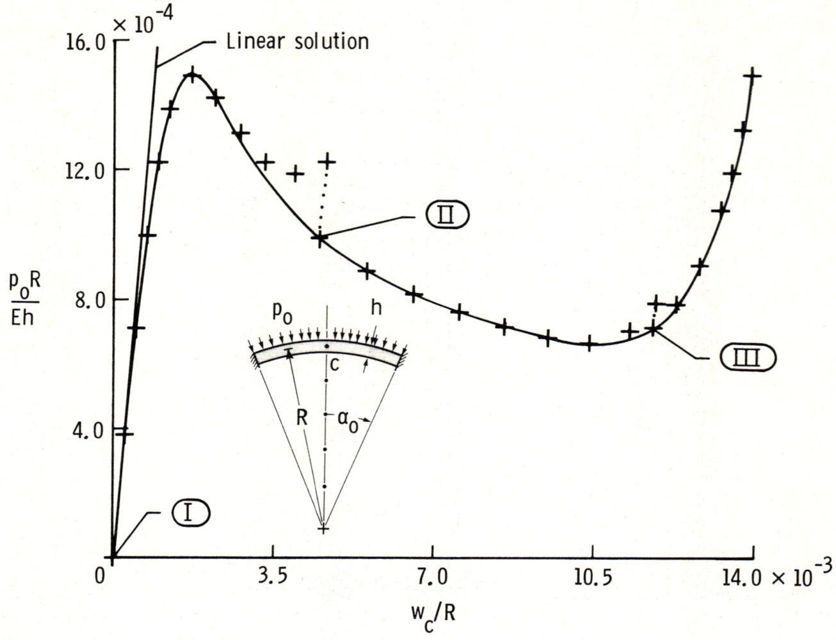

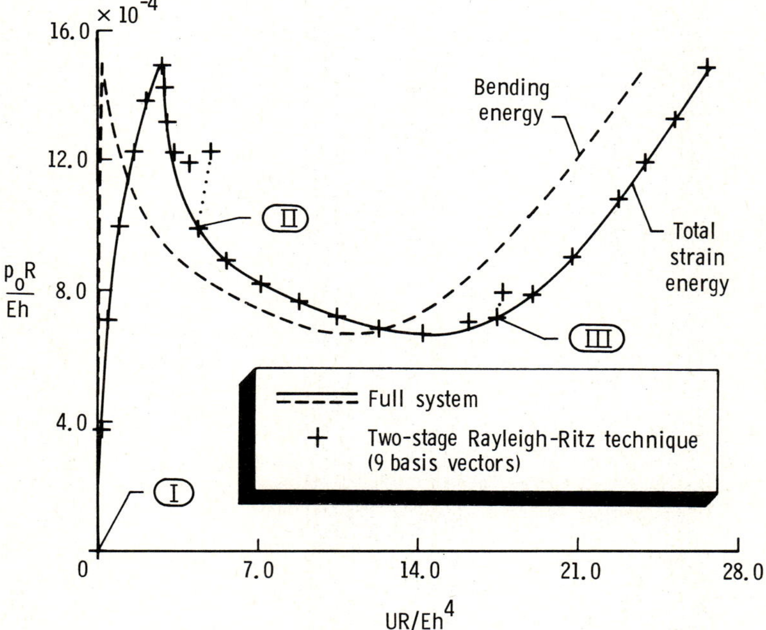

Figure 8. – Accuracy of transverse displacements and strain energies obtained by the two-stage Rayleigh-Ritz technique at various load levels – shallow spherical cap shown in Fig. 5, h = 0.005 m.

Unification of Finite Element Methods
H. Kardestuncer (Editor)
© Elsevier Science Publishers B.V. (North-Holland), 1984

CHAPTER 13

COLLOCATION SOLUTION OF THE TRANSPORT EQUATION USING A LOCALLY ENHANCED ALTERNATING DIRECTION FORMULATION

M.A. Celia & G.F. Pinder

INTRODUCTION

The alternating-direction collocation (ADC) method has been shown to be an attractive technique for the solution of the multi-dimensional transport equation (Celia, 1983; Celia et al., 1980). The method achieves high-order accuracy while benefitting from the computational efficiencies inherent in the alternating direction procedure. In this paper, we present a modification of the ADC procedure designed to achieve enhanced accuracy in the neighborhood of a sharp front. We begin with a review of the ADC method. On this foundation we build the enhancement procedure. Finally, the enhanced ADC method is applied to an example problem. To simplify our presentation, we consider only rectangular subspaces.

THE ADC PROCEDURE

Consider the generalized transport equation

$$(1) \qquad D_t u + \underset{\sim}{B}\cdot\nabla u - C\nabla^2 u = 0,$$

where (x,y) designates a two-dimensional spatial coordinate system, t is time, $D_t(\cdot)$ is differentiation with respect to time, and B and C are assumed spatially constant for simplicity. For a treatment of spatial variability, see Celia (1983).

Let us begin the development of the basic ADC procedure by writing a fractional step algorithm based upon a finite difference approximation of the time derivative, that is

$$(2a) \qquad \frac{u^{n+1/2} - u^n}{\Delta t} + L_x u^{n+1/2} = 0,$$

$$(2b) \qquad \frac{u^{n+1} - u^{n+1/2}}{\Delta t} + L_y u^{n+1} = 0,$$

$$\text{where} \quad L_x(\cdot) \equiv B_x D_x(\cdot) - C D_{xx}(\cdot)$$

$$L_y(\cdot) \equiv B_y D_y(\cdot) - C D_{yy}(\cdot)$$

and B_x and B_y denote vector components in the x and y coordinate directions, respectively, and $D_x(\cdot)$, $D_y(\cdot)$, $D_{xx}(\cdot)$ and $D_{yy}(\cdot)$ are space derivatives. The equivalent one-step equation is obtained by substitution of (2b) into (2a), that is

$$(3) \quad \frac{u^{n+1} - u^n}{\Delta t} + L_x u^{n+1} + L_y u^{n+1} = -(\Delta t) L_x L_y u^{n+1}.$$

Examination of equations (3) and (1) reveals an error of order Δt due to the fractional step procedure presented as equations (2). This error term may be neglected, or a suitable correction can be added to the right hand side of (2). Such a correction is discussed in the Appendix.

One must now choose a suitable approximation for u. If we choose Hermite cubic polynomials as our basis, the approxiation can be written

$$(4) \quad u(x,y,t) \approx \hat{u}(x,y,t) = \sum_{i=1}^{I_o} \sum_{j=1}^{J_o} U_{ij}(t)\phi_i^0(x)\phi_j^0(y)$$

$$+ \frac{\partial U_{ij}}{\partial x}(t)\phi_i^1(x)\phi_j^0(y) + \frac{\partial U_{ij}}{\partial y}(t)\phi_i^0(x)\phi_j^1(y)$$

$$+ \frac{\partial^2 U_{ij}}{\partial x \partial y}\phi_i^1(x)\phi_j^1(y)$$

where $\phi_i^0(x)$, $\phi_j^0(y)$, $\phi_i^1(x)$, $\phi_j^1(y)$ are piecewise Hermite cubic polynomials. The cubic Hermite basis functions defined along x can be written in terms of the local element coordinate

$$\xi = 1 - 2\left(\frac{x_{i+1} - x}{\Delta x}\right), \quad \xi = [-1,1], \quad \text{where} \quad x = [x_i, x_{i+1}],$$

$$\Delta x = x_{i+1} - x_i,$$

and (x_i, y_j) denotes a nodal location, as

$$(5a) \quad \phi^0_{\xi_0}(\xi) = \begin{cases} \frac{1}{4}(\xi^3 - 3\xi + 2) & (\xi_0 = -1) \\[2em] -\frac{1}{4}(\xi^3 - 3\xi - 2) & (\xi_0 = 1) \end{cases}$$

$$(5b) \quad \phi^1_{\xi_0}(\xi) = \begin{cases} (\frac{\Delta x}{2})\frac{1}{4}(\xi^3 - \xi^2 - \xi + 1) & (\xi_0 = -1) \\[2em] (\frac{\Delta x}{2})\frac{1}{4}(\xi^3 + \xi^2 - \xi - 1) & (\xi_0 = 1) \end{cases}$$

The functions $\phi^0_j(y)$ and $\phi^1_j(y)$ are of the same functional form. These functions and the (i,j) notation are presented in figure 1. The parameters I_0 and J_0 denote the number of nodes in the x and y directions respectively. The undetermined coefficients U_{ij}, $\partial U_{ij}/\partial x$, $\partial U_{ij}/\partial y$, $\partial^2 U_{ij}/\partial x \partial y$ are values of $\hat{u}$, $D_x\hat{u}$, $D_y\hat{u}$, and $D_{xy}\hat{u}$ at the nodal locations, respectively.

Substituting $u(x,y,t)$ from (4) into (2a) and regrouping terms, one obtains

$$\sum_{j=1}^{J_0} \phi^0_j(y)\{ \sum_{i=1}^{I_0} U^{n+1/2}_{ij}[\phi^0_i(x) + (\Delta t)L_x\phi^0_i(x)] + \frac{\partial U^{n+1/2}_{ij}}{\partial x}[\phi^1_i(x) + (\Delta t)L_x\phi^1_i(x)]\}$$

$$(6) \quad + \phi^1_j(y)\{ \sum_{i=1}^{I_0} \frac{\partial U^{n+1/2}_{ij}}{\partial y}[\phi^0_i(x) + (\Delta t)L_x\phi^0_i(x)] + \frac{\partial^2 U^{n+1/2}_{ij}}{\partial x \partial y}[\phi^1_i(x) + (\Delta t)L_x\phi^1_i(x)]\}$$

$$= \sum_{j=1}^{J_0} \phi^0_j(y)\{ \sum_{i=1}^{I_0} U^n_{ij}\phi^0_i(x) + \frac{\partial U^n_{ij}}{\partial x}\phi^1_i(x)\} + \phi^1_j(y)\{ \sum_{i=1}^{I_0} \frac{\partial U^n_{ij}}{\partial y}\phi^0_i(x)$$

$$+ \frac{\partial^2 U^n_{ij}}{\partial x \partial y}\phi^1_i(x)\}.$$

Because ϕ^0_j and ϕ^1_j are independent functions of y, and because the quantities in the brackets are functions of x only, equation (7) implies that anywhere along y

$$(7a) \quad \sum_{i=1}^{I_0} U^{n+1/2}_{ij}[\phi^0_i(x) + (\Delta t)L_x\phi^0_i(x)] + \frac{\partial U^{n+1/2}_{ij}}{\partial x}[\phi^1_i(x) + (\Delta t)L_x\phi^1_i(x)]$$

$$= \sum_{i=1}^{I_o} U_{ij}^n \phi_i^0(x) + \frac{\partial U_{ij}^n}{\partial x} \phi_i^1(x)$$

$$(7b) \quad \sum_{i=1}^{I_o} \frac{\partial U_{ij}^{n+1/2}}{\partial y} [\phi_i^0(x) + (\Delta t)L_x\phi_i^0(x)] + \frac{\partial^2 U_{ij}^{n+1/2}}{\partial x \partial y} [\phi_i^1(x) + (\Delta t)L_x\phi_i^1(x)]$$

$$= \sum_{i=1}^{I_o} \frac{\partial U_{ij}^n}{\partial y} \phi_i^0(x) + \frac{\partial^2 U_{ij}^n}{\partial x \partial y} \phi_i^1(x)$$

The solution of (7) for the $4I_o$ unknown parameters requires the evaluation of (7) at $4I_o$ collocation points. Because we employ orthogonal collocation, the collocation point locations correspond to the zeros of the Legendre polynomials. These locations commonly are employed in numerical integration also, wherein they are known as "Gauss points". Each collocation point can be associated with an unknown parameter in equation (7). Thus, for convenience, we identify them, as indicated in figure 1, as u_{IJ}, u_{xIJ}, u_{yIJ} and u_{xyIJ}. The complete set of equations for the x direction ADC sweep can now be written using the notation $\bar{L}_x(\cdot) \equiv (1 + (\Delta t)L_x)(\cdot)$, that is

$$(8a) \quad \sum_{i=1}^{I_o} [U_{ij}^{n+1/2}\bar{L}_x(\phi_i^0(x)) + \frac{\partial U_{ij}^{n+1/2}}{\partial x} \bar{L}_x(\phi_i^1(x)) - U_{ij}^n\phi_i^0(x)$$

$$- \frac{\partial U_{ij}^n}{\partial x} \phi_i^1(x)]\bigg|_{x_\alpha,y_\alpha} = 0 \qquad \begin{array}{l} I = 1,2,\ldots I_o \\[4pt] J = j = 1,2,\ldots J_o \\[4pt] \alpha = u_{IJ}, \ u_{xIJ} \end{array}$$

$$(8b) \quad \sum_{i=1}^{I_o} [\frac{\partial U_{ij}^{n+1/2}}{\partial y} \bar{L}_x(\phi_i^0(x)) + \frac{\partial^2 U_{ij}^{n+1/2}}{\partial x \partial y} \bar{L}_x(\phi_i^1(x)) - \frac{\partial U_{ij}^n}{\partial y} \phi_i^0(x)$$

$$- \frac{\partial^2 U_{ij}^n}{\partial x \partial y} \phi_i^1(x)]\bigg|_{x_\alpha,y_\alpha} = 0 \qquad \begin{array}{l} I = 1,2,\ldots I_o \\[4pt] J = j = 1,2,\ldots J_o \\[4pt] \alpha = u_{yIJ}, \ u_{xyIJ} \end{array}$$

Both (8a) and (8b) require boundary information to be properly posed. The imposition of boundary conditions in the context of splitting schemes must be undertaken with some care; the procedure used herein is given in the Appendix. Once this specification has been made, the resulting systems of $2I_0 - 2$ equations resulting from (8a) or (8b) can be solved directly. This procedure is repeated $2J_0 - 2$ times.

One next proceeds to the second half of the time step. Equations analogous to (8a) and (8b) are written using (2b) and the collocation points (x_α, y_α), $\alpha = u, u_y$, $J = 1, 2, \ldots, J_0$, $I = 1, 2, \ldots, I_0$ and (x_α, y_α), $\alpha = u_x, u_{xy}$, $J = 1, 2, \ldots, J_0$, $I = 1, 2, \ldots, I_0$. These equations are also adjusted for boundary conditions and each set of $2J_0 - 2$ equations is solved for $I = 1, 2, \ldots, I_0$. An entire (x-y) sweep of the mesh is now complete and one proceeds to the next time step.

LOCAL ENHANCEMENT

While the ADC procedure described heretofore constitutes a very efficient and accurate algorithm for the solution of problems in not only two but also three space dimensions, problems exhibiting solutions with steep fronts may require a higher-order approximation in the neighborhood of the front. This local enhancement is achieved by using C^1 quintic polynomials as interpolants in elements where a steep concentration gradient is encountered (see Mohsen and Pinder, 1983). The local use of quintics is particularly attractive when the additional degrees of freedom due to the quintic formulation can be condensed out at the element level. When this local condensation can be achieved, the overall solution can be enhanced without increasing the rank or bandwidth of the global coefficient matrix. Unfortunately, when a standard multi-dimensional collocation approximation is employed, the introduction of midside nodes for local enhancement couples together adjacent elements as illustrated in figure 2. This makes local condensation algorithmically difficult and computationally inefficient. However, as we illustrate shortly, when the ADC procedure is employed, local condensation is readily achieved.

Let us now formulate the approximating equations for an enhanced element. A point of departure is equation (5). The objective in the enhanced procedure is to employ C^1 quintic polynomials in place of cubic Hermites to represent $\hat{u}(x,y,t)$ in elements known to contain solution segments exhibiting a sharp front. While the interpolating polynomial for such elements reads as does (4), the basis functions $\phi_i^0(x)$, $\phi_j^0(y)$, $\phi_i^1(x)$ and $\phi_j^1(y)$ are now defined in the local ξ coordinate system as (see Fig. 2)

$$
(9a) \qquad \phi_{\xi_0}^0(\xi) =
\begin{cases}
-\dfrac{1}{4}(3\xi^5 - 2\xi^4 - 5\xi^3 + 4\xi^2) & (\xi_0 = -1) \\[3mm]
\xi^4 - 2\xi^2 + 1 & (\xi_0 = 0) \\[3mm]
\dfrac{1}{4}(-3\xi^5 - 2\xi^4 + 5\xi^3 + 4\xi^2) & (\xi_0 = 1)
\end{cases}
$$

$$(9b)\qquad \phi^1_{\xi_0}(\xi) = \begin{cases} \dfrac{\Delta x}{8}\left(\xi^5 - \xi^4 - \xi^3 + \xi^2\right) & (\xi_0 = -1) \\[2ex] \dfrac{\Delta x}{2}\left(\xi^5 - 2\xi^3 + \xi\right) & (\xi_0 = 0) \\[2ex] \dfrac{\Delta x}{8}\left(\xi^5 + \xi^4 - \xi^3 - \xi^2\right) & (\xi_0 = 1) \end{cases}$$

One now follows a development for the quintic element that is analogous
to that presented above for the cubics. The outcome is an equation very
similar to (8). The principal difference is that now there are corner,
mid-side, and mid-element nodes with which to contend. Moreover, one
must also employ more collocation points to accommodate the additional
degrees of freedom associated with the quintic polynomial in the
enhanced element. Note that we illustrate in figure 2 how the enhancement
of one element formally leads to the enhancement of the entire row and
column of elements that includes the enhanced element. By employing
this extended approach the x and y sweeps are uniquely defined, and there
is no ambiguity regarding the choice of collocation points to be used in
each sweep.

While we could at this point formulate and solve a global matrix equation
for each row, it is more computationally efficient to reduce the quintic
elements at the local level. To explain this concept, we write the
element collocation matrix equation for a typical (one-dimensional)
quintic element in the x-direction.

$$(10)\quad \begin{bmatrix} L_x(\phi^0_{-1}(\xi))\big|_{u_{x_{I-1,J}}} & L_x(\phi^1_{-1}(\xi))\big|_{u_{x_{I-1,J}}} & L_x(\phi^0_0(\xi))\big|_{u_{x_{I-1,J}}} & L_x(\phi^1_0(\xi))\big|_{u_{x_{I-1,J}}} & L_x(\phi^0_1(\xi))\big|_{u_{x_{I-1,J}}} & L_x(\phi^1_1(\xi))\big|_{u_{x_{I-1,J}}} \\[2ex] L_x(\phi^0_{-1}(\xi))\big|_{u_{IJ}} & L_x(\phi^1_{-1}(\xi))\big|_{u_{IJ}} & L_x(\phi^0_0(\xi))\big|_{u_{IJ}} & L_x(\phi^1_0(\xi))\big|_{u_{IJ}} & L_x(\phi^0_1(\xi))\big|_{u_{IJ}} & L_x(\phi^1_1(\xi))\big|_{u_{IJ}} \\[2ex] L_x(\phi^0_{-1}(\xi))\big|_{u_{xIJ}} & L_x(\phi^1_{-1}(\xi))\big|_{u_{xIJ}} & L_x(\phi^0_0(\xi))\big|_{u_{xIJ}} & L_x(\phi^1_0(\xi))\big|_{u_{xIJ}} & L_x(\phi^0_1(\xi))\big|_{u_{xIJ}} & L_x(\phi^1_1(\xi))\big|_{u_{xIJ}} \\[2ex] L_x(\phi^0_{-1}(\xi))\big|_{u_{I+1,J}} & L_x(\phi^1_{-1}(\xi))\big|_{u_{I+1,J}} & L_x(\phi^0_0(\xi))\big|_{u_{I+1,J}} & L_x(\phi^1_0(\xi))\big|_{u_{I+1,J}} & L_x(\phi^0_1(\xi))\big|_{u_{I+1,J}} & L_x(\phi^1_1(\xi))\big|_{u_{I+1,J}} \end{bmatrix}$$

$$\begin{bmatrix} u^{n+1/2}_{i-1,j} \\[1.5ex] \dfrac{\partial u^{n+1/2}_{i-1,j}}{\partial x} \\[1.5ex] u^{n+1/2}_{ij} \\[1.5ex] \dfrac{\partial u^{n+1/2}_{ij}}{\partial x} \\[1.5ex] u^{n+1/2}_{i+1,j} \\[1.5ex] \dfrac{\partial u^{n+1/2}_{i+1,j}}{\partial x} \end{bmatrix} \quad =$$

$$\left[U^n_{i-1,j}\phi^0_{-1}(\xi) + \frac{\partial U^n_{i-1,j}}{\partial x}\phi^1_{-1}(\xi) + U^n_{ij}\phi^0_0(\xi) + \frac{\partial U^n_{ij}}{\partial x}\phi^1_0(\xi) + U^n_{i+1,j}\phi^0_1(\xi) + \frac{\partial U^n_{i+1,j}}{\partial x}\phi^1_1(\xi) \right]_{u_{xI-1,J}}$$

$$\left[U^n_{i-1,j}\phi^0_{-1}(\xi) + \frac{\partial U^n_{i-1,j}}{\partial x}\phi^1_{-1}(\xi) + U^n_{ij}\phi^0_0(\xi) + \frac{\partial U^n_{ij}}{\partial x}\phi^1_0(\xi) + U^n_{i+1,j}\phi^0_1(\xi) + \frac{\partial U^n_{i+1,j}}{\partial x}\phi^1_1(\xi) \right]_{u_{IJ}}$$

$$\left[U^n_{i-1,j}\phi^0_{-1}(\xi) + \frac{\partial U^n_{i-1,j}}{\partial x}\phi^1_{-1}(\xi) + U^n_{ij}\phi^0_0(\xi) + \frac{\partial U^n_i}{\partial x}\phi^1_0(\xi) + U^n_{i+1,j}\phi^0_1(\xi) + \frac{\partial U^n_{i+1,j}}{\partial x}\phi^1_1(\xi) \right]_{u_{xIJ}}$$

$$\left[U^n_{i-1,j}\phi^0_{-1}(\xi) + \frac{\partial U^n_{i-1,j}}{\partial x}\phi^1_{-1}(\xi) + U^n_{ij}\phi^0_0(\xi) + \frac{\partial U^n_{ij}}{\partial x}\phi^1_0(\xi) + U^n_{i+1,j}\phi^0_1(\xi) + \frac{\partial U^n_{i+1,j}}{\partial x}\phi^1_1(\xi) \right]_{u_{I+1,J}}$$

Equation (10) institutes four equations in six unknown parameters. Because $\phi^0_0(\xi)$ and $\phi^1_0(\xi)$ are non-zero only over the element for which they are defined, the two variables $U^{n+1/2}_{ij}$ and $\partial U^{n+1/2}_{ij}/\partial x$ can be eliminated algebraically from equation (10). Equation (10) then reduces to two equations in four unknowns, the same equation configuration that arises from employing a Hermite cubic interpolation. One can now assemble the resulting two equations into the global coefficient matrix and maintain the same global matrix structure as would be generated using all cubic interpolates. The value of the coefficients derived from quintic elements are, of course, different from those derived from cubic elements. Once the global equations are solved, one can obtain the center node coefficients from the element-level equations.

The enhanced-element concept is a dynamic procedure. As the solution geometry changes, different elements are enhanced. In other words, selected cubic elements become quintic and certain quintic elements revert to cubic. To obtain starting values of u(x,y,t) for a new time step at new center nodes identified with quintic interpolants, the interpolation property of the Hermite cubic polynomial is employed in conjunction with existing nodal coefficients. The entire procedure is presented as a flow chart in figure 3.

EXAMPLE CALCULATIONS

To illustrate the effectiveness of the enhanced ADC procedure, we will now solve a convection-dominated transport equation and investigate the benefits of adding a small number of quintic elements along the principal direction of flow. The equation to be solved is

$$(11) \quad D_t u + 10 D_x u - \nabla^2 u = 0$$

subject to

$$\frac{\partial u}{\partial x} \rightarrow 0 \qquad \text{as} \qquad x \rightarrow \infty$$

$$\frac{\partial u}{\partial y} \rightarrow 0 \qquad \text{as} \qquad y \rightarrow \pm\infty$$

$$u = 1 \text{ at } x = 0, \quad 30 \le y \le 50$$

$$u = 0 \text{ at } x = 0, \quad \text{all other } y$$

$$u = 0 \text{ at } t = 0$$

The solution to this equation is given by

$$(12) \qquad u(x,y,t) = \frac{x}{4\sqrt{\pi}} \, e^{5x} \int_{\tau=0}^{t} \tau^{-3/2} \, e^{-(25\tau + \frac{x^2}{4\tau})}$$

$$\cdot \left\{ - \text{erfc} \left(\frac{50-y}{2\sqrt{\tau}} \right) + \text{ercc} \left(\frac{30-y}{2\sqrt{\tau}} \right) \right\} d\tau.$$

A plot of the solution, along the line $y = 40$, is shown in figure 5.

Because the numerical solution must be solved on a bounded domain, the system illustrated in figure 4a has been solved numerically. To assure that (12) can be used to obtain an error measure, the numerical solution is terminated before the finite boundaries are reflected in the solution.

The numerical solution procedure chooses the n elements to be enhanced by interrogating the solution obtained for the last time step. The n elements in which the largest changes in the solution occur are assigned a quintic formulation. This procedure is clearly dynamic and accommodates the changing solution topology by incorporating enhanced accuracy only in regions where it is needed.

In this example problem, let us consider the grid configuration of figure 4b. Moreover, let us employ in our computational algorithm the numerical correction factor described in (A.9) of the appendix to this paper. We will use as our error measure

$$(13) \qquad E = \sum_{m=1}^{M} e_m(t = 0.5) + e_m(t = 1.0) + e_m(t = 2.0),$$

where $e_m = |u-\hat{u}|$, u is the exact solution, i.e. (12), and $\hat{u}$ is the collocation approximation. The parameter M is the total number of nodes along the lines $y = 40$, $y = 32$, and $y = 28$. (note that the solution is symmetric about $y = 40$). The numerical solution for the cases of $n = 0$ and $n = 2$ are given in figure 5. Figure 6 shows the role of the time truncation error in this approach. As Δt becomes smaller, the spatial error dominates and the improvement in accuracy due to quintic

enhancement becomes evident. The improvement in accuracy is even more
evident for a course mesh. For example, when only four elements are
used along the x-direction, just one quintic reduces the error E from
1.6 for the case of all cubic elements to 0.85, a 47% decrease in error
(for Δt = .01). In comparison, only a 16% error improvement was achieved
using the quintic enhancement and eight elements.

While the error measure described above is certainly of interest,
numerical dispersion, exemplified by overshoot and undershoot, is also
worth investigating. When quintic enhancement is employed, the already
very small overshoot (approximately 2.5%) can be eliminated entirely and
the undershoot (approximately 3.5%) can be reduced to about 1.0%.
Numerical dispersion can be further modified by changing the strategy
used for the selection of the enhanced elements. For example, by
enhancing only the element immediately behind the one exhibiting the
maximum change, the undershoot can be reduced at the expense of a small
(<1%) overshoot.

ACKNOWLEDGEMENTS

This work was supported in part by the Department of Energy under contract
DE-AC02-83ER60170 and also by the National Science Foundation under
contract NSF CEE81-11240. Thanks are given to Dr. Linda M. Abriola for
her helpful discussions concerning the splitting scheme. The contribution
of Dr. Robert Cleary, who supplied the computed code for the analytical
solution used herein, is also appreciated.

REFERENCES

[1] Celia, M.A., Collocation on Deformed Finite Elements and Alternating
     Direction Collocation Methods, Ph.D. Thesis, Department of Civil
     Engineering, Princeton University (Sept. 1983)

[2] Celia, M.A., Pinder, G.F. and Hayes, L.J., Alternating Direction
     Collocation Solution to the Transport Equation, Proc. Third Int.
     Conf. Finite Elements in Water Resources, Wang et al (eds), Univ.
     of Mississippi, (1980), pp. 3.36-3.48.

[3] Gourlay, A.R. and Mitchell, A.R., The Equivalence of Certain
     Alternating Direction and Locally One-Dimensional Methods, S.I.A.M.
     J. Num. Anal.,6 (1) (1969) 37-46.

[4] Mohsen, M.F.N. and Pinder, G.F., Orthogonal Collocation with
     'Adaptive' Finite Elements, submitted to Int. J. Num. Meth. Engrg.,
     (1983).

APPENDIX

In this appendix, the treatment of boundary conditions and of non-zero right hand side forcing functions will be discussed, as will the implementation of the correction term discussed in section 2. To begin, let us consider an equation of the form of (1), written with the definitions of (3), as

$$(A.1) \qquad D_t u + L_x u + L_y u = f(x,y,t)$$

subject to some general Dirichlet or Neumann boundary conditions. The general multi-dimensional, fully-implicit collocation approximation to (A.1) produces an algebraic equation of the form

$$(A.2) \qquad (k_x k_y + (\Delta t)M_x k_y + (\Delta t)M_y k_x)u^{n+1} = k_x k_y u^n$$
$$+ (\Delta t)F + B$$

where K_{x_i}, M_{x_i} correspond to mass and stiffness matrices, respectively, the vector F corresponds to the forcing function f, and B is a vector which accounts for non-zero boundary conditions. When developing split equations, we will require the right hand side of the composite (equivalent one-step) equation to be the same as that of (A.2)

Consider the split equations

$$(A.3a) \qquad \frac{u^{n+1/2} - u^n}{\Delta t} + L_x u^{n+1/2} = F_1/\Delta t$$

$$(A.3b) \qquad \frac{u^{n+1} - u^{n+1/2}}{\Delta t} + L_y u^{n+1} = F_2/\Delta t$$

where F_1 and F_2 are non-zero vectors which are to be determined. Upon application of the collocation method and the elimination of $u^{n+1/2}$ to form the equivalent one-step equation, there results

$$(A.4) \qquad [k_x k_y + \Delta t(M_x k_y + M_y k_x) + (\Delta t)^2 M_x M_y]u^{n+1}$$
$$= k_x k_y u^n + k_y F_1 + (k_x + \Delta t M_x)F_2.$$

Through a comparison of (A.4) and (A.2), it is clear that the condition

$$(A.5) \qquad k_y F_1 + (k_x + \Delta t M_x)F_2 = (\Delta t)F + B$$

should be satisfied. While an infinite number of solutions exist, a simple one is

(A.6a) $F_1 = 0$

(A.6b) $F_2 = (k_x + \Delta t M_x)^{-1} (\Delta t\ F + B)$

or, letting

(A.7) $\bar{F} = (\Delta t)F + B$

one must have

(A.8) $(k_x + \Delta t M_x)F_2 = \bar{F}.$

Notice that the coefficient matrix of (A.8) is exactly the coefficient matrix for the x-direction solutions. Therefore, this matrix needs to be reduced only once to solve both (A.8) and the x-direction equations given by (8). Furthermore, if neither f nor the boundary conditions are time varying, the vector $\bar{F}$ needs to be computed only once. This discussion of non-zero right hand sides for splitting schemes is similar to that presented by Gourlay and Mitchell (1969) where LOD and ADI difference schemes were compared.

The addition of a correction term to account for the perturbation term $(\Delta t)^2\ M_x M_y u^{n+1}$ can be easily achieved through a modification of the definition of $\bar{F}$. Specifically, define the new vector $\bar{F}_c$ as

(A.9a) $\bar{F}_c = (\Delta t)F + B + (\Delta t)^2 C$

where the correction vector C is given by

(A.9b) $C = M_x M_y u^n .$

Notice that the addition of this correction term reduced the error introduced by the perturbation term by one order, since the perturbation term is now of the form $(\Delta t)^2\ M_x M_y (u^{n+1} - u^n)$. With $\bar{F}_c$ replacing $\bar{F}$, the solution procedure for this correction case leads directly to (A.3) via (A.9).

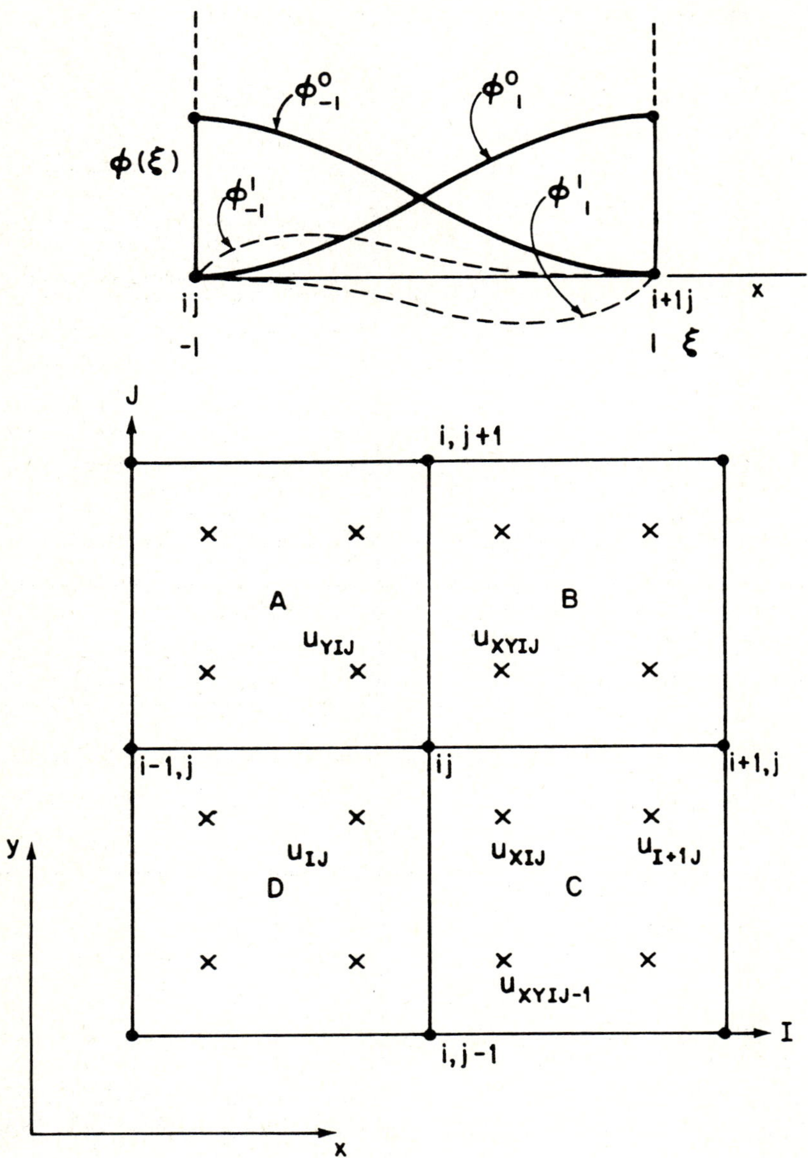

Figure 1: Diagrammatic representation of a collocation-finite element
network illustrating the nodes (·) and collocation points (x).
The nodes carry the subscript (i,j), the collocation points
the subscript α(I,J), and the elements are identified by the
upper case Roman letters (A-D). The cubic bases are also shown.

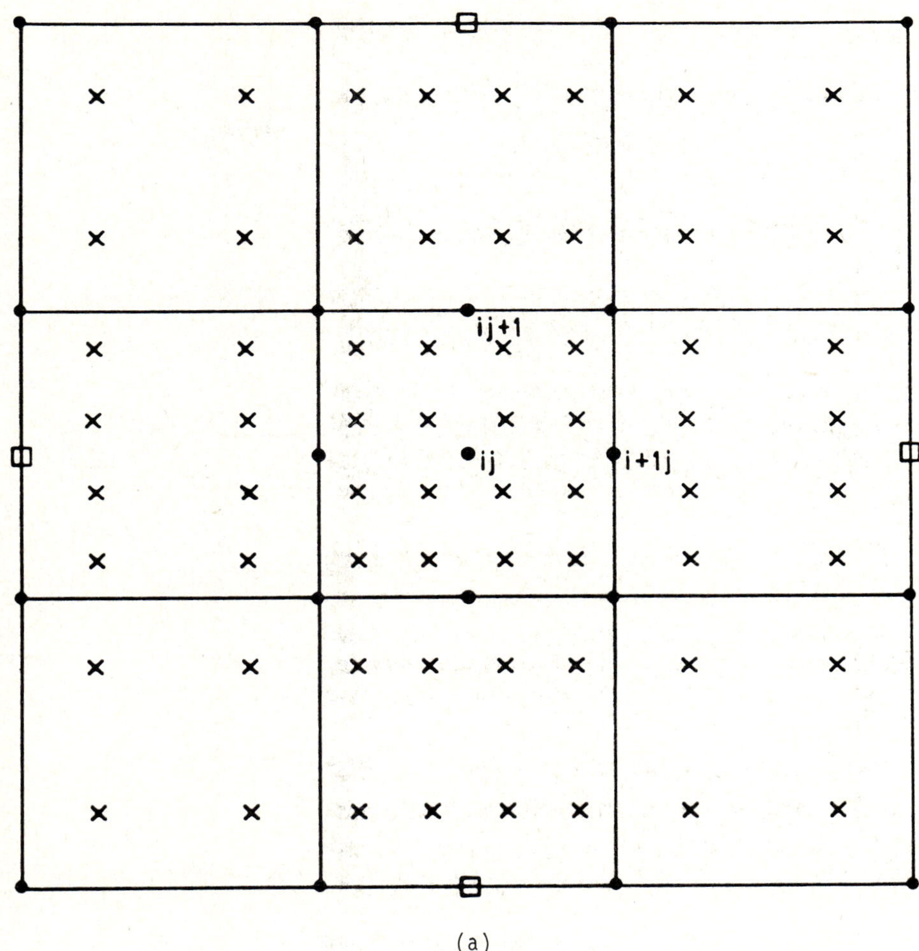

(a)

Figure 2: (a) Diagrammatic representation of the nodes and collocation points associated with local enhancement of the center element. The nodes indicated by ▢ are formally required for the locally enhanced ADC procedure. (b) The quintic bases $\phi_i^0(\xi)$. (c) The quintic functions $\phi_i^1(\xi)$, assuming $\Delta x = 2$.

M.A. Celia & G.F. Pinder

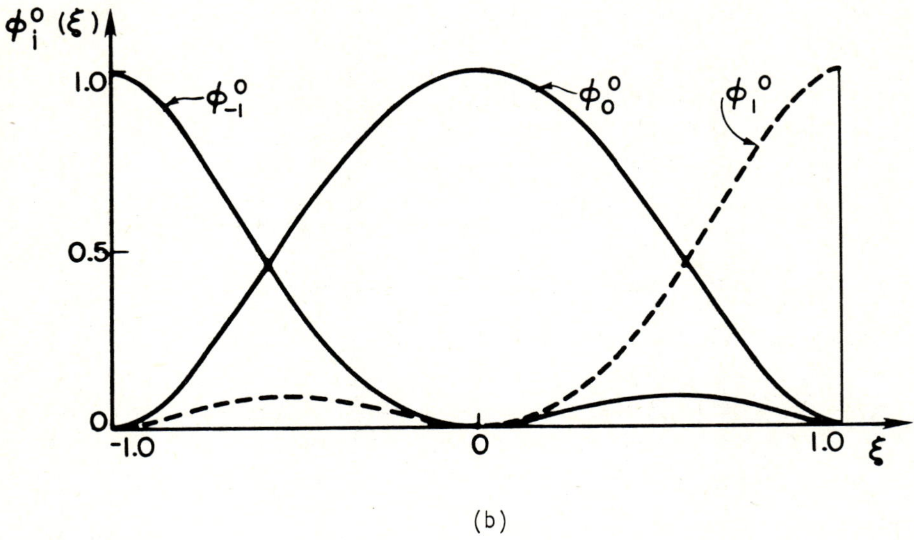

(b)

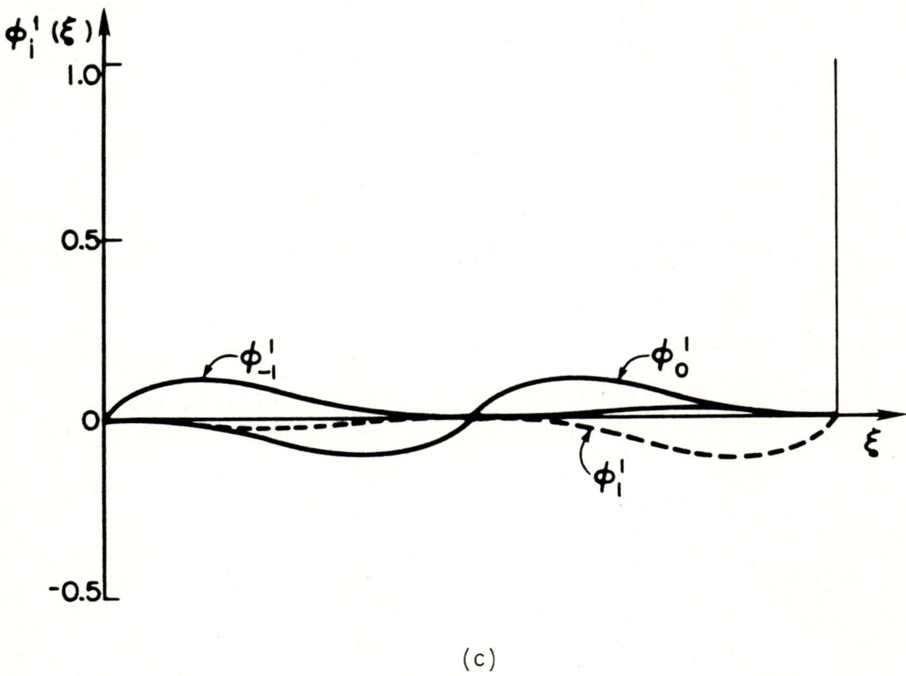

(c)

Figure 2: (continued).

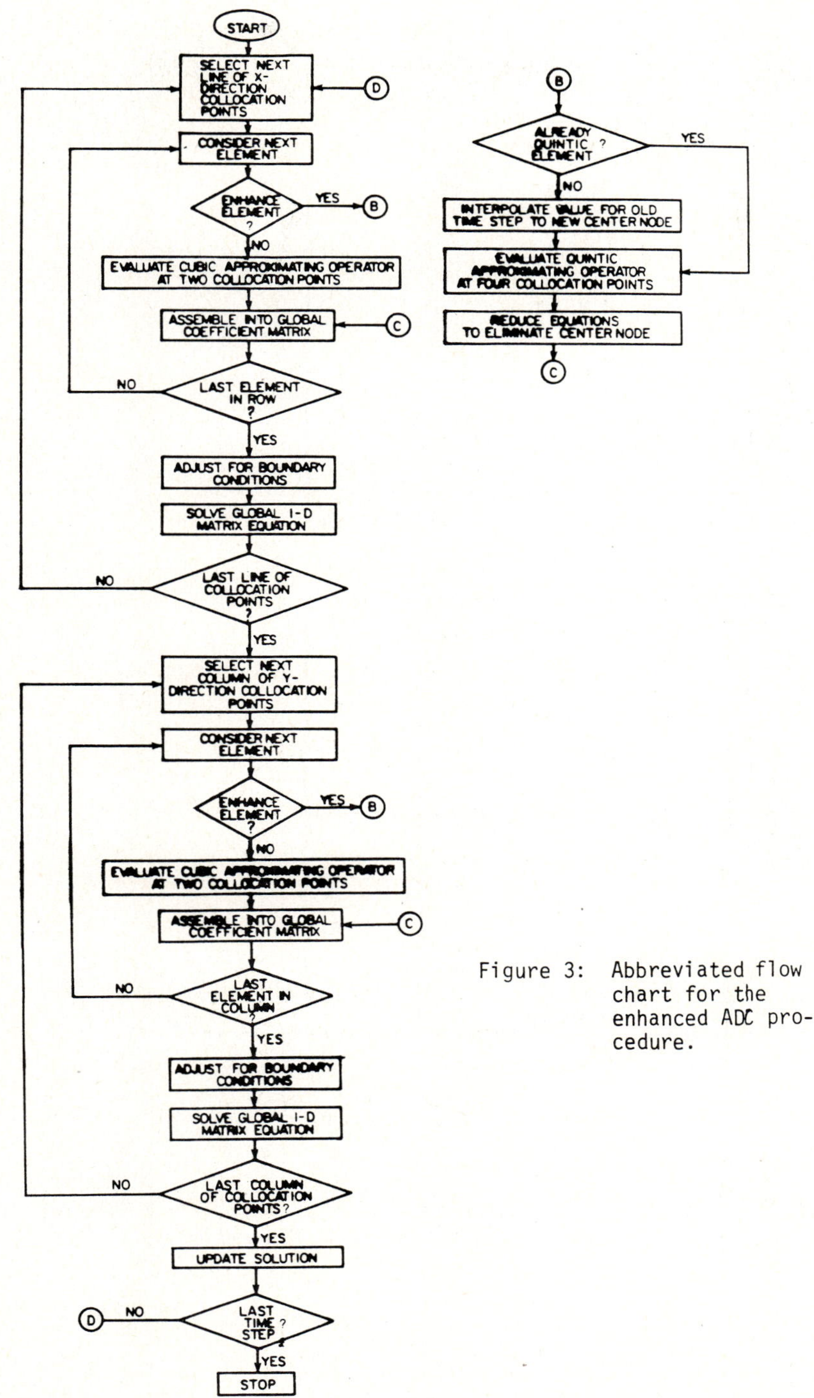

Figure 3: Abbreviated flow chart for the enhanced ADC procedure.

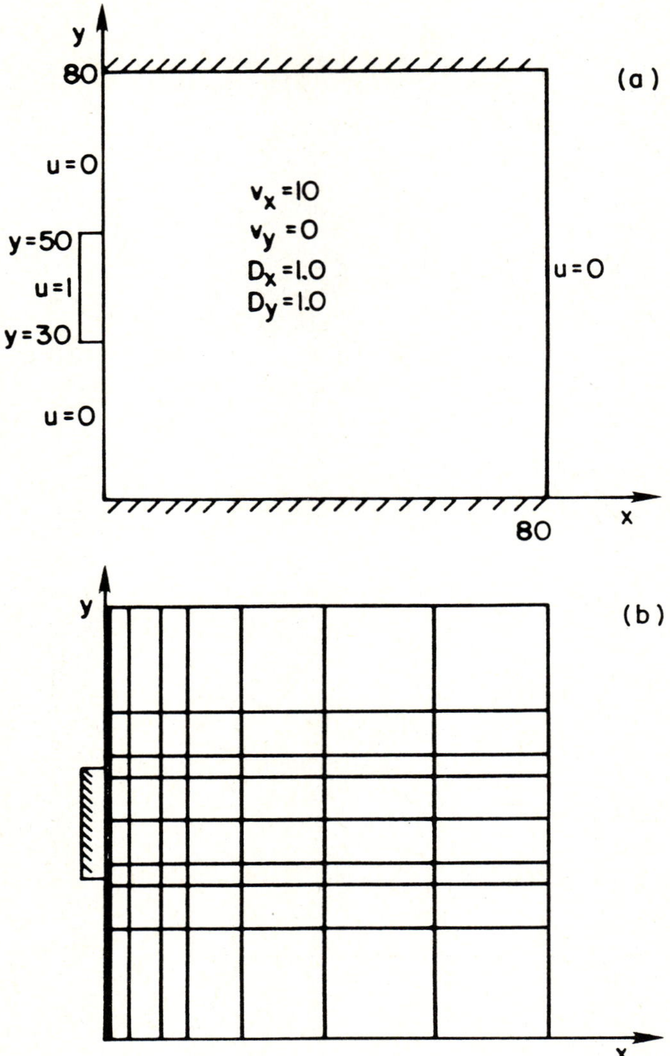

Figure 4:   (a) Domain over which equation (11) is numerically solved,
            (b) Collocation-finite element mesh used to solve (11).

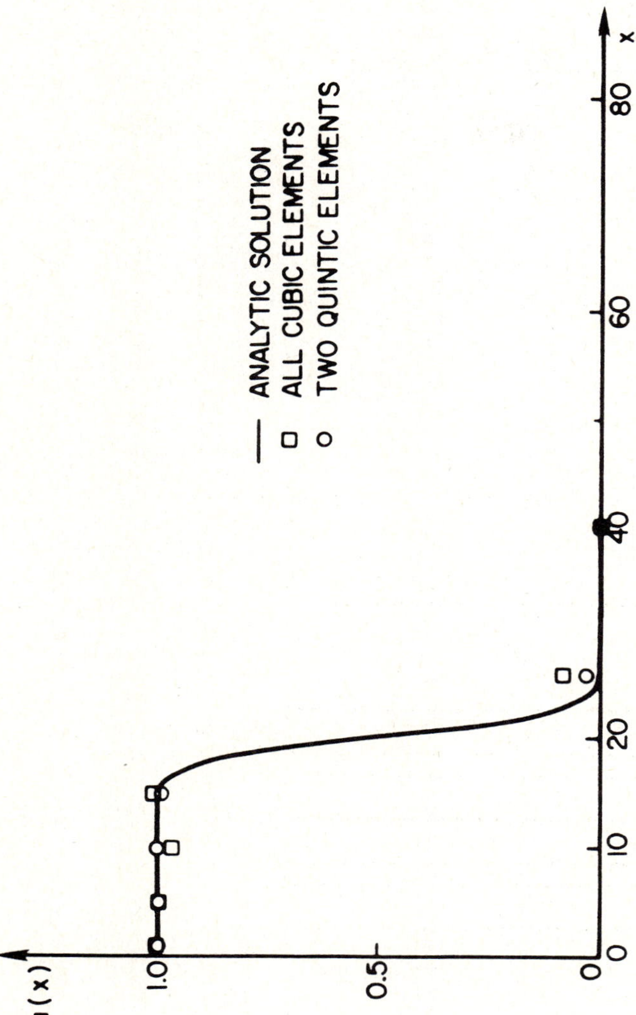

Figure 5: Various solutions to equation (11) evaluated at $y = 40$, $t = 2.0$. Numerical solutions show nodal values only.

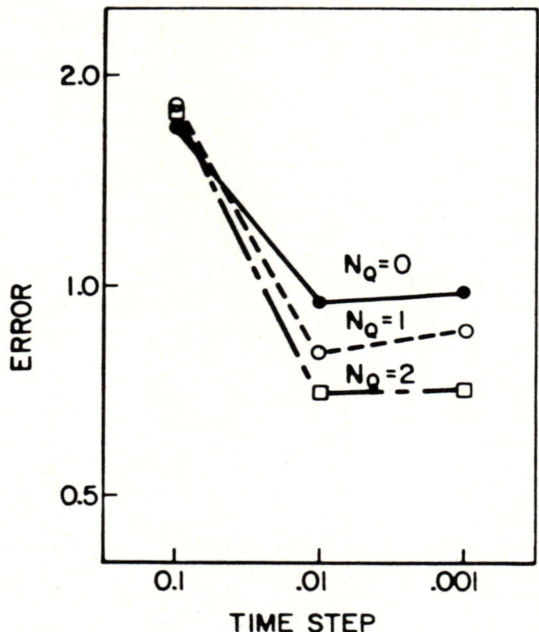

Figure 6:   Error E of (13) vs. time step for the cases of zero, one, and two quintic elements.

Unification of Finite Element Methods
H. Kardestuncer (Editor)
© Elsevier Science Publishers B.V. (North-Holland), 1984   321

CHAPTER 14

NUMERICAL & BIOLOGICAL SHAPE OPTIMIZATION

A. Philpott & G. Strang

This paper describes a shape optimization
code which moves the joints of the structure
toward optimal positions and handles the
possibility of degeneracy in the underlying
linear program. The code is applied to a
design problem arising in biology, attempting
to analyze the framework of internal fibers
in the patella and eventually to predict
their pattern of remodeling after fracture.
The underlying problems also appear in
engineering design applications, involving
optimal bounds on the effective properties
of composite materials and nonconvex
functionals in the calculus of variations.

INTRODUCTION

We report here on an optimization problem which might have
arisen in engineering design, but actually came from biology.
It is a classical problem - under given constraints, to find
a structure of minimum weight. We take the structure to be
a plane truss subject to forces along its boundary. The
forces will roughly approximate those to which the human
patella is actually subjected, and we reduce to two dimensions (plane strain) by considering a vertical cross-section
taken from front to back in the knee. Figure 1 indicates
the tensile forces on the upper and lower edges of the
patella, from the quadriceps and the ligaments, and the
equilibrating compressive force from the support provided
by the femur.

When these forces are concentrated at single points as in
Figure 2, Michell [1] found the solution. It is a truss in
which the members under compression are orthogonal to those

in tension, as is always the case in an optimal design. The
precise geometry depends on the points at which the forces
act, but the design itself is straightforward: the compression members are distributed radially and the tension is
concentrated on a single circumferential member connecting
the two forces. Certainly that is not an accurate description
of the fibers inside the patella. On the other hand it is
not absurd; both stereological data and finite element
analyses [2-4] confirm this general pattern. Photographs of
the trabeculae show a framework too striking to be an accident.
Already in the last century clinical observations led to the
formulation of <u>Wolff's law</u>, that the fibers are in the
directions of principal stress. A more exact statement of
that law, and of the mechanisms that govern the laying down
and resorbing of bone, is badly needed. It is a challenge
that several orthopaedic research groups have accepted, and
we thank W. C. Hayes and his colleagues at Beth Israel
Hospital for patient encouragement to explore the relationship to optimal design. A first step (our immediate goal)
is to refine Figure 2 by admitting a more realistic set of
loads.

This paper discusses the numerical optimization and not the
biology. Even with this limitation it is incomplete: we
could not pursue all possible approaches. Some that we did
pursue were not successful - or rather, their cost increased
so rapidly that a realistic three-dimensional picture (with
the bone allowed to remodel after fracture) was inaccessible.
Other algorithms are more promising. Perhaps we can list
here some of the alternatives, with preliminary comments:

1. <u>Fixed Geometry</u>. Fix the nodes of the truss and allow a
specified subset of edges as candidates in the design optimization. This yields a standard problem in linear programming,
with a sparse coefficient matrix described below. The cost
is effectively the sum $\Sigma\, \ell_j |s_j|$ of edge lengths multiplied
by member forces. The simplex method will yield an optimal
(and statically determinate) truss.

It was hoped that the simplicity of this model would allow
a comparatively large number of nodes and edges, giving the
design enough degrees of freedom to imitate curved fibers.
We now believe that this approach is not efficient. The
approximation of a nonlinear design problem by linear programming can be very slowly convergent. In our experience
the linear programs themselves were also highly degenerate,
with multiple pivot steps before the cost was reduced.

2. <u>Variable Geometry</u>. The positions of the nodes are
optimized as well as the weight of the truss that connects
them. Normally these steps are separate; a linear program
computes the truss for fixed nodes, and a nonlinear calculation moves the nodes. Of the codes we know about (and there
must be many others) we mention three:

(i) The CONMIN code developed by G. Vanderplaats [5]: This
was generously made available to us and was the basis for
the conference paper by Snyder et al [6] on trabecular
orientation in the patella. The code admits a much wider
range of problems in 2 and 3 dimensions, with constraints
on displacements and buckling stiffness and frequency response.

(ii) The Michell structure code developed by McConnel [7]:
This uses a direct search method of Powell to optimize the
nodal positions (with fixed node-edge incidence matrix).
Experimental results on problems of bridge design studied by
Hemp [see 8] look very satisfactory and we hope to test the
code on biological structures.

(iii) A newly developed code: We are completing a special
purpose code to test other algorithms for this class of
optimization problems. This paper describes the use of the
dual variables from the fixed-node program to choose the
directions of movement of the nodes. It seems clear that
this variation of geometry is valuable and probably essential.

3. <u>Michell Displacement Field</u>. It was the great discovery
of Michell that under the presribed load, the optimal design

has principal strains of constant magnitude. (In a region
where no members are required in one or both directions, the
corresponding strains are uniformly below that constant
value.) Lagache [9] has found a beautiful existence proof
for Michell trusses based on a convergent sequence of approximations to this strain field. Although those particular
approximations may be inefficient in computations - they are
based on fixed nodes, as in 1. above - there seems to be
no reason why a program to compute finite element displacements for Michell's strain field should not succeed.

4. <u>Michell stress field</u>. In the stress formulation of the
same problem, the constraint div $\sigma = 0$ can be removed by
deriving the stress σ from Airy's function $\psi(x,y)$:

$$\sigma_{xx} = \frac{\partial^2 \psi}{\partial y^2}, \quad \sigma_{xy} = - \frac{\partial^2 \psi}{\partial x \partial y}, \quad \sigma_{yy} = \frac{\partial^2 \psi}{\partial x^2}.$$

As noted in [10], this yields an ordinary (but nonquadratic)
problem in the calculus of variations. Again a combination
of finite elements and nonlinear optimization would give a
direct approximation of the Michell directions in the optimal
design. The densities and orientations then come directly
from ψ rather than indirectly from a discrete truss.

This paper describes the new variable geometry code mentioned
above. It was prepared in late 1983 and tested on the
biological applications - although it applies equally to
engineering design. The next section of the paper recalls
the steps of the simplex method, modified only to account
for the absolute values in the cost function; members can be
in tension or compression. Section 3 applies the simplex
method to cost (or volume) minimization, for a given placement
of the nodes. Section 4 develops an algorithm for moving
the nodes toward an improved design. There are numerical
difficulties associated with degeneracy, handled by a projection technique. In the last section we briefly discuss
the implementation and the applications.

THE FIXED GEOMETRY DESIGN PROBLEM

A number of authors, starting with Dorn, Greenberg, and Gomory [11], have considered the application of linear programming to the minimum weight problem for a plane truss subjected to forces along its boundary. The problem is to select a set of members connecting a given collection of pin joints which are fixed in number and position. Each member in the optimal truss is in tension or compression at its corresponding limit stress.

The formulation as a linear program is as follows. We consider m pin joints in the plane, and let x_i, y_i, $i = 1, \cdots, m$ be the coordinates (with respect to some origin) of the ith joint. We let $[i,k]$ denote the member connecting joints i and k. To avoid unnecessary subscripts we associate with each $[i,k]$ a unique index $j \in \{1, 2, \cdots, n\}$, and define s_j as the force in member j. Here s_j is positive if j is in tension and negative if j is in compression. The number of members n can be at most $\frac{m(m-1)}{2}$; in practice we restrict our choice of members to a relatively small set, usually those having length below some chosen bound.

The forces s_j must be such that each joint is in static equilibrium. These $2m$ equality constraints may be formulated by resolving each member force in the two coordinate directions and summing the forces in each direction at each node. The right hand sides of these constraints are the x and y components of the applied forces. For joint i, these are denoted by f_{1i} and f_{2i} respectively. For many joints, including those internal to the structure, the applied forces are zero.

If member $j = [i,k]$ makes an angle θ_j with the positive x-axis at joint i, we may write the constraints for node i as

$$\sum_{j \in I_i} - s_j \cos \theta_j = f_{1i} \, ,$$

$$\sum_{j \in I_i} - s_j \sin \theta_j = f_{2i} \, ,$$

where I_i is the set of members connected to the ith joint. Since each member connects exactly two joints, these equilibrium constraints in matrix form become

$$A_1 s = f_1 \, ,$$

$$A_2 s = f_2 \, ,$$

where s is the vector of n member forces, f_1 and f_2 are the x and y components of the boundary forces, and A_1 and A_2 are $m \times n$ matrices with zero entries in column $j = [i,k]$ except in rows i and k - where

$$(A_1)_{ij} = -\cos \theta_j \, ,$$

$$(A_1)_{kj} = \cos \theta_j \, ,$$

$$(A_2)_{ij} = -\sin \theta_j \, ,$$

$$(A_2)_{kj} = \sin \theta_j \, .$$

In what follows we shall refer to the matrix $\begin{bmatrix} A_1 \\ A_2 \end{bmatrix}$ as A and the right hand side vector $\begin{bmatrix} f_1 \\ f_2 \end{bmatrix}$ as f.

An example of a pinjointed truss and its constraint matrix is given in Figure 3.

If ℓ_j is the length of member j, minimizing the volume of a planar truss is equivalent to minimizing

$$W = \sum_{j=1}^{n} \ell_j |s_j|$$

over those vectors s which satisfy the equilibrium condition $As = f$. The derivation of this result requires that the limit stresses in tension (σ_{max}) be the same for

each member, and that the limit stresses under compression
(σ_{min}) be the same for each member. Since W is independent of the values of σ_{max} and σ_{min}, we may assume without
loss of generality that $\sigma_{max} = -\sigma_{min} = \sigma$. When each member
j is fully stressed its cross-sectional area a_j is
proportional to s_j:

$$\frac{s_j}{a_j} = \sigma \text{ , for } j \text{ in tension,}$$

$$= -\sigma \text{, for } j \text{ in compression.}$$

In this way $|s_j|$ measures the thickness required in member
j; the problem of minimizing the weight of a fixed joint
planar truss may thus be formulated as

Problem P: <u>Minimize</u> $\sum_{j=1}^{n} \ell_j |s_j|$ <u>subject to</u> $As = f$. (1)

As posed, this objective function is nonlinear. However it
is easy to reformulate P as a linear program by defining
variables

$$t_j = \text{Maximum } \{s_j, 0\}, \quad u_j = -\text{Minimum } \{s_j, 0\} ,$$

whence $W = \Sigma \ell_j t_j + \Sigma \ell_j u_j$ is linear and the constraints
become

$$A(t - u) = f, \quad t \geq 0, \quad u \geq 0 .$$

We shall write the problem in the formulation P, but make
considerable use of the linear programming formulation and
the large body of theory which this allows us.

We can formulate the problem dual to P as follows:

D: <u>Maximize</u> $v^T f_1 + w^T f_2$ <u>subject to</u> $|(v^T A_1 + w^T A_2)_j| \leq \ell_j$,

$$j = 1, \cdots, n . \tag{2}$$

Multiplying both sides by $|s_j|$, summing on j, and introducing $f_1 = A_1 s$ and $f_2 = A_2 s$ leads to <u>weak duality</u> - the
easy half of duality theory.

<u>Theorem 1</u>: If v and w are feasible for the dual and s
is feasible for the primal then

$$v^T f_1 + w^T f_2 \leq \sum_{j=1}^{n} \ell_j |s_j| . \tag{3}$$

The complementary slackness conditions, which must hold at
the minimizing s and maximizing v and w, are as
follows:

$$\text{if } s_j > 0, \text{ then } (v^T A_1 + w^T A_2)_j = \ell_j \tag{4}$$

$$\text{if } s_j < 0, \text{ then } (v^T A_1 + w^T A_2)_j = -\ell_j . \tag{5}$$

The dual problem is often given the following physical interpretation, with E as Young's modulus. A set of dual
variables v_i, w_i, corresponds to a virtual <u>deformation</u> of
the truss in which each joint i is displaced by $v_i(\frac{\sigma}{E})$ in
the y direction. The elongation e_j of each member j
under such a deformation is given by

$$e^T = (v^T A_1 + w^T A_2)(\frac{\sigma}{E}) .$$

The dual constraints therefore require of each elongation
e_j that

$$- \frac{\sigma}{E} \ell_j \leq e_j \leq \frac{\sigma}{E} \ell_j .$$

In other words the linear strain in each member caused by
the virtual deformation must be at most the value which
corresponds to the member being at its limit stress.

The dual objective function $v^T f_1 + w^T f_2$ can be interpreted
as E/σ times the work done by external forces to produce
this virtual deformation. Since the optimal solutions will
satisfy (4) and (5), the strain in a member with $s_j \neq 0$
produced by the optimal displacements v and w must be

the strain which would occur with j at its limit stress
(σ if $s_j > 0$ and $-\sigma$ if $s_j < 0$). Members with $s_j = 0$
do not appear in the optimal structure - and they are below
the limit stress when subjected to the optimal v and w.

AN ALGORITHM FOR THE FIXED GEOMETRY PROBLEM

Since the fixed geometry design problem P may be formulated
as a linear program, the simplex algorithm will solve it.
This method constructs a sequence of <u>basic feasible solutions</u>
with decreasing weight $\Sigma\, \ell_j |s_j|$. The optimum does occur at
a basic feasible solution (defined below) and the sequence
terminates at an optimal solution after a finite number of
steps.

For a planar truss with m joints we have $2m$ equilibrium
constraints. However the rows of A are not linearly
independent, and for equilibrium the boundary loads must
sum to zero in both coordinate directions. Furthermore the
sum of moments about any point in the truss must be zero.
The number of independent equilibrium constraints is therefore
$2m-3$, and three constraints must be removed before the simplex
algorithm can be applied.

In practice these degrees of freedom are removed by fixing
two or more boundary joints. Then we remove from A_1 and
A_2 the rows which correspond to the anchored joints, and we
remove from the vectors v and w the displacements which
are no longer possible. We shall assume in what follows
that the constraint matrix A has undergone this process,
and take m to be the number of rows of A - equal to the
number of displacement unknowns. This removes at the same
time any restrictions on the boundary loads; they need not be
self-equilibrating since any additional forces necessary for
equilibrium will be taken up by the foundation.

The basic feasible solutions for problem P are found by
choosing $2m$ linearly independent columns of A and
solving (1) uniquely for the member forces s_j corresponding

to these columns. All other member forces are zero. Thus
each basic feasible solution corresponds to a statically
determinate truss. We shall call any such truss a <u>basic
feasible truss</u>. Strictly speaking, the basic feasible
solution to the linear program is given by the nonnegative
variables satisfying s = t - u :

$$t_j = \frac{1}{2} \left(|s_j| + s_j \right) \quad \text{and} \quad u_j = \frac{1}{2} \left(|s_j| - s_j \right).$$

With this understanding we apply the term "basic feasible
solution" to the member forces s and also to the variables
t and u.

If the chosen columns of A are not linearly independent,
the corresponding truss is no longer rigid under an
arbitrary set of boundary loads, and it becomes a mechanism.
On the other hand, if more than 2m columns are chosen
(with 2m linearly independent) then the corresponding
truss is <u>statically indeterminate</u>, and we cannot solve
uniquely for the member forces. A third possibility is
that the 2m columns are linearly independent and yet some
of the member forces in the basic feasible truss are zero.
In this case the basic feasible solution is <u>degenerate</u>.

The algorithm for the fixed geometry optimal design follows
the steps of the simplex method, modified to work with the
formulation P. If we assume that degenerate basic feasible
solutions do not occur, the steps are as follows:

1. Determine a set of 2m members which form a basic
feasible truss. Denote by $B = \begin{bmatrix} B_1 \\ B_2 \end{bmatrix}$ the 2m by 2m
"basis matrix" formed by the corresponding columns of A.

2. Solve Bs = f.

3. For the solution s, construct a complementary slack dual
solution v,w. In other words, solve the linear system

$$(v^T B_1 + w^T B_2)_j = \pm \, \ell_j \text{ (choosing the sign of } s_j).$$

4. Choose the value of j (say $j = q$) which minimizes

$$r_j = \ell_j - |(v^T A_1 + w^T A_2)_j| \, .$$

5. If $r_q \geq 0$ then stop; the current truss is optimal.
If $r_q < 0$ then member q enters the basic feasible truss
- in tension or compression depending on the sign of
$z_q = (v^T A_1 + w^T A_2)_q$.

6. Determine the member to leave the basic feasible truss,
given by the index $i = p$ which minimizes $|\psi_i|$ over
indices i in I. Here

$$\psi_i = \frac{s_i}{(B^{-1}A)_{iq}}$$

$$I = \{i : (B^{-1}A)_{iq} \neq 0 \text{ and sign } \psi_i = \text{sign } z_q\} \, .$$

7. Replace member p in the basic feasible truss (and
the column p in the matrix B) with member q. Return
to Step 2.

It is easy to see that this algorithm follows exactly the
same steps as the simplex method. Step 4 finds a minimum
reduced cost, and ψ_i in Step 6 is the ratio test that
determines which basic variable becomes non-basic after a
pivot operation. In this step, if no quotients of appropriate sign exist then the problem has an unbounded
solution.

Since the constraint matrix A is very sparse, the implementation of the algorithm uses an LU factorization of
the basis matrix B and updates the triangular factors L
and U at each iteration. We employ the method of Forrest
and Tomlin [12]. To take further advantage of sparsity,
the columns of A corresponding to nonbasic variables are
recomputed at Step 4 in each iteration.

The steps described above are followed with no essential
changes when degeneracy is encountered. If the starting

basic feasible truss is degenerate we arbitrarily assign
positive or negative signs to the zero forces. Since by
Step 5 every member entering the basis does so either under
tension or compression, for every subsequent basic feasible
truss we can assign positive and negative signs to the
(possibly zero) member forces. This convention gives a
unique complementary slack dual solution in Step 3.

The choice of member p to leave the basic feasible truss
is however not uniquely determined, since any i with
$s_i = 0$ will render $|t_i|$ a minimum in Step 6. In the code
we choose the variable with lowest index, and although this
rule does not preclude cycling, no instances of cycling have
been observed. (Dantzig has observed only one in long
experience with simplex calculations.) Of course a more
expensive pivot rule could ensure that cycling never occurs.

VARIABLE GEOMETRY

For fixed joint positions the above algorithm gives an optimal
solution - it finds the minimum weight of a plane truss which
supports given boundary loads. Being essentially the simplex
method, the algorithm terminates at the optimum after a finite
number of iterations. Unfortunately, for large n and m
(many joints and many possible members) the time taken to
reach the optimum is prohibitive. If members are permitted
between any two joints then the problem size increases
rapidly with the number of joints. To avoid this problem
we may restrict the allowed members to those whose length
lies below some fixed bound. However, even a small problem
with 100 joints and 800 members can take hours of CPU
time.

One reason for this poor convergence behavior is the high
degree of degeneracy inherent in the linear programs associated
with pin jointed structures. Often a large number of joints
have no external forces, and play no part in the design; the
members incident upon these joints carry zero force. Even
if these joints and the members connected to them are deleted,

it is often the case in a basic feasible truss that other
zero-force members exist. As observed in [11], these cannot
be deleted since the resulting truss would no longer be
rigid. The presence of degenerate basic feasible solutions
causes poor performance in the simplex algorithm, since many
pivot operations occur with no change in the objective
function.

To approximate accurately the trabecular architecture of the
human patella, a large number of internal joints would be
needed in the design process - if their positions are fixed.
This proliferation of joints subject to zero loads exacerbates
the degeneracy problem, and the optimization becomes imprac-
ticable. An alternative is to admit a much smaller number of
joints and then alter their positions in such a way as to
improve the design. We proceed to describe a method for
computing the gradient of the cost W in terms of the joint
positions, thus giving a set of directions of steepest descent;
shifting the joints in these directions produces a truss with
less weight.

To find the gradient, we investigate the effect on the
objective function W of a change in position of the joints
from (x_i, y_i) to $(x_i + \delta x_i, y_i + \delta y_i)$. At an optimal
solution of problem P, the cost is

$$W = \sum_{j=1}^{n} \ell_j |s_j| .$$

The sum is taken over members of the optimal basic feasible
truss. Let this truss correspond to the basis matrix
$B = \begin{bmatrix} B_1 \\ B_2 \end{bmatrix}$ where B_1 and B_2 are $m \times 2m$ matrices. We
assume for the present that this truss is not degenerate.

Changing the positions of the joints will change the lengths
ℓ_j and also affect the entries of B - implying that the
member forces will also change. Suppose that the forces s_j
become $s_j + \delta s_j$, and the changes in angles alter B to
B + δB. Then equation (1) gives

$$(B + \delta B)(s + \delta s) = f,$$

whence to first order

$$B\delta s + (\delta B)s = 0. \tag{6}$$

The first order change in $W = \Sigma \ell_j |s_j|$ given by

$$\delta W = \sum_{s_j > 0} \ell_j \delta s_j - \sum_{s_j < 0} \ell_j \delta s_j + \Sigma \delta \ell_j |s_j| + \sum_{s_j = 0} \ell_j \delta |s_j|. \tag{7}$$

The last term in (7) is absent by the assumption of non-degeneracy. It then follows from the definition of the complementary slack dual variables (the displacements v and w) that

$$\delta W = [v^T w^T] \begin{bmatrix} B_1 \\ B_2 \end{bmatrix} \delta s + \delta \ell^T |s|,$$

and thus from (6) we obtain

$$\delta W = - [v^T w^T](\delta B)s + \delta \ell^T |s|. \tag{8}$$

We now express $\delta \ell$ and δB in terms of the changes in joint positions, δx and δy. If $j = [i,k]$, then the length ℓ_j^2 is $(x_k - x_i)^2 + (y_k - y_i)^2$. Since

$$\frac{\partial \ell_j}{\partial x_k} = \cos \theta_j, \quad \frac{\partial \ell_j}{\partial x_i} = - \cos \theta_j,$$

$$\frac{\partial \ell_j}{\partial y_k} = \sin \theta_j, \quad \frac{\partial \ell_j}{\partial y_i} = - \sin \theta_j,$$

it follows from the definition of B_1 and B_2 that

$$\delta \ell = B_1^T \delta x + B_2^T \delta y. \tag{9}$$

The matrix δB is the first order change in the matrix B and arises from the changes in $\cos \theta_j$ and $\sin \theta_j$ for each member after the positions of some joints are changed. It is easy to show that if $j = [i,k]$,

$$\frac{\partial \cos \theta_j}{\partial x_i} = - \frac{\sin^2 \theta_j}{\ell_j} \quad , \quad \frac{\partial \cos \theta_j}{\partial x_k} = \frac{\sin^2 \theta_j}{\ell_j} \quad ,$$

$$\frac{\partial \cos \theta_j}{\partial y_i} = \frac{\cos \theta_j \sin \theta_j}{\ell_j} \quad , \quad \frac{\partial \cos \theta_j}{\partial y_k} = - \frac{\cos \theta_j \sin \theta_j}{\ell_j} \quad .$$

Therefore the changes in cosine and sine are given to first order by

$$\delta(\cos \theta_j) = \sin^2 \theta_j \left(\frac{\delta x_k - \delta x_i}{\ell_j} \right) - \sin \theta_j \cos \theta_j \left(\frac{\delta y_k - \delta y_i}{\ell_j} \right),$$

$$\delta(\sin \theta_j) = - \sin \theta_j \cos \theta_j \left(\frac{\delta x_k - \delta x_i}{\ell_j} \right) + \cos^2 \theta_j \left(\frac{\delta y_k - \delta y_i}{\ell_j} \right).$$

The change $(\delta B)_j$ in column j of B is zero except in rows i, k, $i + m$, and $k + m$; in these four rows we have

$$(\delta B)_{ij} = -\delta(\cos \theta_j) = \frac{-\sin \theta_j}{\ell_j} [\sin \theta_j (\delta x_k - \delta x_i) + \cos \theta_j (\delta y_i - \delta y_k)]$$

$$(\delta B)_{i+m,j} = -\delta(\sin \theta_j) = \frac{\cos \theta_j}{\ell_j} [\sin \theta_j (\delta x_k - \delta x_i) + \cos \theta_j (\delta y_i - \delta y_k)]$$

with the same formulas and opposite signs for $(\delta B)_{kj}$ and $(\delta B)_{k+m,j}$. The term in brackets is just

$$\left(\begin{bmatrix} B_2 \\ -B_1 \end{bmatrix}^T \right)_j \begin{bmatrix} \delta x \\ \delta y \end{bmatrix} \quad , \text{ so that } (\delta B)_j = \left(\begin{bmatrix} B_2 \\ -B_1 \end{bmatrix} \right)_j \frac{1}{\ell_j} \left(\begin{bmatrix} B_2 \\ -B_1 \end{bmatrix}^T \right)_j \begin{bmatrix} \delta x \\ \delta y \end{bmatrix} \quad . \quad (10)$$

If we let S be the $2m \times 2m$ diagonal matrix with $S_{jj} = \frac{s_j}{\ell_j}$, then δW in (8) can be written as follows:

$$\delta W = |s|^T [B_1^T \; B_2^T] \begin{bmatrix} \delta x \\ \delta y \end{bmatrix} - [v^T \; w^T] \begin{bmatrix} B_2 \\ -B_1 \end{bmatrix} S [B_2^T \; -B_1^T] \begin{bmatrix} \delta x \\ \delta y \end{bmatrix} \quad .$$

We simplify this expression by defining

$$g^T = |s|^T B_1^T - [v^T \ w^T] \begin{bmatrix} B_2 \\ -B_1 \end{bmatrix} S \ B_2^T \ , \qquad (11)$$

$$h^T = |s|^T B_2^T + [v^T \ w^T] \begin{bmatrix} B_2 \\ -B_1 \end{bmatrix} S \ B_1^T \ , \qquad (12)$$

leaving

$$\delta W = g^T \delta x + h^T \delta y \ . \qquad (13)$$

For given loads and a given nondegenerate optimal truss, the
vector $\begin{bmatrix} -g \\ -h \end{bmatrix}$ defined by (11) and (12) is a direction of
steepest descent for W when expressed as a function of
the nodal positions. Therefore it is natural to extend the
algorithm to include a shift of positions in this descent
direction. This variable-geometry algorithm proceeds as
follows:

Step 1.   For a given set of joint coordinates $\overline{x},\overline{y}$, solve
          the fixed geometry problem P

Step 2.   For the optimal basic feasible solution to P,
          calculate g and h from (11) and (12)

Step 3.   For the given boundary loads, find the value of
          the stepsize λ which minimizes $W(\overline{x} - \lambda g, \overline{y} - \lambda h)$

Step 4.   If the change in W is less than some tolerance
          then terminate; otherwise update $\overline{x}$ and $\overline{y}$ and
          return to Step 1.

The crucial step is the line search in Step 3; it is here
that the improvement in geometry is made. There are a
number of ways to carry out this search. One possibility
is to retain the current basic feasible truss and recalculate
W for this truss at each point in the line search (determining
the member forces from equilibrium $Bs = f$). Alternatively,
we may return to the simplex algorithm and find the <u>optimal</u>
basic feasible truss at each point of the search. We tend
to favor the former approach; re-solving the linear

programming problem at each point requires a greater computational effort, a situation which is not improved by the presence of degeneracy.

One might hope that this algorithm gives a reasonably robust method of improving the geometry. Unfortunately this is not the case.

In practice, line searches often terminate at values of λ for which the resulting basic feasible truss is degenerate. This seems reasonable since the cross-sectional area of the degenerate member has a local minimum exactly at the value of λ which renders it degenerate. This contradicts our assumption of nondegenerate basic feasible trusses; even if we begin the line search at a well-behaved truss, we are likely to meet degeneracy.

A possible remedy is to continue as if our assumption were valid, and hope that the sum $\Sigma \, \ell_j \, \delta|s_j|$ over members with $s_j = 0$ (which was neglected in (7)) is small. However it soon becomes obvious that this approach is untenable. In practice, line searches from current degenerate basic feasible trusses often terminate at a steplength λ which is zero to working accuracy, and the algorithm is jammed at a suboptimal solution. Clearly at some point, this neglected fourth term in (7) has a considerable effect.

The solution is to compute a more effective search direction. Since the undesirable behavior is caused by changes in $|s_j|$ for degenerate members, it makes sense to seek a direction which minimizes these changes. We therefore search in a direction which, at least locally, keeps degenerate members degenerate. Then one expects their contribution to δW to be small.

The adjusted search direction is computed as follows. From (6) we have

$$\delta s = -B^{-1}(\delta B)s \, ,$$

whence, using (10),

$$\delta s = -B^{-1} \begin{bmatrix} B_2 \\ -B_1 \end{bmatrix} S \; [B_2^T \; -B_1^T] \begin{bmatrix} \delta x \\ \delta y \end{bmatrix} . \tag{14}$$

Let I_d be the set of indices of columns of B corresponding to degenerate members, and let B_d be the matrix formed by the rows of B^{-1} with these indices. Then if

$D = -B_d \begin{bmatrix} B_2 \\ -B_1 \end{bmatrix} S \; [B_2^T \; -B_1^T]$, any vector $\begin{bmatrix} \delta x \\ \delta y \end{bmatrix}$ in the nullspace of

D will render $(\delta s)_j = 0$ for $j \in I_d$. Thus to obtain a descent direction which keeps the changes in degenerate members as small as possible, we <u>project</u> $[\begin{smallmatrix} g \\ h \end{smallmatrix}]$ onto the nullspace of D :

$$[\tfrac{\overline{g}}{h}] = [\tfrac{g}{h}] - D^T (DD^T)^{-1} D[\tfrac{g}{h}] .$$

Since DD^T is often singular, we use a Gram-Schmidt procedure to construct a matrix Q with orthonormal columns spanning the column space of D^T, whence

$$[\tfrac{\overline{g}}{h}] = [\tfrac{g}{h}] - QQ^T [\tfrac{g}{h}] .$$

Even with this projected search direction, we are not guaranteed a decrease in $W(\overline{x}-\lambda\overline{g}, \overline{y}-\lambda\overline{h})$ for large step sizes λ. Indeed, if a step is so large that some member forces change sign, the current choice of direction may no longer give descent. For this reason, we compute from (14) the values of λ for which a member force changes sign, and evaluate the objection function W at these values in succession. (The equilibrium equation is solved for s.) When an increase in W is detected, this gives an interval in λ over which a golden section line search minimizes W.

CONCLUSION

The algorithm described above was implemented in FORTRAN on a VAX 780, and applied to a model of the human patella. The loads correspond to a $60°$ flexion of the leg (as in

climbing stairs). We worked with 40 joints, and allowed
all edges with length not exceeding one centimeter. An
initial basic feasible truss is shown in Figure 4a. The 14
fixed nodes along the bottom provide reactions to the 5
applied loads. Solid lines denote members in tension and
dashed lines indicate compression. The thickness of the
members is represented by the number of lines in the diagram,
and bars denoted by single lines in Figure 4a are degenerate.
They have zero thickness, as in the two members which meet
at the unloaded node at the top right hand side; they are
present only to avoid a mechanism. The weight of the initial
truss is W = 2.064 grams.

Figure 4b shows the output from the fixed geometry design
algorithm, starting from the truss of Figure 4a. Degenerate
members are now omitted. The optimal truss has a weight of
W = 1.829 grams and was obtained after 25 pivot operations
in the simplex method. For a problem of this size, each
pivot step takes approximately .5 seconds, a value which
increases substantially for larger problems.

Figure 4c shows the output from the variable geometry
algorithm. It is the optimal basic feasible truss after a
succession of changes in nodal positions. The weight is
reduced to W = 1.708 grams after ten line searches, after
each of which the fixed geometry problem was resolved and
new directions of movement were computed (with projection to
control degeneracy). Subsequent line searches failed to give
a significant improvement in the weight of the truss and the
program terminated.

ACKNOWLEDGEMENT

We thank the Army Research Office and the National
Science Foundation for their support under contracts
DAAG 29-83K-0025 and 81-02371. We are also grateful to the
National Institute for Health for the opportunity, under
contract AM30875, to work in the Orthopaedics Research
Laboratory at Beth Israel Hospital.

REFERENCES

[1]   Michell, A.G.M., The limit of economy of material in
      frame structures, Phil. Mag. 8 (1904) 589-597.

[2]   Hayes, W.C., Snyder, B., Levine, B.M. and Ramaswamy, S.,
      Stress-morphology relationships in trabecular bone of the
      patella, in: R. H. Gallagher et al., eds., Finite
      Elements in Biomechanics (John Wiley, New York, 1983).

[3]   Huberti, H.H., Hayes, W.C., Stone, J.L. and Shybut, G.T.,
      Force ratios in the quadriceps tendon and ligamentum
      patellae, J. Biomechanics, to appear.

[4]   Stone, J.L., Three-dimensional stress-morphology rela-
      tionships in trabecular bone, M.Sc. Thesis, M.I.T. (1983).

[5]   Vanderplaats, G., Numerical Optimization Techniques for
      Engineering Design - With Applications (McGraw-Hill,
      New York, 1984).

[6]   Snyder, B., Strang, G., Hayes, W.C. and Norris, G.,
      Application of structural geometry optimization
      techniques in the microstructural remodeling of
      trabecular bone, ASME Annual Meeting, Boston, 1983.

[7]   McConnel, R.E., Least-weight frameworks for load across
      span, ASCE J. Eng. Mech. Div. 100 (1974) 885-901.

[8]   Hemp, W.S., Optimum Structures (Clarendon Press, Oxford,
      1973).

[9]   Lagache, J.-M., Treillis de volume minimal dans une
      région donnée, J. Mécanique 20 (1981) 415-448.

[10] Strang, G. and Kohn, R.V., Hencky-Prandtl nets and
      constrained Michell trusses, Comp. Methods in Appl. Mech.
      Eng. 36 (1983) 207-222.

[11] Dorn, W.S., Gomory, R.E. and Greenberg, H.J., Automatic
      design of optimal structures, J. Mécanique 3 (1964)
      25-52.

[12] Forrest, J.J.H. and Tomlin, J.A., Updated triangular
      factors of the basis to maintain sparsity in the product
      form simplex method, Math. Prog. 2 (1972) 263-278.

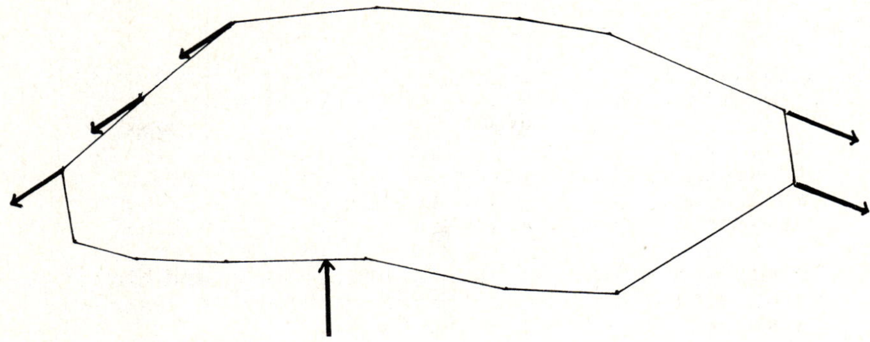

FIGURE 1. HUMAN PATELLA: ROTATED CROSS SECTION, WITH APPLIED
TENSIONS AND SUPPORT REGION.

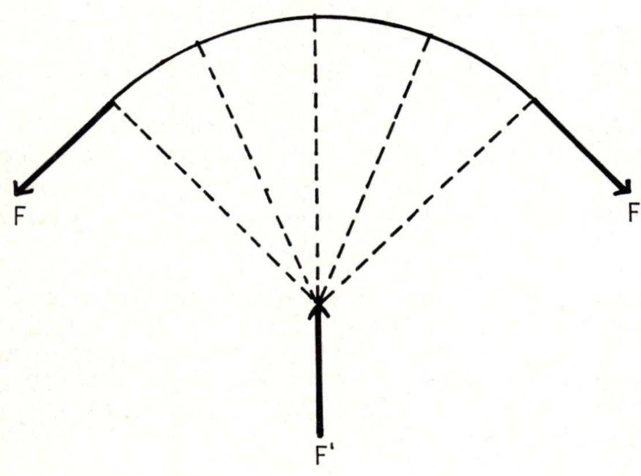

FIGURE 2. MICHELL TRUSS, 3-POINT LOAD.

A. Philpott & G. Strang

$$A = \begin{bmatrix} -1 & -\cdot 87 & 0 \\ 1 & 0 & \cdot 71 \\ 0 & \cdot 87 & -\cdot 71 \\ 0 & -\cdot 5 & 0 \\ 0 & 0 & -\cdot 71 \\ 0 & \cdot 5 & \cdot 71 \end{bmatrix}$$

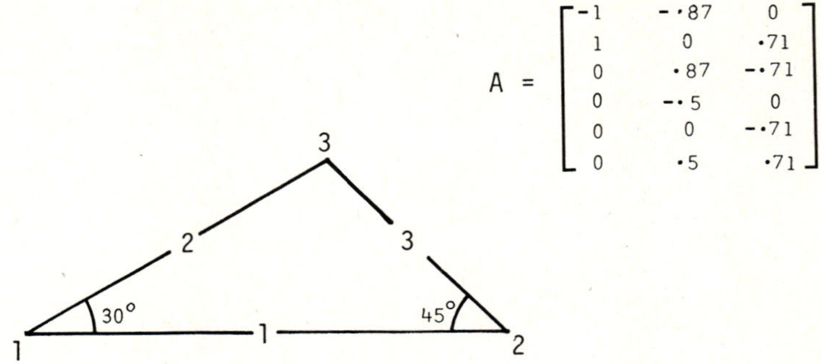

FIGURE 3. A SIMPLE PINJOINTED TRUSS.

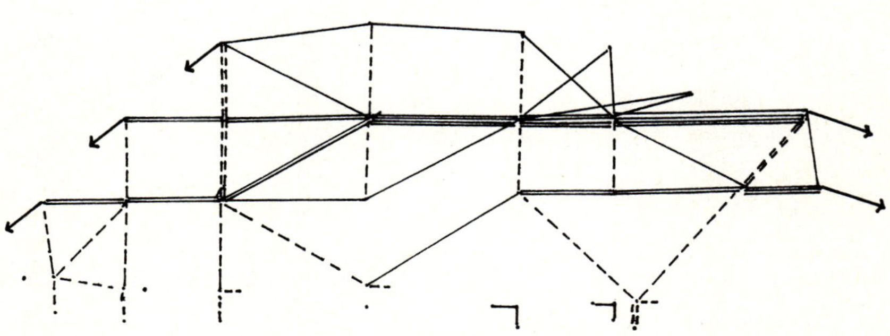

FIGURE 4A. INITIAL TRUSS.

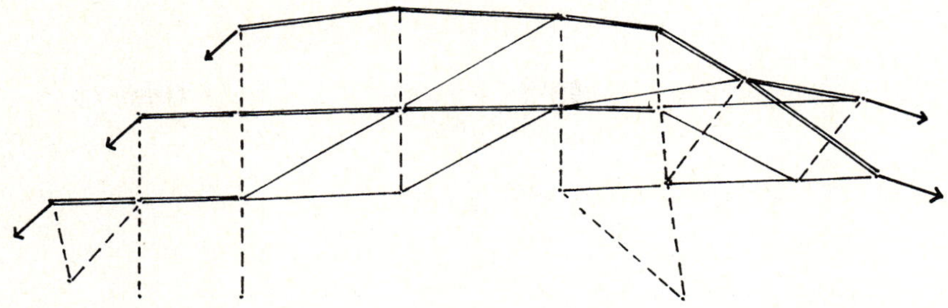

FIGURE 4B. OPTIMAL TRUSS WITH FIXED NODES.

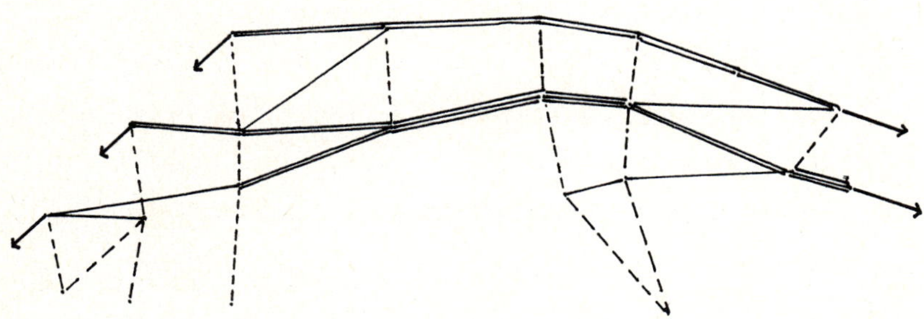

FIGURE 4C. OPTIMAL TRUSS AFTER 10 GEOMETRY CHANGES.

INDEX